INTERNATIONAL GCSE (9–1)

NICK ENGLAND

Physics

for Edexcel International GCSE

HODDER EDUCATION
AN HACHETTE UK COMPANY

In order to ensure that this resource offers high-quality support for the associated Pearson qualification, it has been through a review process by the awarding body. This process confirms that this resource fully covers the teaching and learning content of the specification or part of a specification at which it is aimed. It also confirms that it demonstrates an appropriate balance between the development of subject skills, knowledge and understanding, in addition to preparation for assessment.

Endorsement does not cover any guidance on assessment activities or processes (e.g. practice questions or advice on how to answer assessment questions), included in the resource nor does it prescribe any particular approach to the teaching or delivery of a related course.

While the publishers have made every attempt to ensure that advice on the qualification and its assessment is accurate, the official specification and associated assessment guidance materials are the only authoritative source of information and should always be referred to for definitive guidance.

Pearson examiners have not contributed to any sections in this resource relevant to examination papers for which they have responsibility.

Examiners will not use endorsed resources as a source of material for any assessment set by Pearson. Endorsement of a resource does not mean that the resource is required to achieve this Pearson qualification, nor does it mean that it is the only suitable material available to support the qualification, and any resource lists produced by the awarding body shall include this and other appropriate resources.

The Acknowledgments are listed on page viii.

Orders: please contact Bookpoint Ltd, 130 Milton Park, Abingdon, Oxon OX14 4SB. Telephone: (44) 01235 827720. Fax: (44) 01235 400454. Lines are open 9.00–17.00, Monday to Saturday, with a 24-hour message answering service. Visit our website at www.hoddereducation.co.uk

First published in 2017 by

Hodder Education

An Hachette UK Company,

50 Victoria Embankment,

London EC4Y 0DZ

Impression number 5 4 3

Year 2021 2020 2019 2018

Typeset in ITC Legacy Serif by Elektra Media Ltd.

Printed in Italy

Project managed by Elektra Media Ltd.

A catalogue record for this title is available from the British Library.

ISBN 978 151 040 5189

Contents

Contents

Section 3 Waves

Section 4 Energy resources and energy transfer

Section 5 Solids, liquids and gases

Section 6 Magnetism and electromagnetism

Section 7 Radioactivity and particles

Section 8 Astrophysics

Getting the most from this book

Welcome to the Edexcel International GCSE (9–1) Physics Student Book. This book has been divided into eight Sections, following the structure and order of the Edexcel Specification, which you can find on the Edexcel website for reference.

Each Section has been divided into a number of smaller Chapters to help you manage your learning.

The following features have been included to help you get the most from this book.

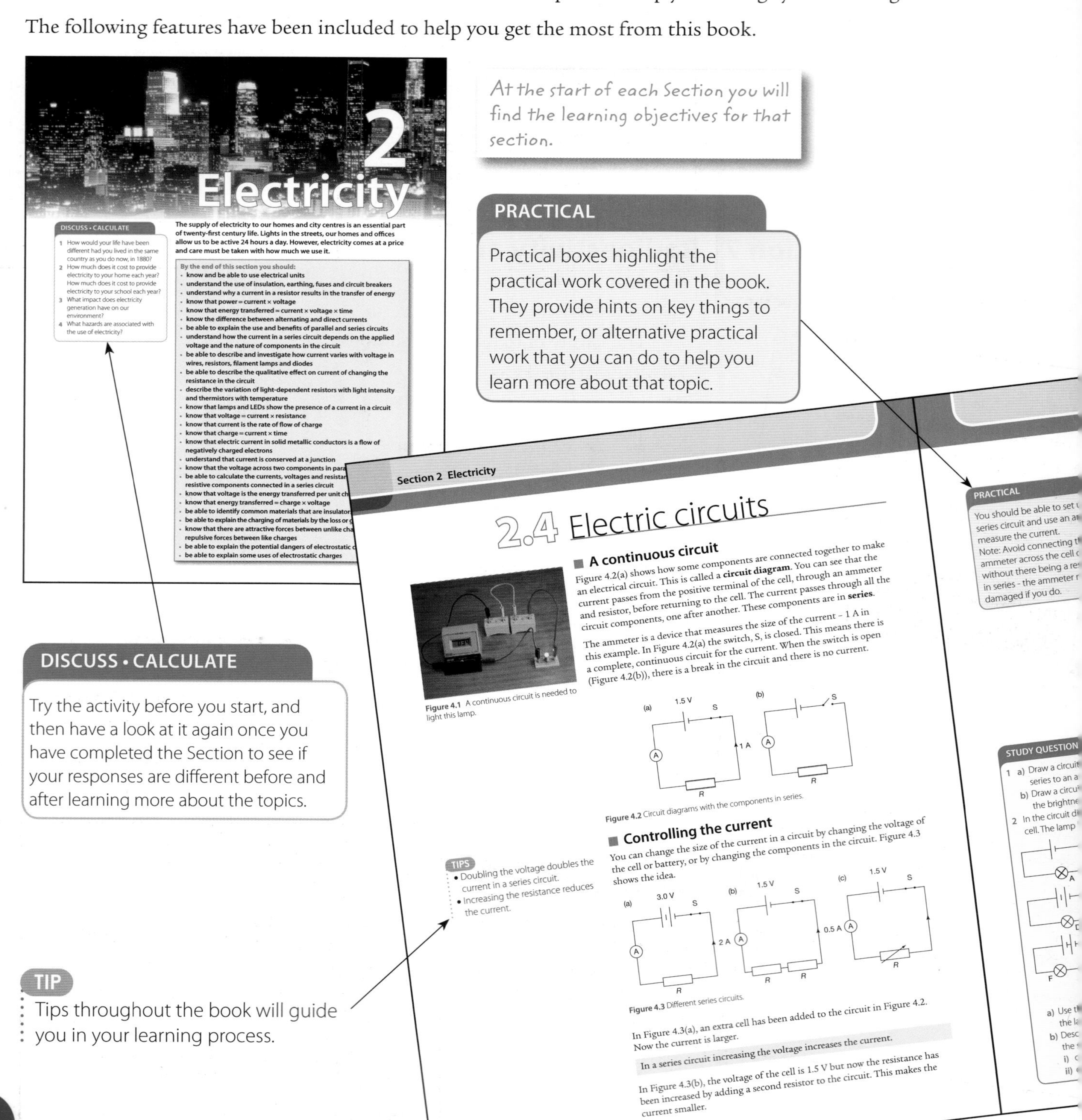

At the start of each Section you will find the learning objectives for that section.
PRACTICAL
Practical boxes highlight the practical work covered in the book. They provide hints on key things to remember, or alternative practical work that you can do to help you learn more about that topic.
DISCUSS • CALCULATE
Try the activity before you start, and then have a look at it again once you have completed the Section to see if your responses are different before and after learning more about the topics.
TIP
Tips throughout the book will guide you in your learning process.
2
Electricity
Section 2 Electricity
2.4 Electric circuits
A continuous circuit
Controlling the current
TIPS
STUDY QUESTION
58

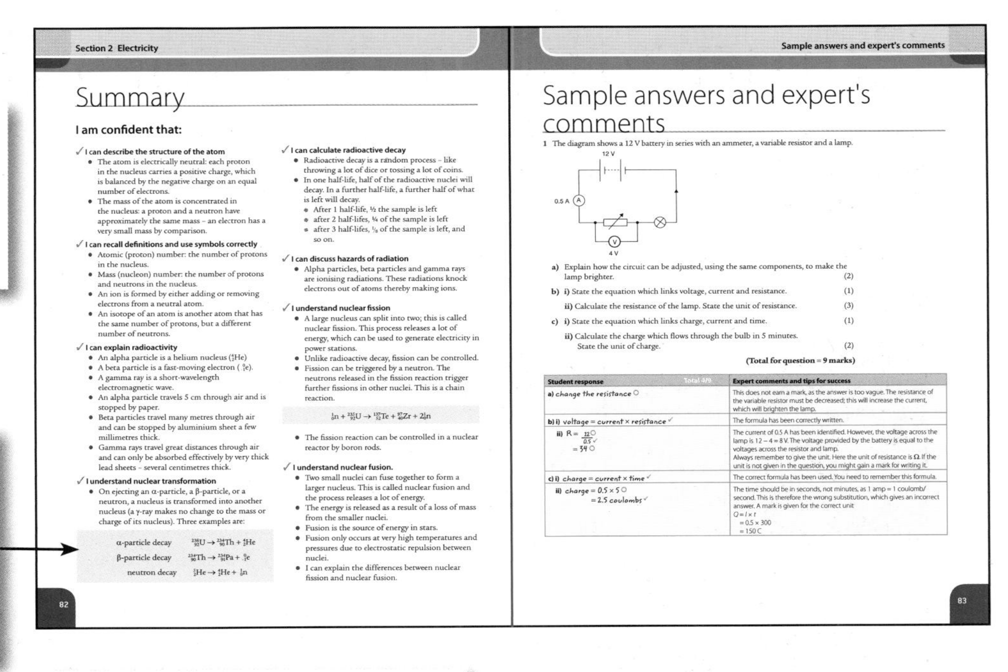

You will find Exam-style questions at the end of each Section covering the content of that Section and the different types of questions you will find in an examination.

Formulae and laws have been highlighted so that you can easily find them as you work through the book. Remember that in your exam you will be given some formulae; others you have to memorise.

Before you try the exam-style questions, look at the sample answers and expert's comments to see how marks are awarded and common mistakes to avoid.

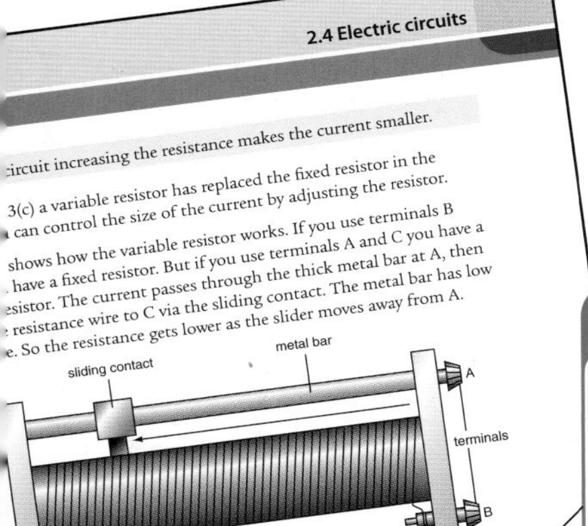

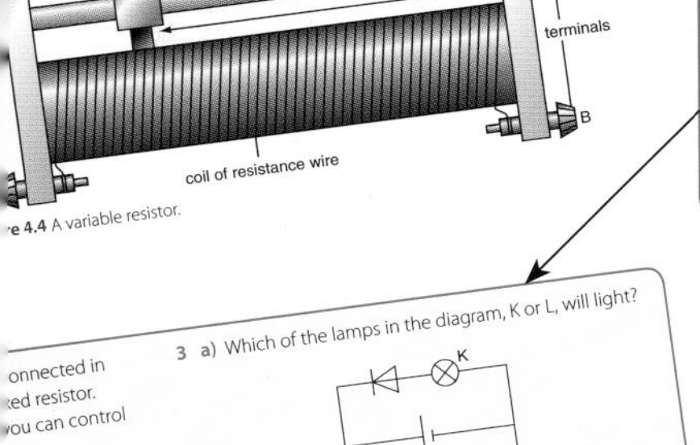

STUDY QUESTIONS

At the end of each Chapter you will find Study Questions. Work through these in class or on your own for homework. Answers are available online.

ANSWERS

Answers for all questions and activities in this book can be found online at www.hoddereducation.co.uk/igcsephysics

EXTEND AND CHALLENGE

When you have completed all the Exam-style questions for the Section, try the Extend and Challenge questions.

EXTEND AND CHALLENGE

1 a) The table shows the current in three different electrical appliances when connected to a 230 V a.c. supply.

appliance	current / A
kettle	11.5
lamp	0.05
toaster	4.2

i) Which appliance has the greatest resistance? How does the data show this?

ii) The lamp is connected to the mains supply using thin, twin-cored cable, consisting of live and neutral connections. State **two** reasons why this cable should not be used to connect the kettle to the mains supply.

b) i) Calculate the power rating of the kettle when it is operated from the 230 V a.c. mains supply.

ii) Calculate the current flowing through the kettle when it is connected to a 115 V mains supply.

iii) The kettle is filled with water. The water takes 90 s to boil when working from the 230 V supply. Explain how the time it takes to boil changes when the kettle operates on the 115 V supply.

2 The graph shows how the current through a type of filament lamp depends on the voltage applied to it. Three of these lamps are connected into the circuit shown.

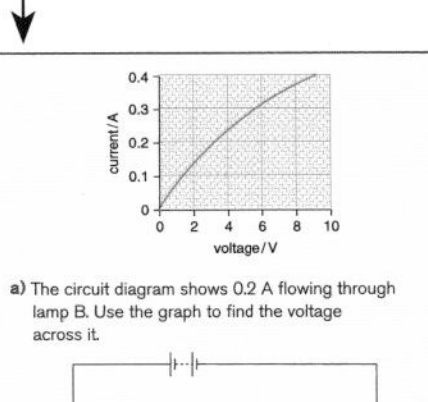

a) The circuit diagram shows 0.2 A flowing through lamp B. Use the graph to find the voltage across it.

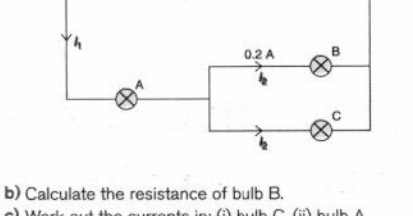

b) Calculate the resistance of bulb B.

c) Work out the currents in: (i) bulb C, (ii) bulb A.

d) Use your answer to part (c) (i) to calculate the voltage across bulb A.

e) Now calculate the voltage of the battery.

The Publisher would like to thank the following for permission to reproduce copyright photographs:

p.1 © Philippe Devanne / Fotolia; p.2 © OLIVIER MORIN / Staff/ Getty Images; p.5 t © Carolina K Smith MD - Fotolia.com, b © jpmatz - Fotolia.com; p.8 © GUSTOIMAGES/SCIENCE PHOTO LIBRARY; p.11 t © Daniel Vorley / Stringer / Getty Images, b © Toutenphoton - Fotolia.com; p.15 © NASA; p.18 © Look and Learn / The Bridgeman Art Library; p.19 © Anton Podoshvin / 123RF; p.20 © Stephen Finn – Fotolia; p.22 © MSPhotographic – Fotolia; p.26 © bytesurfer - Fotolia.com; p.27 © Volvo Car UK Ltd; p.28 © Peter Ginter / Peter Ginter / Superstock; p.29 © SondraP/ iStockphoto.com; p.33 © baranq - Fotolia.com; p.36 © Wally Stemberger; p.49 © Larry Brownstein/ Getty Images; p.50 t © SeanPavonePhoto - Fotolia.com, b © nikkytok - Fotolia.com; p.51 t © photobyjimshme - Fotloia.com, c © ermess - Fotolia.com, b © Aaron Kohr - Fotolia.com; p.52 © Nicky Rhodes - Fotolia.com; p.53 t © Reidos - Fotolia.com, b © Sanguis - Fotolia.com; p.54 © ia_64 - Fotolia.com; p.56 © Dariusz Kopestynski - Fotolia.com; p.58 © TREVOR CLIFFORD PHOTOGRAPHY/SCIENCE PHOTO LIBRARY; p.62 © Calek - Fotolia.com; p.66 © adisa - Fotolia.com; p.69 l © xalanx - Fotolia.com, r © Ricardo Reitmeyer / 123RF; p.75 ©TopFoto/ImageWorks; p.79 © JEAN-LOUP CHARMET/SCIENCE PHOTO LIBRARY; p.85 l © lucielang - Fotolia.com, r © Bondarau - Fotolia.com; p.92 © pljvv / 123RF; p.93 t © Junjie - Fotolia.com, b © MARTIN DOHRN/SCIENCE PHOTO LIBRARY; p.94 © Art Konovalov / Shutterstock; p.101 © Freefly - Fotolia.com; p.102 t © BABAK TAFRESHI/SCIENCE PHOTO LIBRARY, b © Jinx Photography Animals / Alamy; p.103 t © EUROPEAN SPACE AGENCY/SCIENCE PHOTO LIBRARY, c © FRANCK FIFE / Staff / Getty Images, b © Csák István - Fotolia.com; p.105 © Flying Colours Ltd / Getty Images; p.107 © GIPhotoStock/SCIENCE PHOTO LIBRARY; p.109 © Nick England; p.110 © Igor Kali - Fotolia.com; p.112 © BSIP SA / Alamy; p.115 © Grafvision - Fotolia.com; p.126 © ASHLEY COOPER/SCIENCE PHOTO LIBRARY; p.127 t © Zoe - Fotolia.com, b © pixeldigits / istock; p.131 t © Rob Wilkinson / Alamy Stock Photo, b © Alexander Erdbeer - Fotolia.com; p.134 © Nick England; p.136 © nevodka / 123RF; p.138 © AVAVA - Fotolia.com; p.141 t © Image Source Plus / Alamy, c © MGrushin / Megapixl.com, b © Stefan Scheer / Creative Commons; p.142 © Lovrencg - Fotolia.com; p.143 © Mikael Damkier - Fotolia.com; p.147 © Nigel Hicks / Alamy; p.148 © China photos / Contributor / Getty Images; p.149 t © Laurence Gaugh - Fotolia.com, b © Stephen Hill / 123RF; p.150 t © Saskia Massink / 123RF, b © kilukilu /Shutterstock; p.158 © Fotonanny - Fotolia.com; p.159 © DigiMagic Editorial / Alamy; p.160 © ISO400 - Fotolia.com; p.162 tr © Valuykin S. - Fotolia.com, tl © azure / Shutterstock, l © Minik - Fotolia.com, br © Nick England, bl © Gina Sanders - Fotolia.com; p.164 © frantisek hojdysz – Fotolia; p.167 l © Bradford Calkins / 123RF, c © Richard Lindie / 123RF, r © Dmitry Pichugin / 123RF; p.171 © OJPHOTOS / Alamy Live News; p.175 © Wajan - Fotolia.com; p.187 © Jay Pasachoff / Getty Images; p.188 © Stockbyte / Getty Images; p.189 © ALFRED PASIEKA/SCIENCE PHOTO LIBRARY; p.192 © Science and Society Picture Library / Getty; p.193 © sciencephotos / Alamy; p.195 l © Leslie Garland Picture Library / Alamy, r © Martin Bond/Science Photo Library; p.196 © SCIENCE PHOTO LIBRARY; p.197 © Nick England; p.199 © mario beauregard – Fotolia; p.201 YVES SOULABAILLE/LOOK AT SCIENCES/SCIENCE PHOTO LIBRARY; p.204 © Artem Merzlenko – Fotolia; p.207 © TebNad – Fotolia; p.218 © Mark Clifford / Barcroft Media / Getty Images; p.219 © Jack C. Schultz; p.221 © ullsteinbild / TopFoto; p.224 © Science Photo Library; p.226 © NASA / Harrison Schmitt; p.227 © Pamela Maxwell / 123RF; p.228 t © James King-Holmes/Science Photo Library, b © Ton Keone/Visuals Unlimited/Science Photo Library; p.230 t © Centre Oscar Lambret / Phanie / Rex Features, b © Martyn F. Chillmaid/Science Photo Library; p.232 © PUBLIC HEALTH ENGLAND/SCIENCE PHOTO LIBRARY; p.235 © Tim Wright/ Getty; p.238 © EFDA-JET/SCIENCE PHOTO LIBRARY; p.248 © neutronman – Fotolia; p.249 t © Dieter Willasch / NASA, b © Galaxy Picture Library / Alamy Stock Photo; p.250 l © David Brewster/Minneapolis Star Tribune/ZumaPress/Corbis, c © NASA / ESA / Hubble Heritage Team, r © NASA, b © Royal Observatory, Edinburgh/AATB/Science Photo Library; p.251 © Brent Hofacker – Fotolia; p.253 © European Southern Observatory (ESO) / Science Photo Library; p.255 © EUROPEAN SOUTHERN OBSERVATORY/Science Photo Library; p.256 l © European Southern Observatory (ESO), c © NASA / ESA, r © NASA; p.258 © NASA, ESA, J. Hester, A. Loll (ASU); p.259 © NASA/ESA/STScI/HUBBLE SM4 ERO TEAM/SCIENCE PHOTO LIBRARY; p.262 © NASA/ESA/STSCI/N.BENITEZ, JHU/ SCIENCE PHOTO LIBRARY, b © NASA / Bjarn Hindler; p.263 t © NASA, c © NASA / ESA & Hubble Heritage Team, b © NASA / ESA & Hubble Heritage Team; p.264 © HALE OBSERVATORIES/SCIENCE PHOTO LIBRARY; p.274 NASA / ESA & Hubble Heritage Team

1 Forces and motion

CONSIDER • DISCUSS

1 Can you suggest how many forces are acting on the surfer in the photograph?
2 How do they balance to keep him upright and moving forwards?

When moving in a straight line at a constant speed, balanced forces must act on this surfer.

By the end of this section you should:

- be able to plot and explain distance-time graphs
- know the relationship between average speed, distance and time
- know the relationship between acceleration, change in velocity and time taken
- be able to plot and explain velocity-time graphs, and use them to calculate an acceleration and the distance travelled
- be able to use the relationship between final speed, initial speed, acceleration and distanced moved
- describe the effects of forces that act on bodies
- be able to identify different types of force
- understand how vector quantities differ from scalar quantities
- be able to calculate the resultant of forces that act along a line
- know that friction is a force that opposes motion
- know and be able to use the relationship between unbalanced force, mass and acceleration
- know and be able to use the relationship between weight, mass and gravitational field strength
- know that the stopping distance of a car is the sum of the thinking and braking distances
- be able to describe the factors which affect vehicle stopping distance
- be able to describe the forces that act on falling objects, and explain why a falling object reaches a terminal velocity
- know Hooke's Law and be able to describe elastic behaviour
- know and be able to use the relationship between momentum, mass and velocity
- be able to use the idea of momentum to explain safety features
- be able to use the conservation of momentum to calculate mass, velocity and momentum
- be able to use the relationship between force, change in momentum and time taken
- understand Newton's Third Law
- know that moment = force × perpendicular distance from the pivot
- know that the weight of an object acts through its centre of gravity
- be able to use the principle of moments to analyse the forces acting in one plane.

1.1 How fast do things move?

Average speed

When you travel in a fast car you finish a journey in a shorter time than when you travel in a slower car. If the speed of a car is 100 kilometres per hour (100 km/h), it will travel a distance of 100 kilometres in one hour.

Figure 1.1 Hannah, Jenny and Natalia finish the 1500 m in 4 minutes and 5 seconds. What was their average speed?

Often, however, we use the equation to calculate an average speed because the speed of the car changes during the journey. When you travel along a motorway, your speed does not remain exactly the same. You slow down when you get stuck behind a lorry and speed up when you pull out to overtake a car.

The average speed of an object can be calculated using the equation:

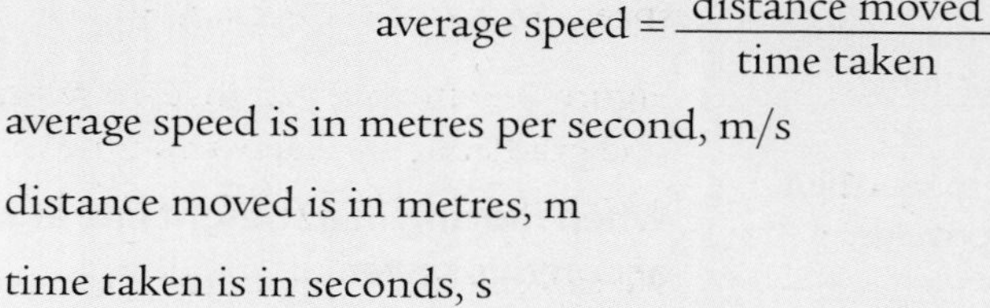

$$\text{average speed} = \frac{\text{distance moved}}{\text{time taken}}$$

average speed is in metres per second, m/s

distance moved is in metres, m

time taken is in seconds, s

Example. A train travels 440 km in 3 hours. Calculate its average speed in m/s.

To solve this problem, you need to remember that 1 km = 1000 m and that 1 hour = 3600 s.

$$\text{average speed} = \frac{\text{distance moved}}{\text{time taken}}$$

$$= \frac{440 \times 1000}{3 \times 3600}$$

$$= 41 \text{ m/s}$$

Speed and velocity

The word speed is defined in the previous paragraph. When we use the word **velocity** we are stating a speed in a certain direction – for example a car has a velocity of 20 m/s in an easterly direction. Often you will find that speed and velocity are used interchangeably. However, in International GCSE Physics we must be careful because sometimes direction is important.

Definition: **velocity** is a speed in a defined direction

Distance–time graphs

When an object moves along a straight line, we can represent how far it has moved using a distance–time graph. Figure 1.2 is a distance–time graph for a runner. He sets off slowly and travels 20 m in the first 10 seconds. He then speeds up and travels the next 20 m in 5 seconds.

We can calculate the speed of the runner using the gradient of the graph.

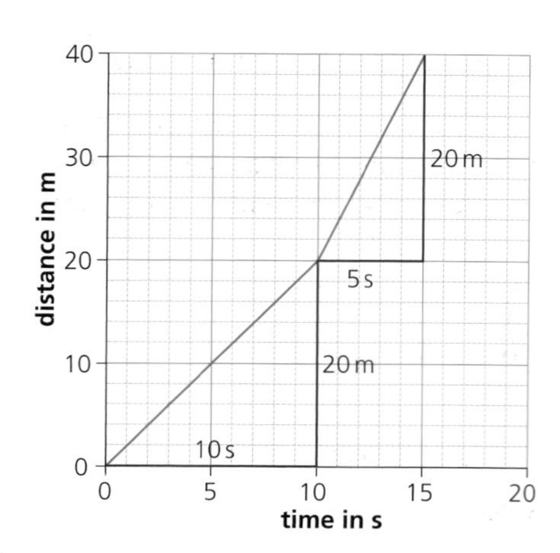

Figure 1.2

Example. Calculate the speed of the runner using the distance–time graph (Figure 1.2)

a) over the first 10 seconds,

b) over the time interval 10 s to 15 s.

$$\text{a) speed} = \frac{\text{distance}}{\text{time}}$$
$$= \frac{20\,\text{m}}{10\,\text{s}} = 2\,\text{m/s}$$

$$\text{b) speed} = \frac{40-20}{5}$$
$$= 4\,\text{m/s}$$

You can see from the graph that these speeds are the gradients of each part of the graph.

When you set off on a bicycle ride, it takes time for you to reach your top speed. You accelerate gradually.

Figure 1.3 shows a distance–time graph for a cyclist at the start of a ride. The gradient gets steeper as time increases. This tells us that her speed is increasing. So the cyclist is accelerating.

We can calculate the speed of the cyclist at any point by drawing a tangent to the curve, and then measuring the gradient.

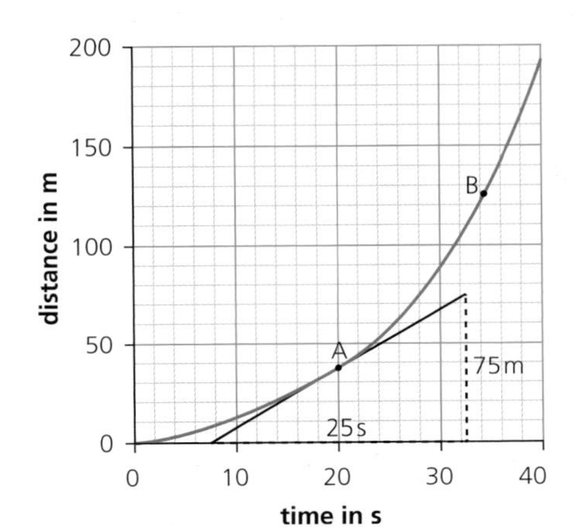

Figure 1.3

Example. On the graph in Figure 1.3, a gradient has been drawn at point A, 20 seconds after the start of the ride. Calculate the speed at this time.

$$\text{speed} = \text{gradient}$$
$$\text{speed} = \frac{75}{25}$$
$$= 3\,\text{m/s}$$

Investigating motion

The purpose of this experiment is for you to be able to determine the speed of an object or a person, by measuring how long it takes something (or someone) to cover a measured distance.

You can design your own experiments, or base yours on the ideas below.

Go outside and mark places at a separation of 2 m. A pupil walks slowly for 10 m, then walks quickly or runs for 10 m. His motion is recorded by ten students with stopwatches; they start their watches as the walker starts, and they record the time that the walker passes.

TIP

You should be familiar with an experiment to investigate the motion of everyday objects.

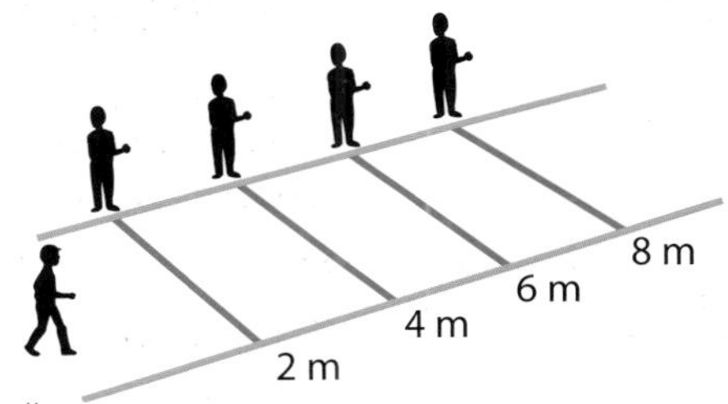

Figure 1.4 Timing a walker

Student	A	B	C	D	E	F	G	H	I	J
distance in m	2.0	4.0	6.0	8.0	10.0	12.0	14.0	16.0	18.0	20.0
time in s	1.75	3.60	5.33	7.08	8.80	10.13	11.70	12.79	14.16	15.45

1 Plot a graph of distance against time, draw the best straight lines through the points.
2 Which student was slow to react as the walker passed?
3 Use the graph to determine the walker's speed:
   a) over the first 10 m
   b) over the second 10 m.
4 Explain the significance of the gradient of the graph.

STUDY QUESTIONS

1 A helicopter flies from London to Paris in 2 hours, covering a distance of 300 km. Calculate the helicopter's speed in km/h.

2 Curtis cycles to school. Figure 1.5 shows the distance–time graph for his journey.

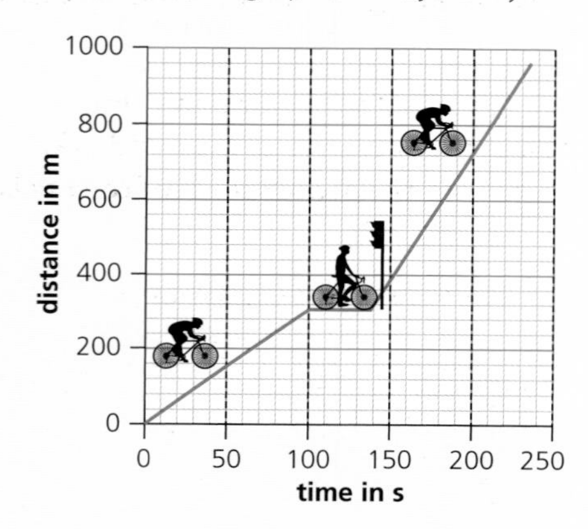

Figure 1.5

a) How long did Curtis stop at the traffic lights?
b) During which part of the journey was Curtis travelling fastest?

3 A car travels 100 m in a time of 5 s at a constant speed. Sketch a distance–time graph to show the motion of the car.

4 The table below shows average speeds and times recorded by top male athletes in several track events. Copy and complete the table.

Event	Average speed in m/s	Time
100 m		9.6 s
200 m	10.3	
400 m	8.9	
	7.1	3 m 30 s
10 000 m		29 m 10 s
	5.5	2 h 7 m 52 s

5 Ravi, Paul and Tina enter a 30 km road race. Figure 1.6 shows Ravi's and Paul's progress through the race.

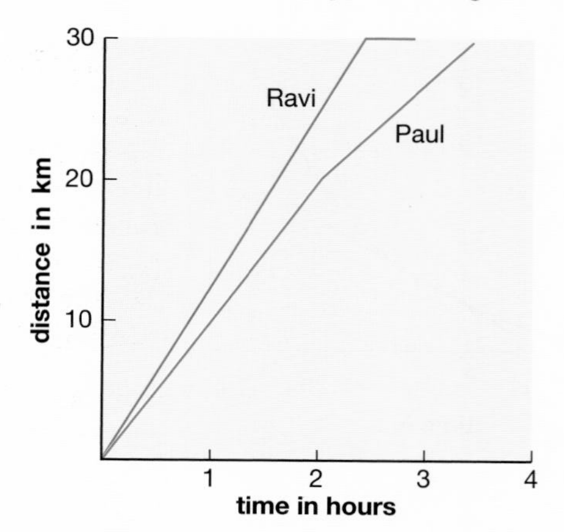

Figure 1.6

a) Which runner ran at a constant speed? Explain your answer.
b) Calculate Paul's average speed for the 30 km run.
c) What happened to Paul's speed after 2 hours?
   Tina was one hour late starting the race. During the race she ran at a constant speed of 15 km/h.
d) Copy the graph and add to it a line to show how Tina ran.
e) Determine how far Tina had run when she overtook Paul.

6 Sketch a graph of distance travelled (*y*-axis) against time (*x*-axis) for a train coming into a station. The train stops for a while at the station and then starts again.

7 Determine which of the answers below is the closest to the speed at point B of Figure 1.3.
   a) 4 m/s
   b) 10 m/s
   c) 15 m/s

1.2 Acceleration

Starting from the grid, a Formula 1 car reaches a speed of 30 m/s after 2 s. A flea can reach a speed of 1 m/s after 0.001 s. Which accelerates faster?

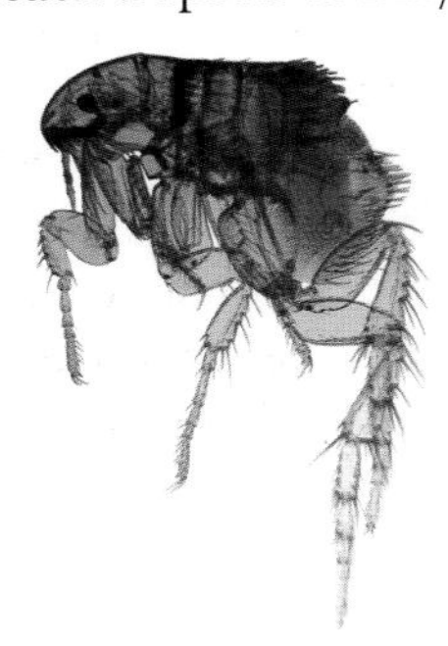

Figure 2.1

Figure 2.2

Speeding up and slowing down

When a car is speeding up, we say it is accelerating. When it is slowing down we say it is decelerating.

A car that accelerates rapidly reaches a high speed in a short time. For example, a car might speed up to 12.5 m/s in 5 seconds. A van could take twice as long, 10 s, to reach the same speed. So the acceleration of the car is twice as big as the van's acceleration.

You can calculate the acceleration of an object using the equation:

$$\text{acceleration} = \frac{\text{change in velocity}}{\text{time taken}}$$

$$a = \frac{v - u}{t}$$

where acceleration, a, is in metres per second squared, m/s^2

change in velocity ($v - u$) is in metres per second, m/s

v is the final velocity after something has accelerated and u is the velocity before the acceleration.

time, t, taken is in seconds, s

TIP
The units of acceleration are m/s^2

$$\text{acceleration of car} = \frac{12.5}{5} = 2.5\ m/s^2$$

$$\text{acceleration of van} = \frac{12.5}{10} = 1.25\ m/s^2$$

Velocity–time graphs

It can be helpful to plot graphs of velocity against time.

Figure 2.3 shows the velocity–time graph for a cyclist as she goes on a short journey.

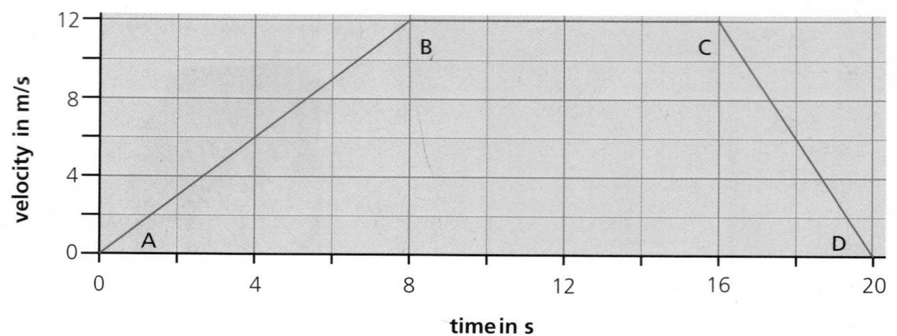

Figure 2.3

- In the first 8 seconds, she accelerates up to a speed of 12 m/s (section AB of the graph).
- For the next 8 seconds, she cycles at a constant speed (section BC of the graph).
- Then for the last 4 seconds of the journey, she decelerates to a stop (section CD of the graph).

The gradient of the graph gives us the acceleration. In section AB she increases her speed by 12 m/s in 8 seconds.

$$\text{acceleration} = \frac{\text{change of velocity}}{\text{time}}$$

$$= \frac{12\,\text{m/s}}{8\,\text{s}}$$

$$= 1.5\,\text{m/s}^2$$

You can also work out the distance travelled by calculating the area under the velocity–time graph. The area under section BC gives the distance travelled because:

$$\text{distance} = \text{average speed} \times \text{time}$$

$$= 12\,\text{m/s} \times 8\,\text{s}$$

$$= 96\,\text{m}$$

TIP

The area under a velocity-time graph is the distance travelled.

The distance travelled in the first 8 seconds, the region AB, is calculated using the formula:

$$\text{area of triangle} = \tfrac{1}{2} \times \text{base} \times \text{height}$$

$$= \tfrac{1}{2} \times 8\,\text{s} \times 12\,\text{m/s}$$

$$= 48\,\text{m}$$

TIP

The gradient of a velocity-time graph is the acceleration.

STUDY QUESTIONS

1 Which is the correct unit for acceleration?

ms^2 ms m/s m/s^2

2 Explain the meaning of the words:

a) acceleration

b) deceleration.

3 a) Write down the equation which links acceleration to a change of velocity and time.

b) Use the information at the top of page 5 to calculate the acceleration of:

i) the F1 car

ii) the flea

4 This question refers to the journey shown in Figure 2.3.

a) Calculate the cyclist's deceleration over region CD of the graph.

b) Use the area under the graph to calculate the distance covered during the whole journey.

c) Calculate the average speed over the whole journey.

5 The table shows how the speed, in m/s, of a Formula 1 racing car changes as it accelerates away from the starting grid at the beginning of a Grand Prix.

a) Plot a graph of speed (*y*-axis) against time (*x*-axis).

b) Use your graph to calculate the acceleration of the car at:

i) 16 s

ii) 1 s.

c) Calculate how far the car has travelled after 16 s.

Table 1

Speed in m/s	0	10	20	36	49	57	64	69	72	72
Time in s	0	1	2	4	6	8	10	12	14	16

6 Drag cars are designed to cover distances of 400 m in about 6 seconds. During this time the cars accelerate very rapidly from a standing start. At the end of 6 seconds, a drag car reaches a speed of 150 m/s.

a) Calculate the drag car's average speed.

b) Calculate its average acceleration.

7 Copy the table and fill in the missing values.

Table 2

	Starting speed in m/s	**Final speed in m/s**	**Time taken in s**	**Acceleration in m/s²**
Cheetah	0		5	6
Train	13	25	120	
Aircraft taking off	0		30	2
Car crash	30	0		−150

1.3 Observing and calculating motion

Figure 3.1 This high-speed photograph of the athlete's long jump enables the coach to work on style, strength and distance, and helps the athlete to improve technique.

An athlete's trainer used to watch them and time their motion with a stopwatch. Now motion can be tracked with a GPS tracker in great detail. Movement can also be analysed by high speed multiframed photography.

The following practicals show two further ways of analysing motion.

Light gates

The speed of a moving object can also be measured using light gates. Figure 3.2 shows an experiment to determine the acceleration of a rolling ball as it passes between two light gates. When the ball passes through a light gate, it cuts a beam of light. This allows the computer to measure the time taken by the ball to pass through the gate. By knowing the diameter of the ball, the speed of the ball at each gate can be calculated. You can tell if the ball is accelerating if it speeds up between light gates A and B. Follow the questions below to show how the gates can be programmed to do the work for you. The measurements taken in an experiment are shown here.

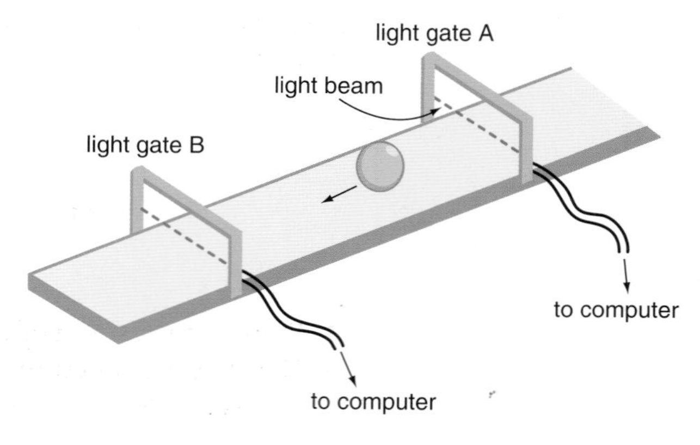

Figure 3.2

Diameter of the ball	6.2 cm
Time for the ball to go through gate A	0.12 s
Time for the ball to go through gate B	0.09 s
Time taken for the ball to travel from gate A to gate B	0.23 s

1 Explain why it is important to adjust the light gates to the correct height.
2 Calculate the speed of the ball as it goes through:
   a) gate A
   b) gate B.
3 Calculate the ball's acceleration as it moves from gate A to gate B.

Ticker timer

Changing speeds and accelerations of objects in the laboratory can be measured directly using light gates, data loggers and computers. However, motion is still studied using the ticker timer (Figure 3.3), because it collects data in a clear way, which can be usefully analysed. A ticker timer has a small hammer that vibrates up and down 50 times per second. The hammer hits a piece of carbon paper, which leaves a mark on a length of tape.

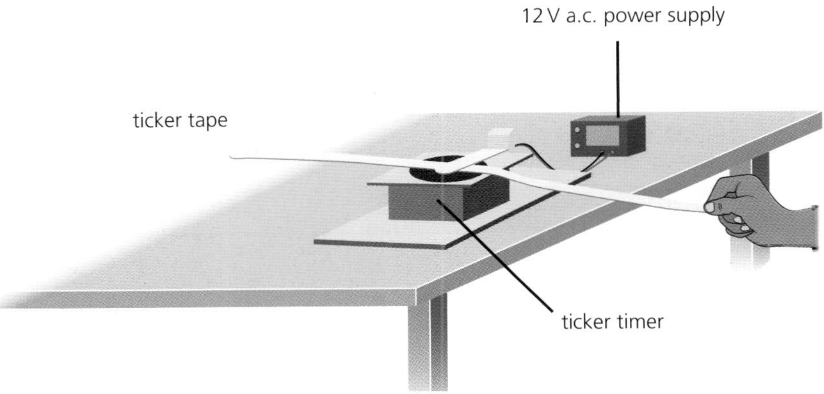

Figure 3.3

0.5 cm
1 cm
P Q R S

Figure 3.4

Figure 3.4 shows you a tape that has been pulled through the timer. You can see that the dots are close together over the region PQ. Then the dots get further apart, so the object moved faster over QR. The movement slowed down again over the last part of the tape, RS. Since the timer produces 50 dots per second, the time between dots is 1/50 s or 0.02 s. So we can work out the speed:

$$\text{speed} = \frac{\text{distance between dots}}{\text{time between dots}}$$

$$\text{Between P and Q: speed} = \frac{0.5\ \text{cm}}{0.02\ \text{s}}$$

$$= 25\ \text{cm/s or } 0.25\ \text{m/s}$$

1 Work out the speed of the tape in the region QR.
2 Which is the closest to the speed in the region RS?

0.1 m/s   0.3 m/s   0.8 m/s

Equation of Motion

When an object accelerates in a straight line, the final speed, initial speed, the acceleration and the distance travelled may be connected by the following equation.

(final speed)^2 = (initial speed)^2 + 2 × acceleration × distance moved

$$v^2 = u^2 + 2as$$

where v is the final velocity in metres per second, m/s

u is the initial (starting) velocity in metres per second, m/s

a is the acceleration in metres per second squared, m/s^2

s is the distance in metres, m.

Provided the value of three of the quantities is known, the equation can be rearranged to calculate the unknown quantity.

Example. The second stage of a rocket accelerates at 3 m/s^2. This causes the velocity of the rocket to increase from 450 m/s to 750 m/s. Calculate the distance the rocket travels while it is accelerating.

Answer.

$v = 750\,\text{m/s},\ u = 450\,\text{m/s},\ a = 3\,\text{m/s}^2$

$v^2 - u^2 = 2as$

$750^2 - 450^2 = 2 \times 3 \times s$

$562\,500 - 202\,500 = 6 \times s$

$s = \dfrac{360\,000}{6} = 60\,000\,\text{m}$

STUDY QUESTIONS

1 A model car is rolled down a slope in a laboratory. Describe apparatus and explain how you would use that apparatus to show that the car is accelerating.

2 The graph shows how the velocity of a jet aircraft increases as it takes off from the deck of an aircraft carrier.

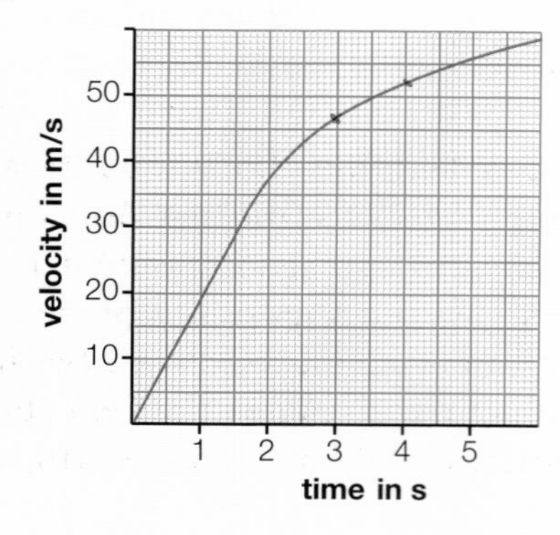

Calculate its acceleration

a) during the first second

b) between 3 and 4 seconds.

3 An aeroplane can accelerate at a rate of 2.5 m/s^2. The plane takes off when it speed reaches 60 m/s. Calculate the minimum length of runway that the plane needs to take off.

1.4 Introducing forces

Figure 4.1

Figure 4.2

A force is a push or a pull. The shot putter pushes the shot and the archer pulls the string on the bow. Forces are measured in newtons, N. Where does the name come from?

What is a force?

A force is a push or a pull. Whenever you push or pull something you are exerting a force on it. The forces that you exert can cause three things:

- **You can change the shape of an object.** You can stretch or squash a spring. You can bend or break a ruler.
- **You can change the speed of an object.** You can increase the speed of a ball when you throw it. You decrease its speed when you catch it.
- **A force can also change the direction in which something is travelling.** We use a steering wheel to turn a car.

The forces described so far are called **contact forces**. Your hand touches something to exert a force. There are also **non-contact forces**. Gravitational, magnetic and electric forces are non-contact forces. These forces can act over large distances without two objects touching. The Earth pulls you down whether or not your feet are on the ground. Although the Earth is 150 million km away from the Sun, the Sun's gravitational pull keeps us in orbit around it. Magnets also exert forces on each other without coming into contact. Electrostatic forces act between charged objects.

The size of forces

The unit we use to measure force is the **newton** (N). The box in the margin will help you to get the feel of the size of several forces.

Large forces can be measured in kilonewtons, kN.

$$1000 \text{ N} = 1 \text{ kN}$$

TIP

The size of some forces

- The pull of gravity on a fly = 0.001 N
- The pull of gravity on an apple = 1 N
- The frictional force slowing a rolling football = 2 N
- The force required to squash an egg = 50 N
- The pull of gravity on a 50 kg student = 500 N
- The tension in a rope, towing a car = 1000 N
- The braking force on a car = 5000 N

Vectors and scalars

Vector quantities have both size and direction. Scalar quantities only have size. Speed is a scalar because we only define how fast something is moving. Velocity is a vector quantity because we should define both a size and a direction. Figure 4.3 shows the importance of direction: a helicopter can fly at 150 km/h, it can reach one of three cities in two hours, depending on the direction of travel. When the helicopter flies to Brussels from London, its velocity is 150 km/h on a bearing of 110°.

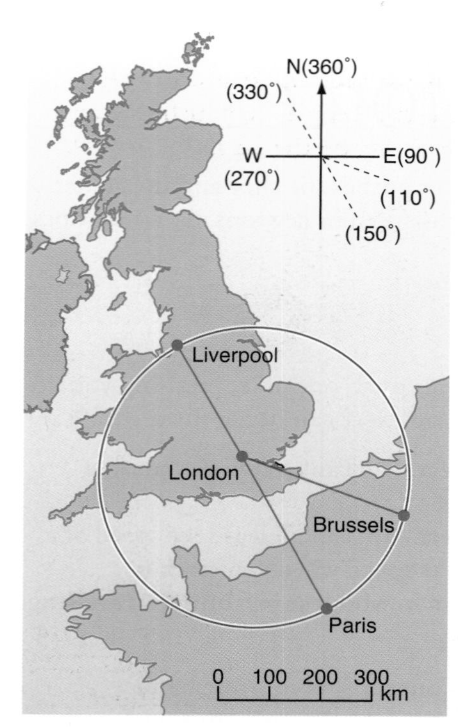

Figure 4.3 At the same speed, in 2 hours a helicopter from London can reach Liverpool, Brussels or Paris.

Force is an example of a **vector** quantity. Vector quantities have both size and direction. Other examples of vectors are: velocity (the wind blows at 50 km/h from the North); displacement (a car travels 20 km due East). A quantity that has only size is a **scalar** quantity. Some examples of scalar quantities are: mass (3 kg of potatoes); temperature (20 °C); energy (100 joules).

Some important forces

Weight is the name that we give to the pull of gravity on an object. Near the Earth's surface the pull of gravity is approximately 10 N on each kilogram.

Tension is the name given to a force that acts through a stretched rope; when two teams pull on a rope it is under tension.

Friction is the contact force that slows down moving things. Friction can also prevent stationary things from starting to move when other forces act on them.

Drawing forces

Figure 4.4 shows two examples of forces acting on Michael: (a) his weight (the pull of gravity on him) is 800 N; (b) a rope with a tension of 150 N pulls him forwards.

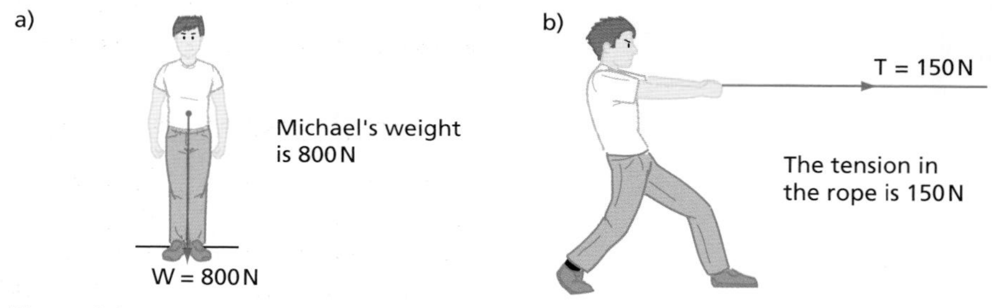

Figure 4.4

It is usual for more than one force to act on something. Then we must show all the forces acting. When Michael is pulled by the rope (Figure 4.5), his weight still acts on him, and the floor supports him too – if the floor did not exert an upwards force on him equal to his weight, he would be falling downwards. The force is called the floor's **normal contact force**, R. The floor will also exert a frictional force on him, in the opposite direction to that in which he is moving. All these forces are shown together in Figure 4.5.

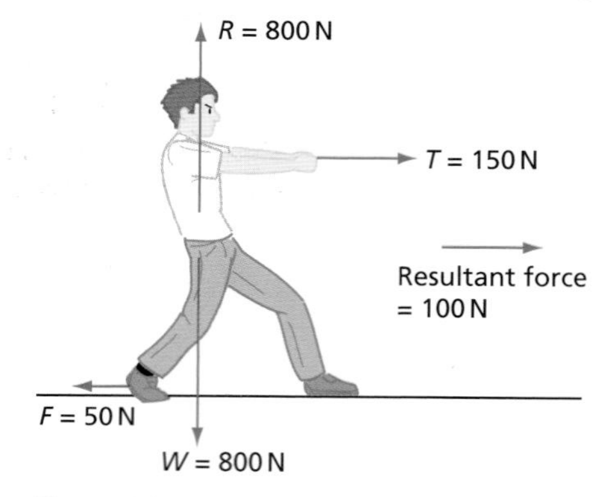

Figure 4.5

Adding forces

When two forces act in the same direction, they add up to give a larger **resultant** force. In Figure 4.6, for example, two people each push the car with a force of 300 N. The resultant force acting on the car is now 600 N.

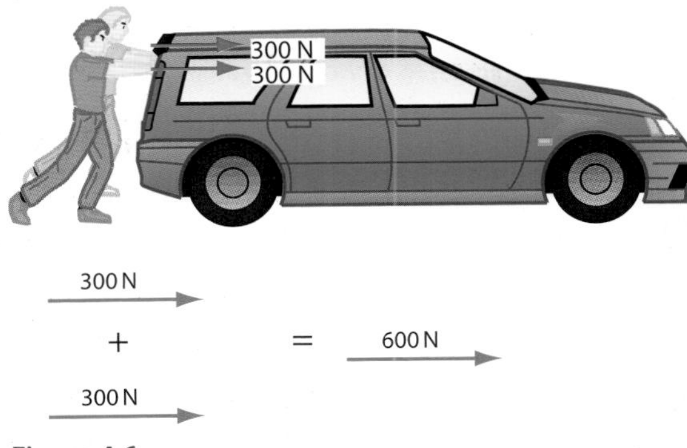

Figure 4.6

If forces act in opposite directions they may cancel each other out. In Figure 4.5 Michael's weight, which pulls him downwards, is balanced by the upwards force from the floor. The resultant force is zero; 800 N - 800 N. We say that these forces are **balanced**. Michael therefore stays on the floor.

There are other forces that act on Michael; the pull from the rope to the right is 150 N, but the frictional force to the left is 50 N. The resultant horizontal force on Michael is therefore 150 N – 50 N = 100 N, to the right.

Resultant force and state of motion

The diagrams in Figure 4.7 show two more examples of how forces add up along a line. In Figure 4.7(a) the dolphin has adjusted his buoyancy so that the upthrust from the water balances his weight; he can now stay at the same depth. In Figure 4.7(b), a car is moving forwards and increases its speed. The push from the engine (900 N) is greater than the wind resistance (700 N), so there is an unbalanced force of 200 N forwards. The car accelerates forwards.

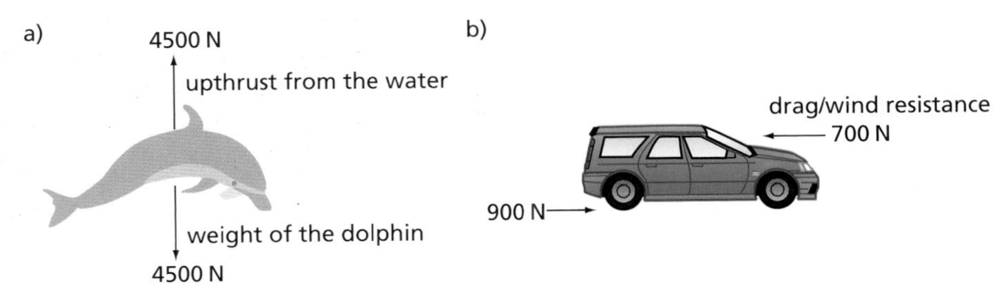

Figure 4.7 **(a)** The forces balance so the resultant force on the dolphin is zero.
(b) The resultant force on the car is 200 N to the right. This means the car increases its speed.

STUDY QUESTIONS

1 a) Give three examples of forces that are pulls, and three examples that are pushes.

b) For each of the examples of forces you have given, state an approximate value for the size of the force.

2 A fisherman has caught a large fish and has to use two balances to weigh it. Look at the diagram to calculate its weight.

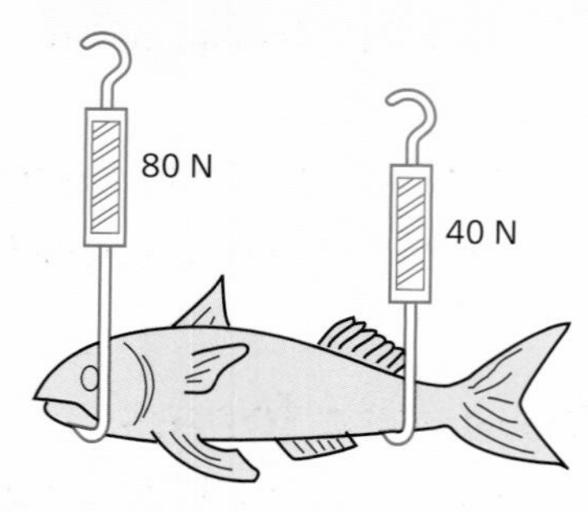

3 A golf ball is hit off its tee, 200 m down the fairway. Draw diagrams to show all the forces acting on the ball:

a) when the ball rests on the tee

b) while the club strikes the ball on the tee

c) as the ball is in flight.

4 For each of the diagrams state the size and direction of the resultant force.

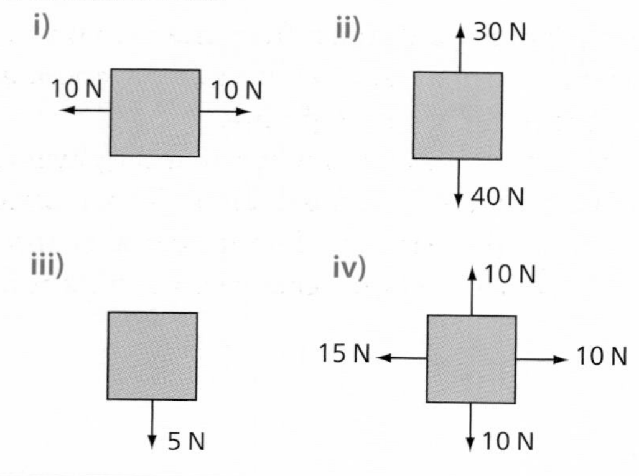

5 Frank's weight is 800 N and his bike's weight is 2500 N. Determine the upwards force exerted by the road on his rear wheel.

6 Calculate the resultant force on this rocket.

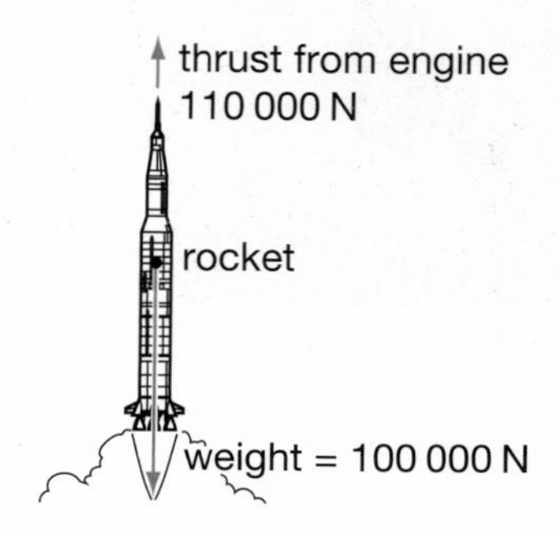

1.5 Forces, acceleration and Newton's laws of motion

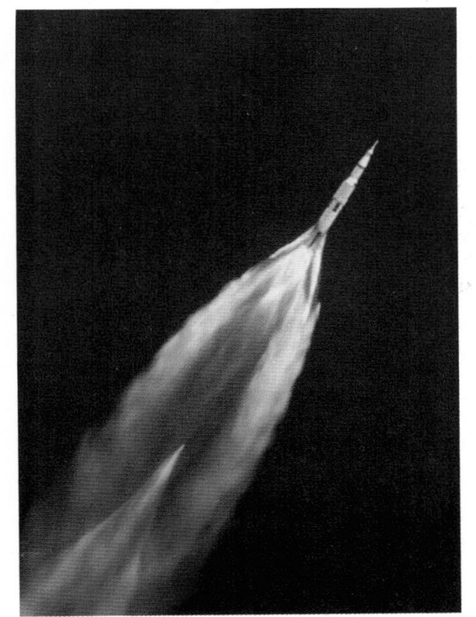

Figure 5.1 The *Saturn V* rocket on its way into space.

Newton's laws of motion allowed scientists to calculate how to land a spacecraft on the Moon.

Can you state three equations that they used in their calculations?

What important device do you use every day that was developed as a result of the NASA space programme?

Newton's first law: balanced forces

When the resultant force acting on an object is zero, the forces are balanced and the object does not accelerate. It remains stationary, or continues to move in a straight line at a constant speed.

Figure 5.2 and 5.3 show some examples where the resultant force is zero.

- A person is standing still. Two forces act on him: his weight downwards and the normal contact force from the floor upwards. The forces balance; he remains stationary.
- A car moves along the road. The forwards push from the road on the car is balanced by the air resistance on the car. The forces are balanced and so the car moves with a constant speed in a straight line.
- A spacecraft is in outer space, so far away from any star that the gravitational force is zero. There are no frictional forces. So the resultant force is zero. The spacecraft is either at rest or moving in a straight line at a constant speed.

The speed and/or direction of an object will only change if a resultant force acts on the object.

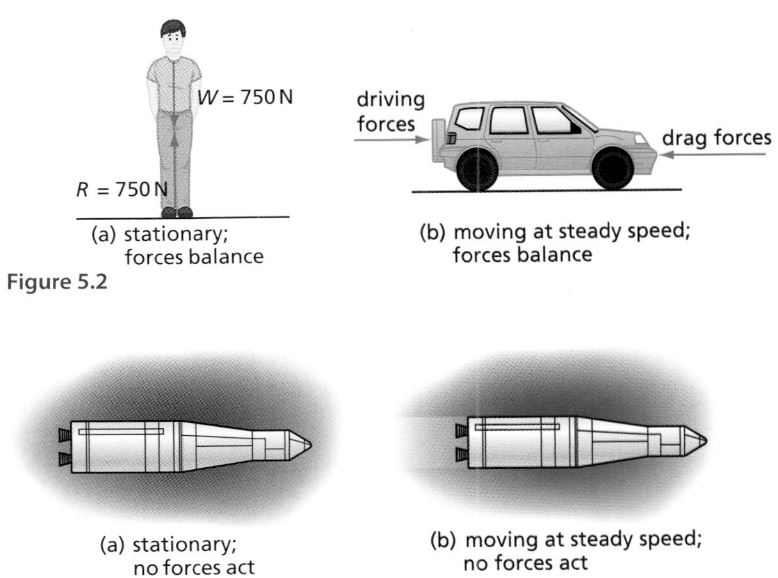

Figure 5.2

Figure 5.3

Newton's second law: unbalanced forces

When an unbalanced force acts on an object it accelerates. The object could speed up, slow down or change direction. Figure 5.4 shows two examples of unbalanced forces acting on a body.

- The spacecraft has turned its engine on. There is a force pushing the craft forwards, so it accelerates.
- The driver has taken his foot off the accelerator while the car is moving forwards. There is an air resistance force that acts to decelerate the car.

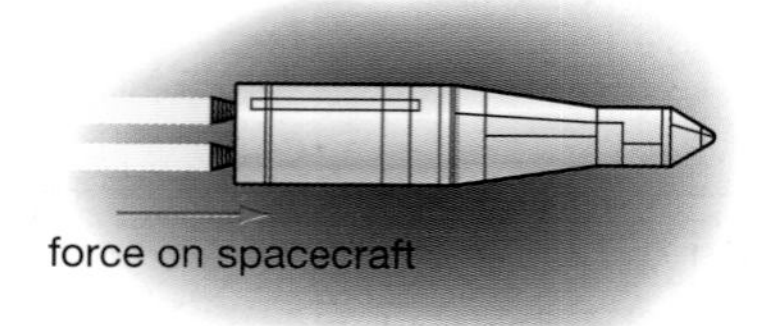

a) acceleration

b) deceleration

Figure 5.4

Force, mass and acceleration

You may have seen people pushing a car with a flat battery. When one person tries to push a car, the acceleration is very slow. When three people give the car a push, it accelerates more quickly.

You will know from experience that large objects are difficult to get moving. When you throw a ball you can accelerate your arm more quickly if the ball has a small mass. You can throw a tennis ball much faster than you put a shot. A shot has a mass of about 7 kg so your arm cannot apply a force large enough to accelerate it as rapidly as a tennis ball.

Newton's second law states that

- acceleration is proportional to the resultant force
  $a \propto F$
- acceleration is inversely proportional to the mass
  $a \propto \frac{1}{m}$

This can be written as an equation:

$$F = ma$$

resultant force = mass × acceleration

where force is in newtons, N

mass is in kilograms, kg

acceleration is in metres per second squared, m/s^2

TIP

When using $F = ma$, F means resultant force.

Example. The mass of the car in Figure 5.5 is 1200 kg. Calculate the acceleration.

$$F = ma$$

$$1000 - 640 = 1200 \times a$$

$$a = \frac{360}{1200}$$

$$= 0.3\ m/s^2$$

Figure 5.5

TIP

When you use the equation $F = ma$, the force must be in N, the mass in kg and the acceleration in m/s^2

Note

- First we had to work out the resultant force.
- The acceleration is in the same direction as the resultant fo... speeds up or accelerates.

STUDY QUESTIONS

1 a) A car is stationary on some ice on a road. Explain why the car might have difficulty starting to move on ice.

b) A sprinter fixes blocks into the ground to help him get a good start. Explain how these assist him.

2 You leave a parcel on the seat of a car. When you brake suddenly, the parcel falls onto the floor. Explain why.

3 The diagram shows the direction of a force on a model car. Which of the following is a possible state of motion for the car? Explain your answers.

a) Staying at rest.

b) Beginning to move backwards.

c) Moving backwards at a constant speed.

d) Slowing down while moving forwards.

4 Draw all the forces that act on a plane

a) just after take off

b) while flying at a constant speed at a constant height.

5 Trains accelerate slowly out of stations. Why is their acceleration slower than that of a car?

6 You are in a spacecraft in a region where there is no gravitational pull. You have two biscuit tins, one of which has no biscuits (because you have eaten them) and the other is full. How can you tell which is full, without opening the lid?

7 a) Calculate the acceleration of a mass of 3 kg, which experiences a resultant force of 15 N.

b) Calculate the mass of an object that experiences an acceleration of 4 m/s^2 when a resultant force of 10 N acts on it.

8 The figure shows an experimental arrangement for measuring the acceleration of a trolley and then investigating how the acceleration depends on the applied force.

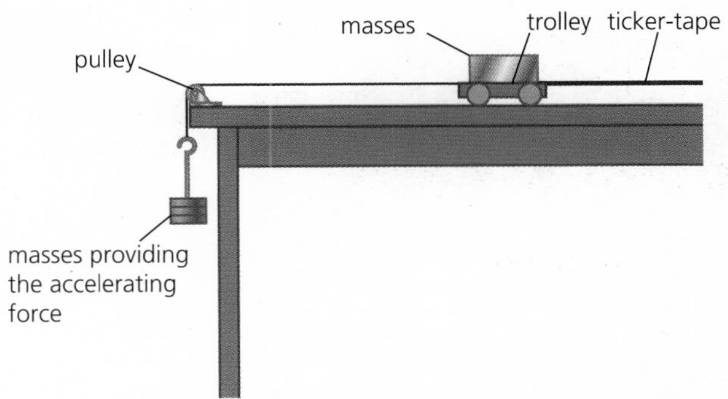

a) The slope of the table is adjusted to compensate for friction. How is this done? Explain why this is important.

In the diagram, the trolley has a mass of 0.7 kg and the three masses each have a mass of 0.1 kg.

b) i) State the accelerating force on the trolley and masses on the string.

ii) State the total mass being accelerated by that force.

iii) Calculate the acceleration you would be expect the masses and the trolley to have when released.

When a student accelerates the trolley, she uses a ticker tape to analyse its motion. She discovers that the trolley accelerates over a distance of 29.4 cm in a time of 0.44 s.

c) i) Calculate the average speed of the trolley over the distance travelled.

ii) State the final speed of the trolley after it has travelled 29.4 cm. Why is this twice the average speed?

iii) Now calculate the acceleration of the trolley.

d) i) Explain how you would use the weights on the trolley to vary the accelerating force, while keeping the mass accelerated constant.

ii) Design a second experiment to investigate how the mass of an object affects its acceleration for a constant applied force.

1.6 Applying forces

Figure 6.1 Galileo demonstrating to the crowds from the Leaning Tower of Pisa.

When you drop something it accelerates downwards, moving faster and faster until it hits the ground. In 1589 Galileo demonstrated to the crowds in Pisa that objects of different masses fall to the ground at the same rate. He dropped a large iron cannon ball and a small one. Both balls hit the ground at the same time. They accelerated at the same rate of about 10 m/s^2. Aristotle, the Greek philosopher, had previously taught that heavier objects fall more quickly.

Weight and mass

The size of the Earth's gravitational pull on an object is proportional to its mass. The Earth pulls a 1 kg mass with a force of 10 N and a 2 kg mass with a force of 20 N. We say that the strength of the **Earth's gravitational field**, g, is 10 N/kg.

The **weight**, W, of an object is the force that gravity exerts on it, which is equal to the object's mass × the pull of gravity on each kilogram.

$$W = mg$$

The value of g is roughly the same everywhere on the Earth, but away from the Earth it has different values. The Moon is smaller than the Earth and pulls things towards it less strongly. On the Moon's surface the value of g is 1.6 N/kg. In space, far away from all stars and planets, there are no gravitational pulls, so g is zero, and therefore everything is weightless.

Example. Calculate the weight of a 70 kg man on the Moon.

$$\begin{aligned} W &= mg \\ &= 70\text{ kg} \times 1.6\text{ N/kg} \\ &= 112\text{ N} \end{aligned}$$

The size of g also gives us the **gravitational acceleration**, because:

$$\text{acceleration} = \frac{\text{force}}{\text{mass}} \text{ or } g = \frac{W}{m}$$

The difference between mass and weight

Weight is the pull of gravity on an object. Weight is a force measured in newtons, N. The weight of an object depends on the gravitational field strength of a planet.

The mass of an object is always the same anywhere and is not affected by a planet's gravitational pull; it depends on the amount of matter in the object. Mass is measured in kilograms, kg.

Mass is also a measure of how difficult it is to accelerate an object.

Since acceleration = $\frac{\text{force}}{\text{mass}}$ a large mass accelerates more slowly than a small mass, when the same force is applied to each.

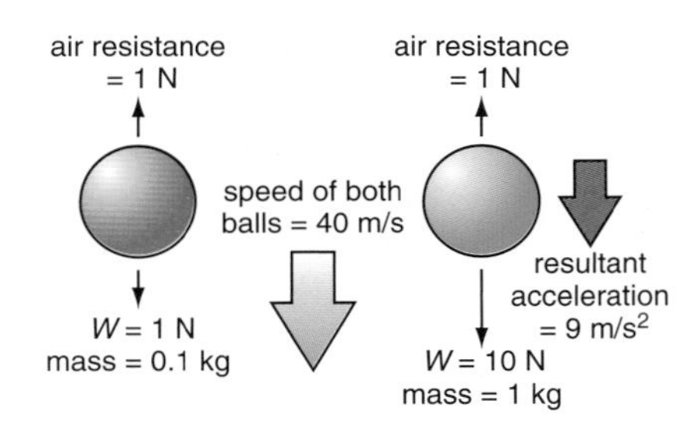

Figure 6.2 At this instant both balls have a speed of 40 m/s. At this speed the weight of the red ball is balanced by air resistance, but the heavier blue ball is still accelerating.

Falling and parachuting

You have read above that everything accelerates towards the ground at the same rate. But that is only true if the effects of air resistance are small. If you drop a feather you know that it will flutter slowly towards the ground. That is because the size of the air resistance on the feather is only slightly less than the downwards pull of gravity.

The size of the air resistance on an object depends on the area of the object and its speed:

- The larger the area, the larger the air resistance.
- The larger the speed, the larger the air resistance.

- mass
- area

Figure 6.2 shows the effect of air resistance on two balls, which are the same size and shape, but the red ball has a mass of 0.1 kg and the blue ball a mass of 1 kg. The balls are moving at the same speed, so the air resistance is the same, 1 N, on each. The pull of gravity on the red ball is balanced by air resistance, so it now moves at a constant speed. It will not go any faster and we say it has reached **terminal velocity**. For the blue ball, however, the pull of gravity is greater than the air resistance so it continues to accelerate.

Figure 6.3 The sky diver has reached terminal velocity. How could the skydiver slow down before opening the parachute?

Figure 6.4 shows how the speed of a sky diver changes as she falls towards the ground. The graph has five distinct parts:

1. OA. She accelerates at about 10 m/s^2 just after leaving the aeroplane.
2. AB. The effects of air resistance mean that her acceleration gets less as there is now a force acting in the opposite direction to her weight.
3. BC. The air resistance force is the same as her weight. She now moves at a constant speed because the resultant force acting on her is zero.
4. CD. She opens her parachute at C. There is now a very large air resistance force so she decelerates rapidly.
5. DE. The air resistance force on her parachute is the same size as her weight, so she falls at constant speed until she hits the ground at E.

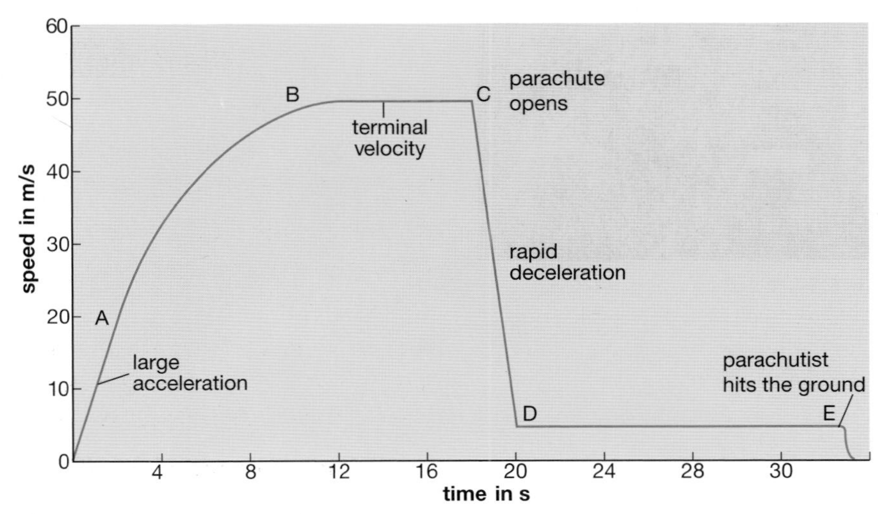

Figure 6.4 Speed–time graph for a parachutist.

Driving safely – 1

When you learn to drive, the most important thing you must understand is not how to start, but how to stop. Coming to a halt safely is vital for your own and others' well-being.

Stopping distance

When you are driving you should be aware of your **stopping distance** for a particular speed. The stopping distance is the sum of your **thinking distance** and **braking distance**. The thinking distance is the distance the car travels while you react to a hazard ahead – it takes time for you to take your foot off the accelerator and apply the brake.

$$\text{thinking distance} = \text{speed} \times \text{reaction time}$$

If you are travelling at 13 m/s, which is just less than 50 km/h (a typical speed in a city) and your reaction time is 0.7 s, your thinking distance will be:

$$\begin{aligned}\text{thinking distance} &= 13\ \text{m/s} \times 0.7\ \text{s} \\ &= 9.1\ \text{m}\end{aligned}$$

You will have travelled about 9 m before you begin to brake.

What affects your reaction time?

You will react more slowly if you are tired, so you should take a break from driving every few hours. Some medicines might make you drowsy, so you should not drive if you are taking these medicines. It is illegal to drive under the influence of alcohol or drugs, both of which slow your reactions and impair your judgement. Your reaction time is longer if you are distracted by using your phone or changing your music.

Figure 6.5 Why are variable speed limits imposed on motorways during rush hours?

Braking distance

Your braking distance depends on your speed. Table 1 shows typical braking distances for a saloon car. There is a gap in the second column, which you will be able to fill in when you answer Study Question 7.

Table 1

Speed in m/s	Braking distance in m
5	2
10	8
15	17
20	31
25	
30	69
35	94

Certain factors can increase your braking distance:

- Icy or wet roads can reduce the grip on the tyres.
- Gravel, mud or oil on the road can reduce the grip on the tyres.
- Worn tyres can reduce their grip on the road.

- Worn brake pads can reduce the braking force.
- A heavily laden car means there is more mass to slow, so the deceleration is less and the braking distance increased.
- Your braking distance will be longer if you are going downhill, as gravity is pulling the car forwards.

Visibility

It is dangerous to drive fast when the visibility is restricted by heavy rain, or mist or fog. You need to make sure that you are able to stop within the distance you can see ahead.

STUDY QUESTIONS

1 A student wrote 'my weight is 67 kg'. What is wrong with this statement, and what do you think his weight really is?

2 A hammer has a mass of 1 kg. What is its weight **a)** on Earth, **b)** on the Moon, **c)** in outer space?

3 Explain this observation: 'when a sheet of paper is dropped it flutters down to the ground, but when the same sheet of paper is screwed up into a ball it accelerates rapidly downwards when dropped'.

4 Refer to Figure 6.2 and explain the following.
  a) Why is the red ball falling at a constant speed?
  b) Why does the blue ball fall with an acceleration of 9 m/s^2?

5 This question refers to the speed–time graph in Figure 6.4.
  a) Determine the speed of the sky diver when she hit the ground.
  b) Why is her acceleration over the part AB less than it was at the beginning of her fall?
  c) Use the graph to estimate roughly how far she fell during her dive. Was it nearer 100 m, 1000 m, or 10 000 m?
  d) Use the information in the graph at point E to make a rough estimate of the acceleration on landing.

6 The graph (right) shows how the force of air resistance on our sky diver's parachute changes with her speed of fall.
  a) State the resistive force acting on her when she is travelling at a constant speed of 5 m/s.
  b) Explain why your answer to part **a)** must be the same size as her weight.
  c) Use the graph to predict the terminal velocity of the following people using the same parachute:
    i) a boy of weight 400 N
    ii) a man of weight 1000 N.
  d) Make a copy of the graph and add to it a sketch to show how you think the air resistance force would vary on a parachute with twice the area of the one used by our sky diver.

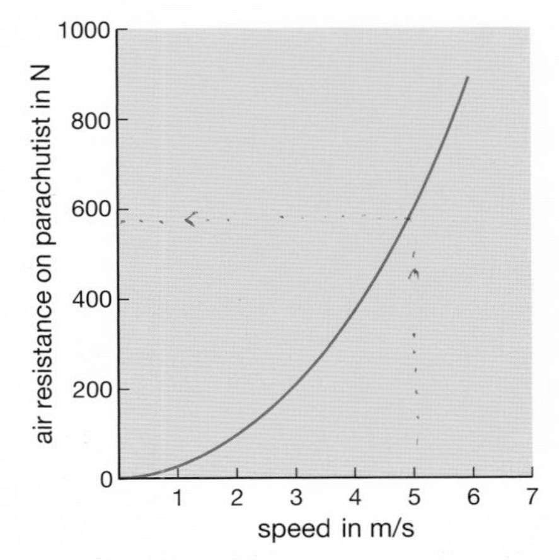

7 List three features that affect:
  a) The thinking distance for a driver
  b) The breaking distance of a car.

8 Copy Table 1 on page 20 and add to it two further columns headed 'thinking distance' and 'stopping distance'.
  a) Calculate the thinking distances for the speeds listed, assuming your reaction time is 0.6 s.
  b) Now fill in the stopping distances where you can.
  c) Plot a graph to show the stopping distances (*y*-axis) against the speed (*x*-axis).
  d) Use your graph to predict the stopping distance for a speed of 25 m/s. Now also fill in the braking distance for a speed of 25 m/s.
  e) Calculate the maximum safe speed if the visibility is limited to 50 m.

1.7 Stretching

Our muscles exert forces when they are stretched. An exercise rope helps this woman stretch and tone up her muscles. On most occasions our muscles behave elastically. What does this mean? Describe what happens to a muscle when too much force is applied to it.

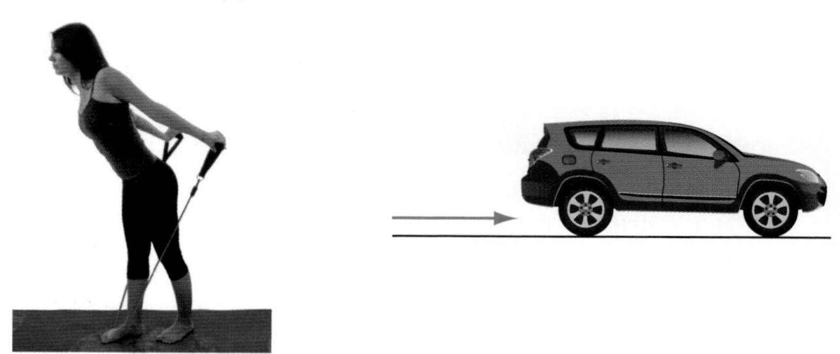

Figure 7.1

Figure 7.2 This car accelerates.

If one force only is applied to an object, for example a car, then it will change speed or direction. If we want to change the shape of an object, we have to apply more than one force to it.

Figure 7.3 shows some examples of how balanced forces can change the shape of some objects. Because the forces balance, the objects remain stationary.

- Two balanced forces can stretch a spring.
- Two balanced forces can compress a beam.
- Three balanced forces cause a beam to bend.

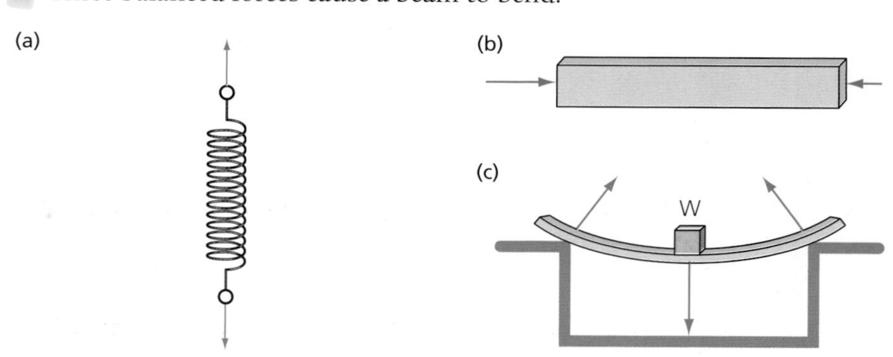

Figure 7.3 Balanced forces can change the shape of some objects

Sometimes when an object has been stretched, it returns to its original length after the forces are removed. If this happens, the object experiences **elastic deformation**.

Sometimes an object that has been stretched does not return to its original length when the forces are removed. If the object remains permanently stretched, the object experiences **inelastic deformation**.

Elastic and inelastic deformations can be shown easily by stretching a spring in the laboratory. When small forces are applied and then removed, the spring returns to its original length and shape. When large forces are applied and then removed, the spring does not return to its original length.

You can also explore elastic and inelastic behaviour with an empty drinks can. When you squeeze the can gently, it springs back to its original shape when you remove your fingers. However, by applying larger forces you can change the can's shape permanently.

Stretching a spring

For a spring that is elastically deformed, the force exerted on a spring and the **extension** of the spring are linked by the equation:

$$F = k \times e$$

force = spring constant × extension

Where: force is in newtons, N

spring constant is in newtons per metre, N/m

extension is in metres, m

The spring constant is a measure of how stiff a spring is. If k is large, the spring is stiff and difficult to stretch. When a spring has a spring constant of 180 N/m, this means that a force of 180 N must be applied to stretch the spring 1 m. The equation, $F = k \times e$, can also be applied to the compression of a stiff spring by two forces. In this case, e is the distance by which the spring has been compressed (squashed).

Example. George uses a spring to weigh a fish he has just caught. The spring stretches 8 cm. George knows that the spring constant is 300 N/m. Calculate the weight of the fish.

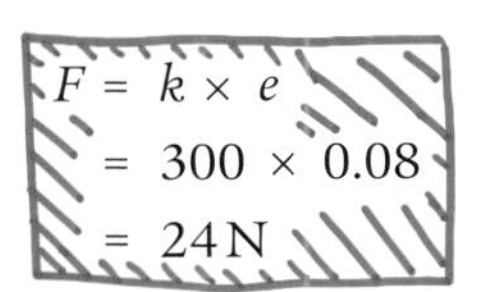

$$F = k \times e$$
$$= 300 \times 0.08$$
$$= 24\,\text{N}$$

Remember: you must convert the 8 cm to 0.08 m, because the spring constant is measured in N/m.

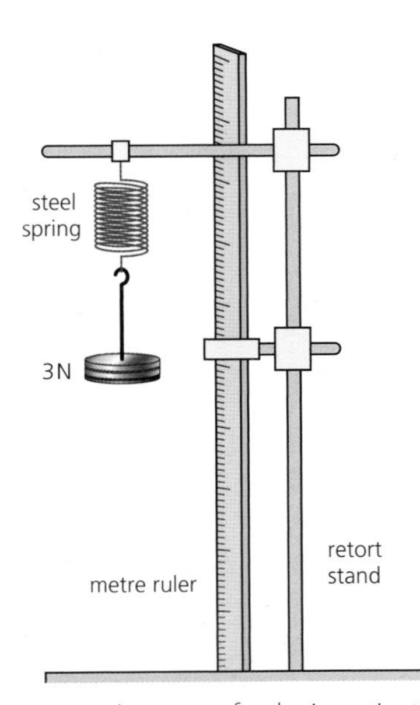

Figure 7.4 Apparatus for the investigation

Investigating the relationship between force and the extension of a spring

Method

1. Set up the apparatus as shown in Figure 7.4. Make sure that the ruler is vertical and close to the spring.
2. Measure the position of the bottom of the spring on the metre rule. This is l_0.
3. Extend the spring by placing a 100 g mass on the bottom. This exerts a force of 1.0 N. Measure the new position of the bottom of the spring. This is l_1.
4. Calculate the extension of the spring $l_1 - l_0$.

5 Now construct a suitable table to record the applied force, original length, extended length and the extension.
6 Add a further weight so that the spring is stretched by a force of 2 N. Record the new length l_2, and new extension $l_2 - l_0$.
7 Repeat the procedure up to a weight of 6 N.

Analysing the results

1 Plot a graph of the extension of the spring against the force applied to it.
2 Draw a straight line of best fit through your results.
3 What conclusion can you draw from your results?

PRACTICAL

Safety: during the investigation you should wear safety glasses.

Advisory notes

1 Before starting the experiment, check that 6 N is an appropriate force to stretch the spring.
2 Sometimes new springs are compressed. Under those circumstances begin with one weight on the spring, and use that length as l_0.
3 Make sure the retort stand is clamped to the bench or weighted down so that it cannot topple over.

Limit of proportionality

The graph in Figure 7.5 shows some typical results as you increase the weights to stretch the spring. In this case you can see that the extension is proportional to the force up to a force of 7 N. This is called the **limit of proportionality**. Beyond this point there is a greater extension for each weight added to the spring.

The spring constant for the spring can be calculated from the linear section of the graph by dividing the force by the extension:

$$k = \frac{F}{e}$$

The fact that the initial linear extension of a spring is proportional to the Force is known as **Hooke's Law**.

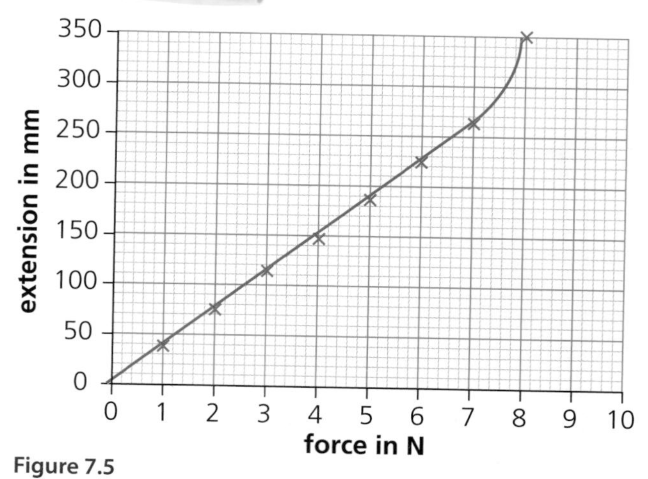

Figure 7.5

Further investigations

1 You can use the same apparatus (in Figure 7.4) and procedure for measuring the extension of a rubber band as different forces are applied to it. Question 4 refers to the extension–force graph for rubber.

2 If you want to investigate the extension of a metal wire – copper for example – you will find it best to use a long piece of wire stretched along a bench as shown in Figure 7.6.
Your teacher will provide a wire of suitable thickness for the investigation. A wire diameter of about 0.2 mm or 0.3 mm works well.
If the wire snaps the mass may damage the floor as it falls – place some padding beneath it just in case.

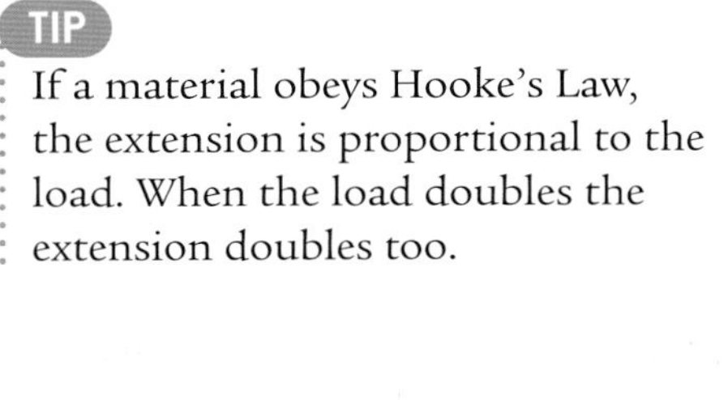

TIP

If a material obeys Hooke's Law, the extension is proportional to the load. When the load doubles the extension doubles too.

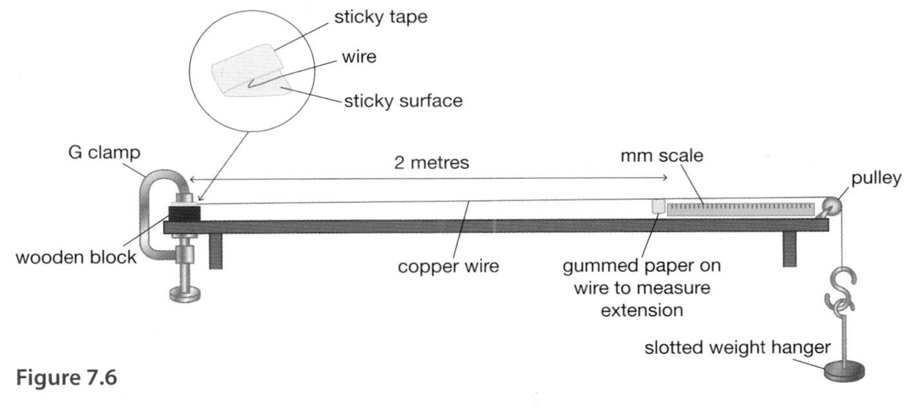

Figure 7.6

STUDY QUESTIONS

1 Explain what is meant by:
 a) elastic deformation
 b) inelastic deformation.

2 This question refers to the graph of the extension of a spring shown in Figure 7.5.
 a) Up to what extension does the spring obey Hooke's Law?
 b) Use the graph to determine the spring constant for the spring.

3 A force of 600 N is applied to compress a car suspension spring, that has a spring constant of 20 000 N/m. Calculate the compression of the spring.

4 Figure 7.7 shows an extension–force graph for a rubber band.

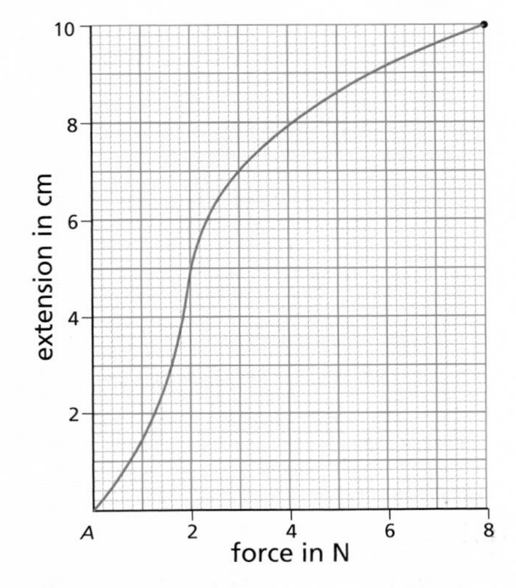

Figure 7.7

 a) Explain whether the rubber band obeys Hooke's Law.
 b) Describe the relationship between extension and the applied force.

5 Figure 7.8 shows an extension/force graph for a long thin copper wire. When the wire is stretched beyond A, it does not return to its original length.

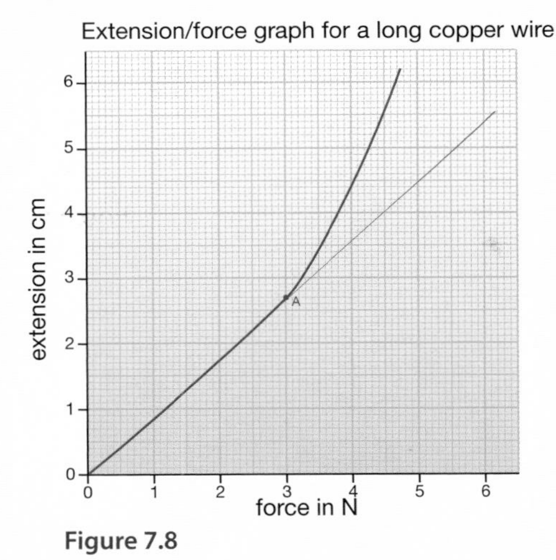

Figure 7.8

 a) Describe the relationship between the extension of the wire and the applied force. In your answer include the words elastic, inelastic and limit of proportionality.
 b) Discuss whether a copper wire obeys Hooke's Law.

1.8 Momentum

Figure 8.1 Why does a shot gun recoil when it is fired?

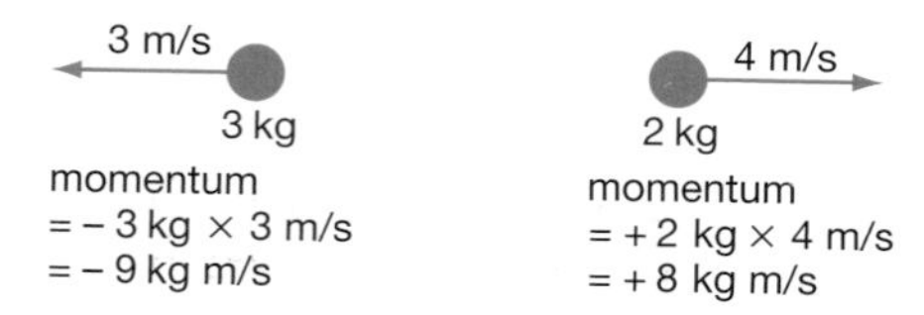

Figure 8.2 In this diagram we have given anything moving to the right positive momentum, and anything moving to the left negative momentum.

Momentum is defined as the product of mass and velocity.

$$\textbf{momentum} = \textbf{mass} \times \textbf{velocity}$$

The units of momentum are kilogram metres per second, kg m/s.

Velocity is a vector quantity and therefore momentum is too. You must give a size and a direction when you talk about momentum. Look at Figure 8.2, where you can see two objects moving in opposite directions. One has positive momentum and the other negative momentum.

Forces and change of momentum

When a force, F, pushes a mass, m, we can relate it to the mass and acceleration, using the equation:

$$\text{force} = \text{mass} \times \text{acceleration}$$
$$F = ma$$

The acceleration is defined by the equation:

$$\text{acceleration} = \frac{\text{change of velocity}}{\text{time}}$$

These two equations can be combined to give:

$$\text{force} = \frac{\text{mass} \times \text{change of velocity}}{\text{time}}$$

Because (mass × change of velocity) is the change of momentum, the force can be written as:

$$\text{force} = \frac{\text{change of momentum}}{\text{time}}$$

Example. A mass of 7 kg is accelerated from a velocity of 2 m/s to 6 m/s in 0.5 s. Calculate the force applied.

$$\text{force} = \frac{\text{change of momentum}}{\text{time}}$$
$$= \frac{7\text{ kg} \times 6\text{ m/s} - 7\text{ kg} \times 2\text{ m/s}}{0.5\text{ s}}$$
$$= 56\text{ N}$$

Momentum and safety

The equation relating force and change of momentum is important when it comes to considering a number of safety features in our lives. If you are moving you have momentum. To stop moving a force must be applied. But the equation shows us that if we stop over a long period of time, the force to slow us is smaller. This way we get hurt less.

Here are some examples of how we protect ourselves by increasing the time we or another object has to stop.

- When we catch a fast ball we move our hands backwards with the ball.
- We wear shin pads in hockey so that the ball has more time to stop when it hits us.
- When we jump off a wall we bend our legs when we land, so that we stop in a longer time.
- Children's playgrounds have soft, rubberised matting under climbing frames, so that a faller takes longer to stop.
- In a bungee jump the elastic rope slows your fall gradually.

TIP

Make sure you understand and can explain these examples.

Driving safely – 2

Figure 8.3 shows two cars being tested in a trial crash. In the centre of the cars are rigid passenger cells, which are designed not to buckle in a crash. However, the front and back of the cars *are* designed to buckle in a crash – these are the **crumple zones**. These zones absorb energy and also reduce the deceleration by *increasing the time and the distance* to slow the passenger. The force acting on the passenger is less.

Figure 8.3 Most of the energy of an impact in a car crash is absorbed by the crumple zones.

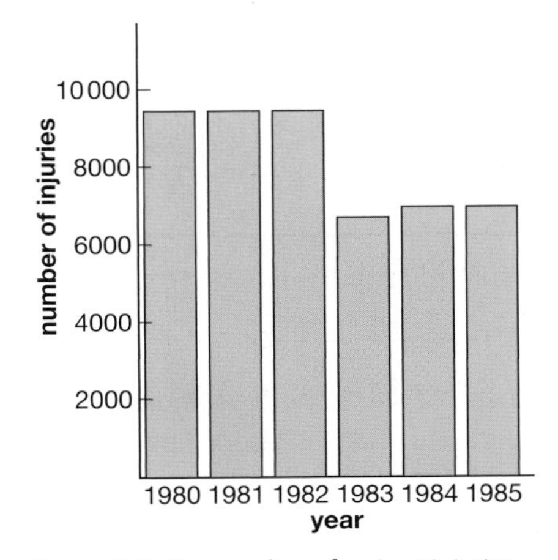

Figure 8.4 The number of serious injuries to front-seat passengers and drivers in cars and light vans in Great Britain. In January 1983 it became compulsory to wear seat belts in the front seats. The statistics speak for themselves. Will you drive a car without a seat belt?

Seat belts

Another vital safety feature in a car is a seat belt – you must wear one by law in most countries. Figure 8.4 shows some statistics that illustrate how wearing seat belts has reduced the rate of serious injuries in car crashes.

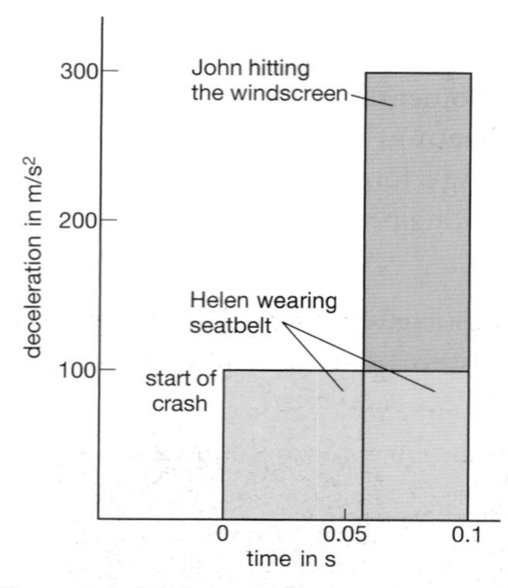

Figure 8.5 Graph to show approximate decelerations in a car crash.

There is some 'give' in a seat belt, which allows the passenger to slow over a longer distance. The seat belt allows you to take advantage of the time it takes the crumple zone to buckle. The graph in Figure 8.5 shows approximate decelerations of two passengers in a car crash. Helen was wearing a seat belt and John was not. Helen decelerated over a period of 0.1 s while the front of the car crumpled. John kept moving forwards until he hit the windscreen. His deceleration was much larger as he stopped in a shorter time. John received very serious injuries.

Figure 8.6 Wearing a seat belt could save your life in an accident. Only dummies forget to put their seat belts on.

STUDY QUESTIONS

1. A car of mass 1200 kg moves at a speed of 30 m/s. Calculate its momentum.
2. Using your knowledge of momentum, explain:
   a) why you bend your knees when you land after a jump
   b) why cars are designed to have crumple zones.
3. Explain why passengers should wear seat belts in cars and coaches.
4. Explain why it hurts more if you fall over on concrete than it does if you fall over on grass.
5. This question refers to Figure 8.5. John's mass is 80 kg and Helen's mass is 60 kg.
   a) Use the graph to calculate the force that acted on each of them as they decelerated. Explain why John received more serious injuries than Helen.
   b) Use the graph (Figure 8.5) and the equation:
   $$\text{acceleration} = \frac{\text{change of speed}}{\text{time}}$$
   to calculate the speed of the car before the crash.
   c) Helen was driving the car and was responsible for the crash. She was breathalysed by the police and found to be over the drink-drive limit. What measures should governments take to deter drink-drivers?
6. In the UK in 1966, 7985 people were killed in road accidents and 65 000 seriously injured. In 2013, 1713 people were killed in road accidents and 20 800 seriously injured. Suggest six factors that have improved road safety during this period of time.

1.9 Collisions and explosions

When the firework explodes, stored chemical energy is transferred to kinetic energy and thermal energy. But the momentum after the explosion is the same as it was before the explosion. Can you explain why?

Figure 9.1 Why does this firework explode symmetrically in all directions?

Conservation of momentum

When it comes to calculating what happens in collisions and explosions, momentum is a very useful quantity. Momentum is always conserved.

When two bodies collide, the total momentum they have is the same after the collision as it was before the collision.

Figure 9.2 shows two ice hockey players colliding on the ice. When they meet they push each other – the blue player's momentum decreases and the red player's momentum increases by the same amount.

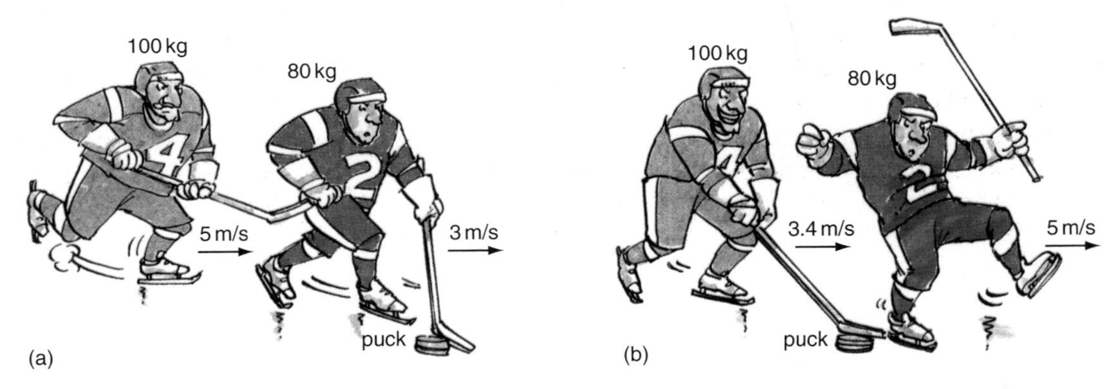

Figure 9.2 The total momentum of the players is the same before and after the collision.

Before the collision the total momentum was:

$$\text{momentum of blue player} = 100\ \text{kg} \times 5\ \text{m/s} = 500\ \text{kg m/s}$$

$$\text{momentum of red player} = 80\ \text{kg} \times 3\ \text{m/s} = 240\ \text{kg m/s}$$

TIP

In a collision between two bodies, where no external force acts on then, momentum is always conserved

This makes a total of 740 kg m/s.

After the collision the total momentum was:

$$\text{momentum of blue player} = 100\ \text{kg} \times 3.4\ \text{m/s} = 340\ \text{kg m/s}$$

$$\text{momentum of red player} = 80\ \text{kg} \times 5\ \text{m/s} = 400\ \text{kg m/s}$$

This makes a total of 740 kg m/s – the same as before.

Momentum as a vector

Figure 9.3 shows a large estate car colliding head-on with a smaller saloon car. They both come to a halt. How is momentum conserved here?

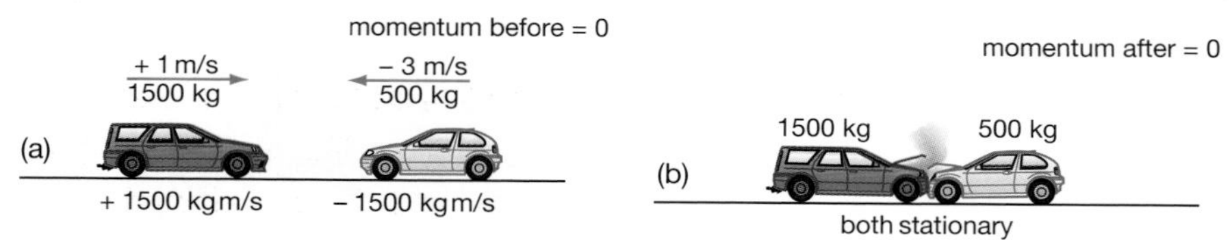

Figure 9.3

Momentum is a vector quantity. In Figure 9.3(a) you can see that the red car has a momentum of + 1500 kg m/s and the blue car has momentum of – 1500 kg m/s. So the total momentum is zero before the collision – there is as much positive momentum as negative momentum. After the collision they have both stopped moving and the total momentum remains zero.

Investigating collisions

Figure 9.4 shows how you can investigate conservation of momentum using laboratory trolleys. One trolley is pushed into a second stationary trolley and the speeds measured before and after the collision. These speeds can be measured with light gates and data loggers, or with ticker tape.

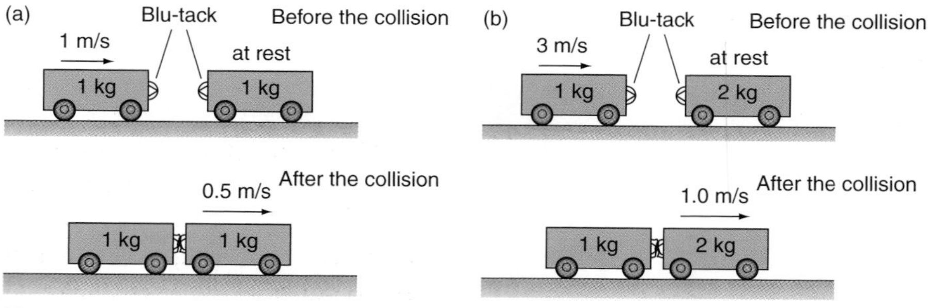

Figure 9.4

In Figure 9.4(a) a trolley of mass 1 kg, travelling at 1 m/s, collides with an identical trolley, which is at rest. They stick together. The momentum before the collision is:

$$1\ \text{kg} \times 1\ \text{m/s} = 1\ \text{kg m/s}$$

The trolleys move off after the collision with the same total momentum of 1 kg m/s, but this is shared by the two trolleys, so they move with a speed of 0.5 m/s.

What happens in Figure 9.4(b)?

The total momentum before the collision is:

$$1\ \text{kg} \times 3\ \text{m/s} = 3\ \text{kg m/s}$$

The momentum after the collision = 3 kg m/s (momentum is conserved). The trolleys stick together to make a new mass of 3 kg, which now carries the momentum. So:

$$3\ \text{kg m/s} = 3\ \text{kg} \times v$$

The unknown speed, v, is therefore 1 m/s.

Newton's third law

Newton's third law of motion states that: to every force there is an equal and opposite force. This law sounds easy to apply, but it requires clear thinking. First, it is important to appreciate that the pair of forces mentioned in the law act on different bodies. Some examples are given in Figure 9.5.

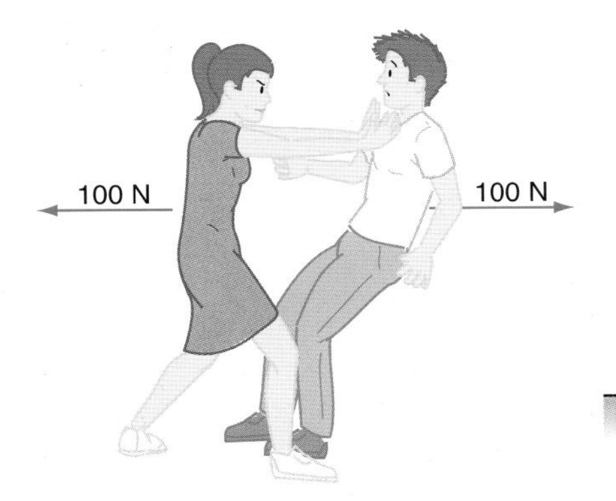

(a) Each person experiences the same force when contact is made.

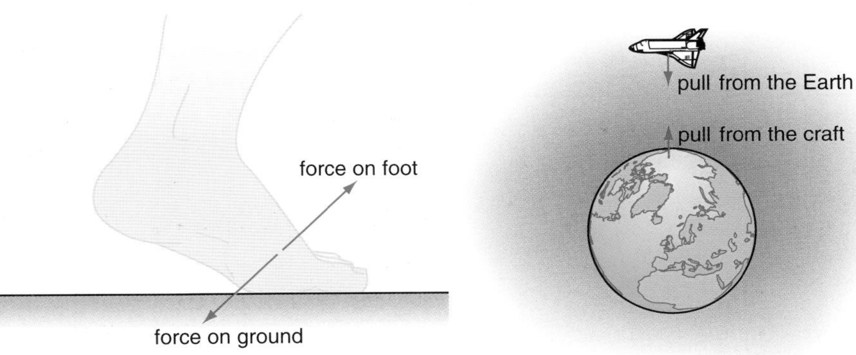

(b) A foot pushes the ground backwards; the ground pushes the foot forwards.

(c) There is an equal and opposite gravitational pull on the craft and on the Earth.

Figure 9.5

a) If I push you with a force of 100 N, you push me back with a force of 100 N in the opposite direction.
b) When you walk, you push the ground backwards; the ground pushes you forwards.
c) A spacecraft orbiting the Earth is pulled downwards by Earth's gravity; the spacecraft exerts an equal and opposite gravitational pull on the Earth. This means that as the spacecraft moves, the Earth moves too. But the Earth is so big that it only moves a tiny amount – far too little for us to notice.

STUDY QUESTIONS

1 Explain why momentum is a vector quantity.
2 Explain what is meant by the conservation of momentum.
3 In Figure 9.3 two cars collided head-on. Each driver had a mass of 70 kg.
   a) Calculate the change of momentum for each driver.
   * b) The cars stopped in 0.25 s. Calculate the average force that acted on each driver.
   c) Explain which driver is more likely to be seriously injured.
4 In each of the following experiments shown in the diagrams, the two trolleys collide and stick together, Work out the speeds of the trolleys after their collisions.

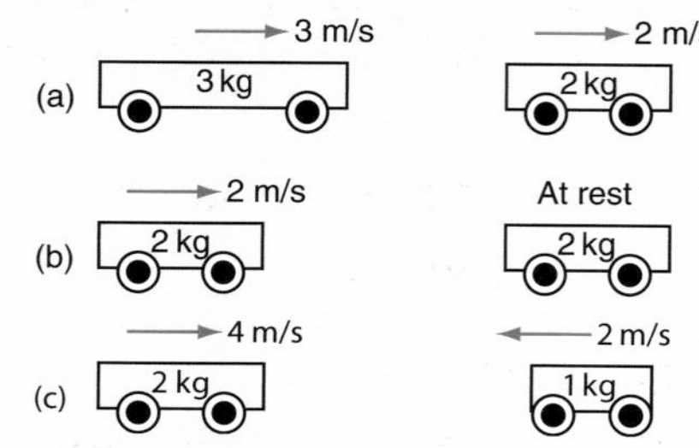

5 A lorry of mass 18 000 kg collides with a stationary car of mass 2000 kg.

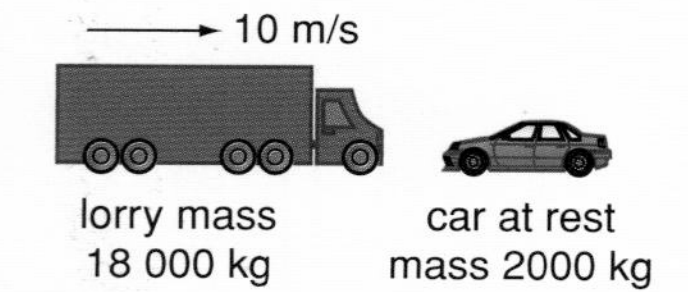

   a) Calculate the momentum of the lorry before the collision.
   b) After the collision the lorry and car move off together. State their combined momentum after the collision.
   c) Calculate the speed of the lorry and car after the collision.
   d) Each of the drivers of the car and lorry has a mass of 80 kg. Calculate their changes of momentum during the collision.
   * e) The collision last for 0.2 s. Calculate the force that acts on each driver during the collision. Explain which driver is more likely to be injured.
   f) Explain why a crumple zone protects the drivers in a collision.
6 Apply Newton's third law of motion to explain the following:
   a) A gun recoils when it is fired.
   b) When you lean against a wall you do not fall over.
   c) You have to push water backwards so that you can swim forwards.
7 A field gun of mass 1000 kg, which is free to move, fires a shell of mass 10 kg at a speed of 200 m/s.
   a) Calculate the momentum of the shell after firing.
   b) State the momentum of the gun just after firing.
   c) Calculate the recoil velocity of the gun.
   * d) Why do you think very large guns are mounted on railway trucks?

1.10 Turning moments

Figure 10.1

If you have ever tried changing the wheel of a car, you will know that you are not strong enough to undo the nuts with your fingers. You need a spanner to get a large turning effect. You need a long spanner and a large force.

The size of the turning effect a force exerts about a point is called a **turning moment**. The turning moment is increased by applying a large force and by using a long lever.

turning moment = force applied × perpendicular distance of the line of force from pivot

Perpendicular means 'at right angles'. Figure 10.2 shows you why this distance is important. In Figure 10.2(a) you get no turning effect at all as the force acts straight through the pivot, *P*. In Figure 10.2(b) you can calculate the turning moment.

$$\text{turning moment} = 100\text{ N} \times 0.3\text{ m}$$
$$= 30\text{ Nm}$$

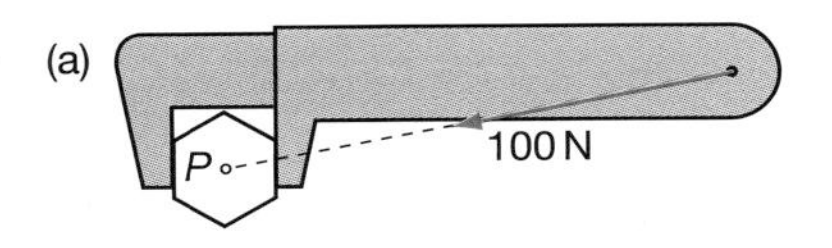

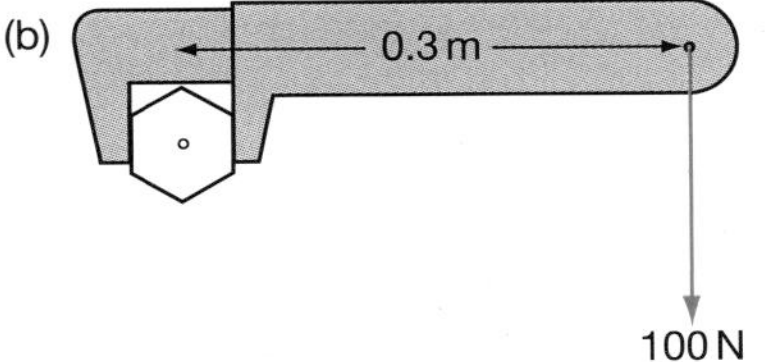

Figure 10.2

Turning forces in action

Figures 10.3 to 10.6 show four examples of how tools can be used to give us large turning moments.

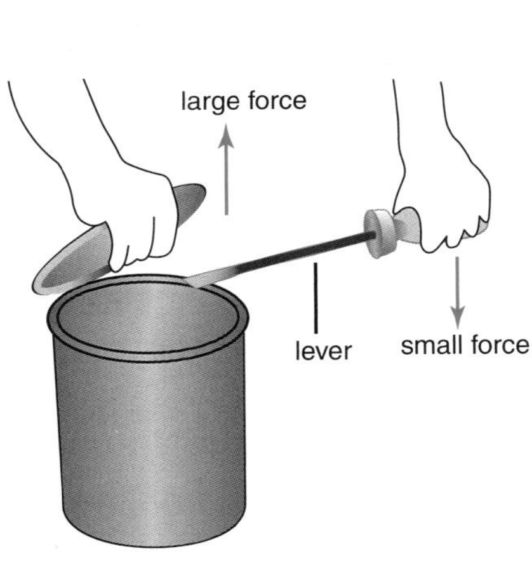

Figure 10.3 A screwdriver can be used to lever off the lid of a paint tin.

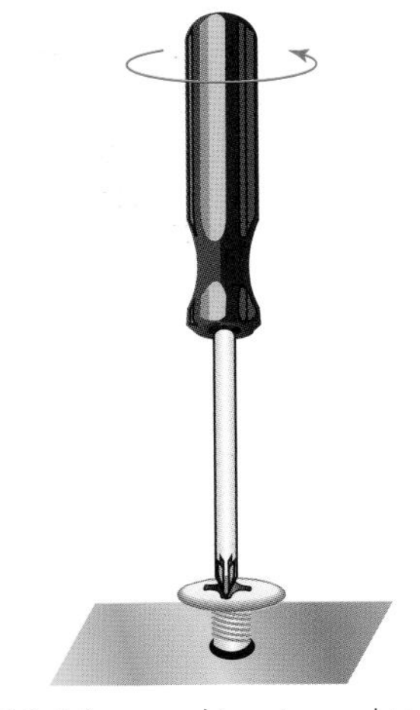

Figure 10.4 A screwdriver is used to insert a screw – a broad handle allows us a good grip and leverage.

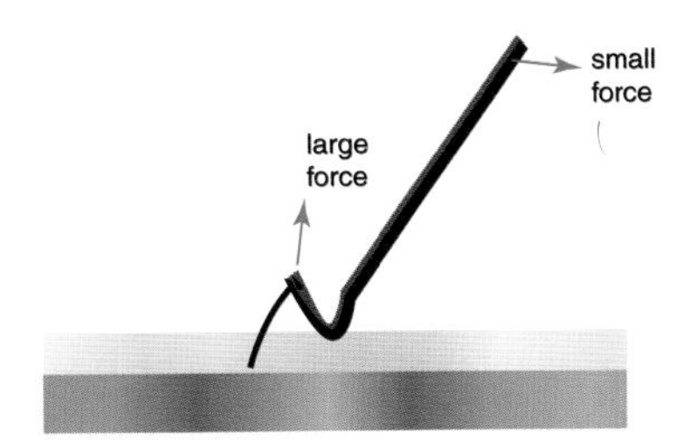

Figure 10.5 A crowbar with a long handle can be used to pull out old nails.

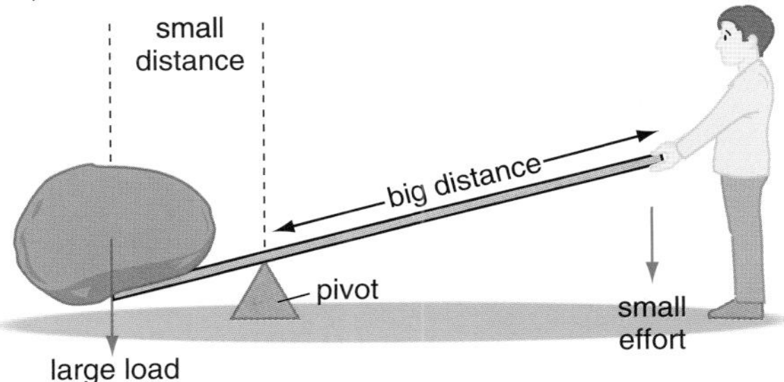

Figure 10.6 A simple lever. The rock can be lifted when its turning effect is balanced by the turning effect of the man's push. With a long lever a small effort can lift a large load.

Lifting loads

Turning moments need to be considered when lifting heavy loads with a mobile crane (Figure 10.7). If the turning effect of the load is too large the crane will tip over. So, inside his cab, the crane operator has a table to tell him the greatest load that the crane can lift for a particular **working radius**.

TIP

The unit of a turning moment is N m or N cm.

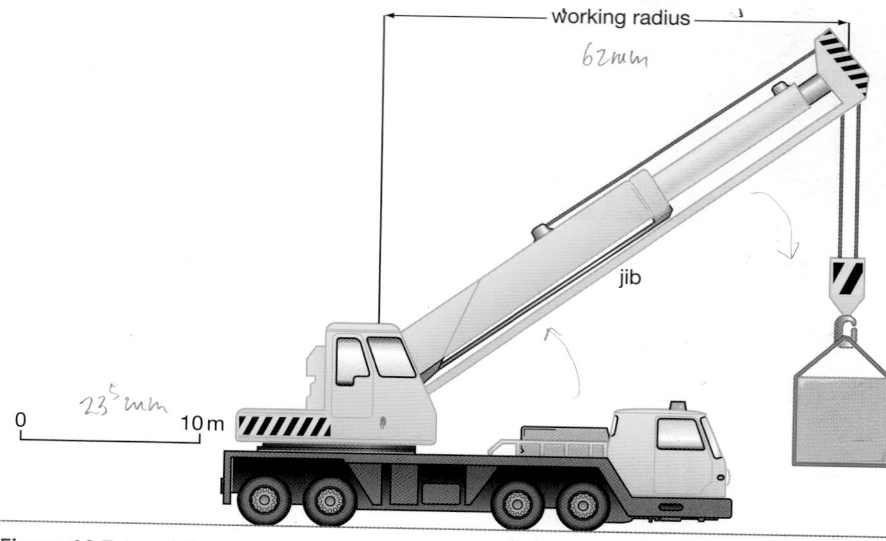

Figure 10.7 A mobile crane.

Table 1 shows you how this works. For example, the crane can lift a load of 60 tonnes safely with a working radius of 16 m. If the crane is working at a radius of 32 m, it can only lift 30 tonnes. You get the same turning effect by doubling the working radius and lifting half the load.

Table 1 A load table for a crane operator

Working radius in m	Maximum safe load in tonnes	load × radius in tonne × m
12	80	960
16	60	960
20	48	960
24	40	960
28	34	952
32	30	960
36	27	972

Centre of gravity

Figure 10.9(a) on the next page shows a see-saw that is balanced about its midpoint. Gravity has the same turning effect on the right-hand side of the see-saw, as it has on the left-hand side. The resultant turning effect is zero.

The action of the weight of the see-saw is the same as a single force, *W*, which acts downwards through the pivot (Figure 10.9(b)). This force has no turning moment about the pivot. As the see-saw is stationary, the pivot must exert an upwards force *R* on it, which is equal to *W*.

When the see-saw is not pivoted about its midpoint, the weight will act to turn it (Figure 10.9(c)).

Figure 10.8 This rocking toy is stable because its centre of gravity is below the pivot.

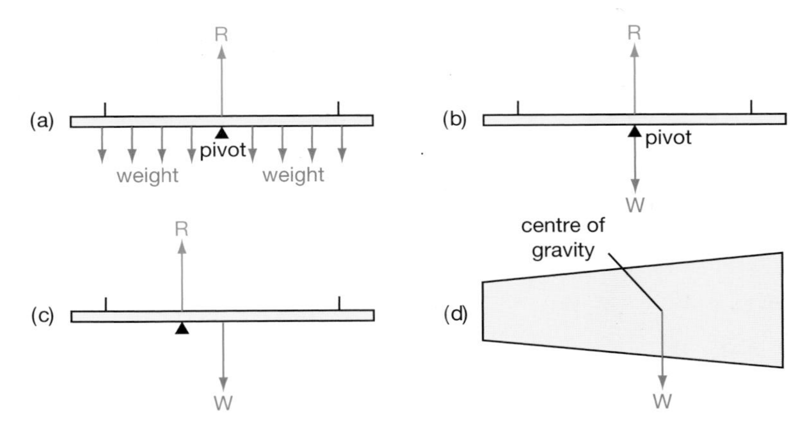

Figure 10.9

The point that the weight acts through is called the **centre of gravity**. The centre of gravity of the see-saw lies at its midpoint because it has a regular shape. In Figure 10.9(d) the centre of gravity lies nearer the thick end of the shape.

STUDY QUESTIONS

1 **a)** When you cannot undo a tight screw, you use a screwdriver with a large handle. Explain why.
b) Explain why door handles are not put near hinges.

2 Look at the crane in Figure 10.7.
a) What is meant by working radius?
b) Use the scale in Figure 10.7 to show that the working radius of the crane is 26 m.
c) Use the data in Table 1 to calculate the greatest load that the crane can lift safely in this position.
d) Laura, a student engineer, makes this comment: 'You can see from the driver's table that the crane can lift 960 tonnes, when the working radius is 1 m.' Do you agree with her?

3 **a)** A crowbar is used to lever a nail out of a block of wood. Use the information in the diagram to calculate the size of the turning moment exerted.

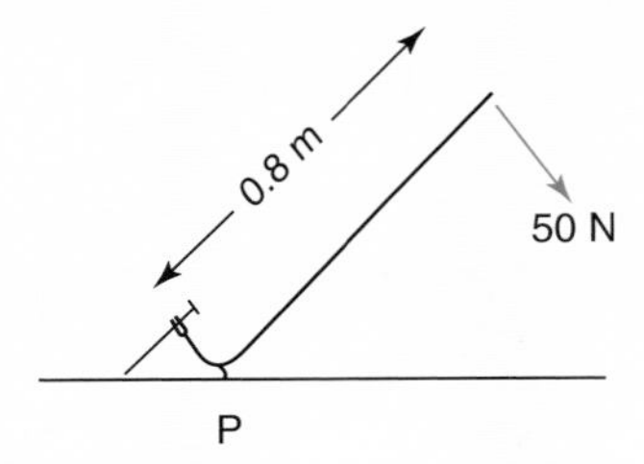

b) A student finds that the minimum force required to lift up the lid of a computer is 6.4 N.

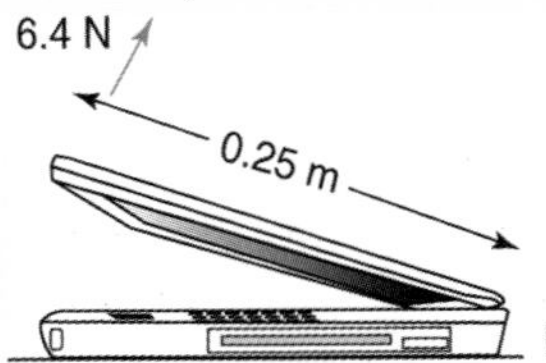

i) Calculate the moment of the force that opens the computer.
ii) Explain why the minimum force required to close the computer lid is likely to be less than 6.4 N.

4 A passenger in an airport has a case with wheels.

She decides it will be easier to pull the case if she packs all the heavy items of luggage nearest the wheels. Explain why this is a good idea. Include ideas about turning moments in your answer.

5 The diagram shows some apparatus being used to find the centre of gravity of a piece of card. Three small holes have been made in the card. There is a retort stand that supports a cork with a pin it. There is also a weight attached to a thin piece of cotton.

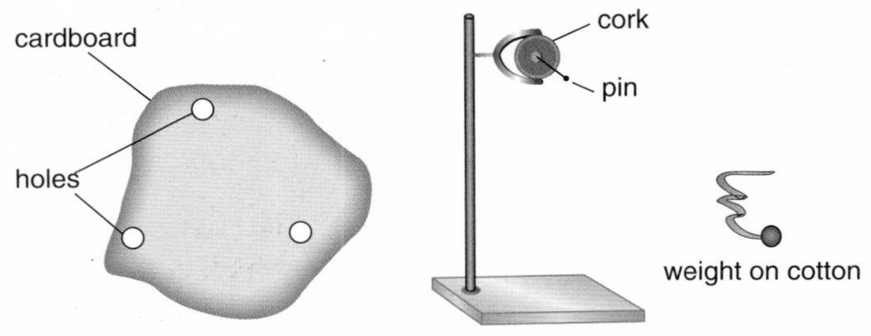

Explain carefully how you would use this apparatus to find the centre of gravity of the card.

1.11 Balancing forces

Figure 11.1

This tightrope walker is helped if he has a head for heights and an understanding of physics. What conditions must exist for him to remain balanced? How does the pole help him?

Balancing

In Figure 11.2 you can see Jaipal and Mandy sitting on a see-saw, which is balanced. It balances because the turning moment produced by Jaipal's weight exactly balances the turning moment produced by Mandy's weight in the opposite direction.

$$\text{Jaipal's anti-clockwise turning moment} = 450\text{ N} \times 1\text{ m} = 450\text{ Nm}$$

$$\text{Mandy's clockwise turning moment} = 300\text{ N} \times 1.5\text{ m} = 450\text{ Nm}$$

The turning moments balance. The seesaw is in equilibrium.

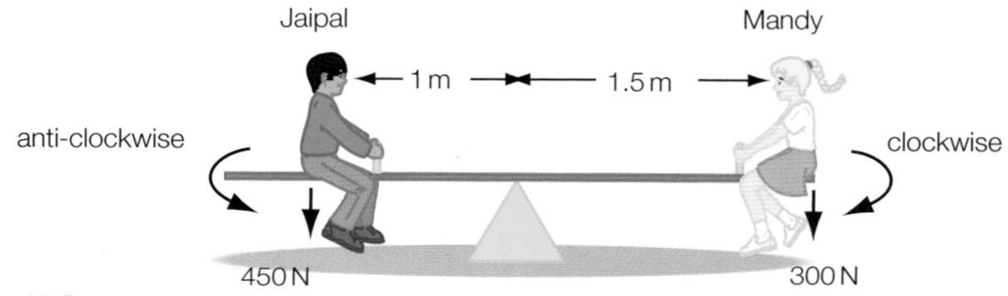

Figure 11.2

TIP

When an object is balanced (or in equilibrium) the turning forces in one direction balance the turning forces in the opposite direction.

In equilibrium:

- the sum of the anti-clockwise moments = the sum of the clockwise moments
- the sum of the forces balance.

Moments in action

In Figure 11.3 a small crane is being used to load a boat. A counterbalance weight on the left-hand side is used to stabilise the crane and to stop it from toppling over.

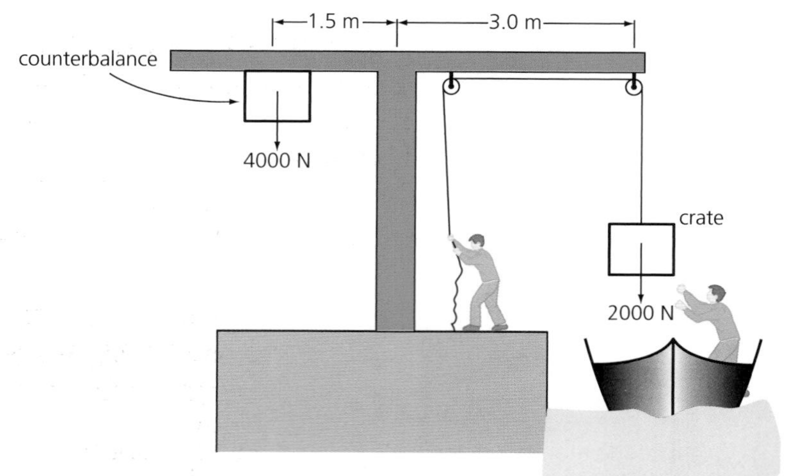

Figure 11.3 Forces acting on a small crane.

The clockwise turning moment from the crate being lifted = 2000 × 3
= 6000 Nm

The anti-clockwise turning moment from the counterbalance = 4000 × 1.5
= 6000 Nm

These turning moments keep the beam in equilibrium.

A window cleaner is carrying his ladder (Figure 11.4). The weight of the ladder and the bucket exert a downwards turning moment behind him. He must balance this by pulling down on the ladder in front of him.

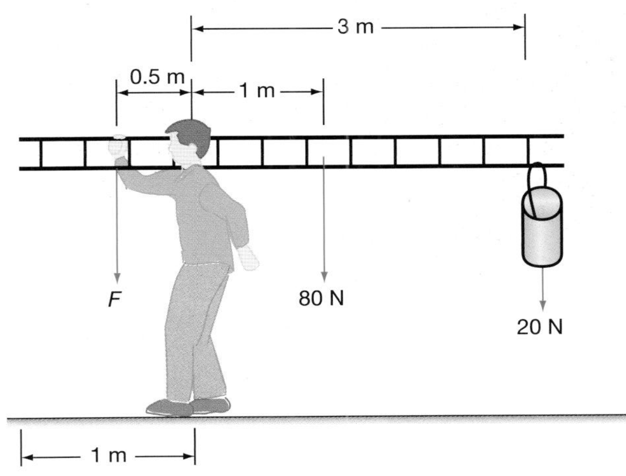

Figure 11.4 Forces on a window cleaner's ladder which is balanced on his shoulder.

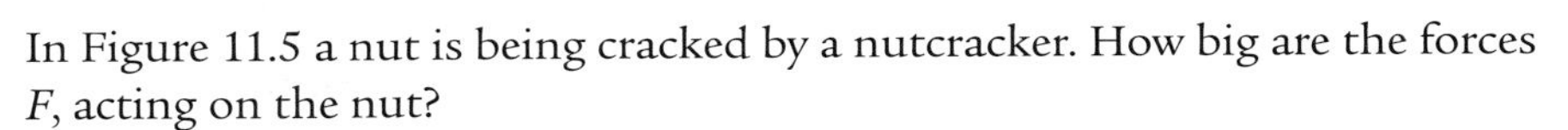

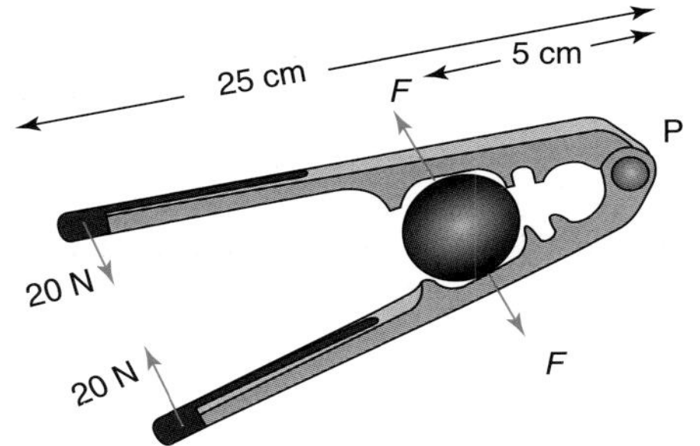

Figure 11.5 Forces acting on a nutcracker.

In Figure 11.5 a nut is being cracked by a nutcracker. How big are the forces *F*, acting on the nut?

Just before the nut cracks, the turning moments of the forces, *F*, balance the turning moments of the 20 N forces.

$$F \times 5\text{ cm} = 20\text{ N} \times 25\text{ cm}$$

$$F = 100\text{ N}$$

This is the force which the nut exerts on the nutcracker. Newton's 3rd law tells us that the nutcracker exerts the same force on the nut, in the opposite direction.

Forces on beams

In Figure 11.6 a man is running with a constant velocity across a light bridge (which has a negligible weight). How do the forces at A and at B change as he crosses?

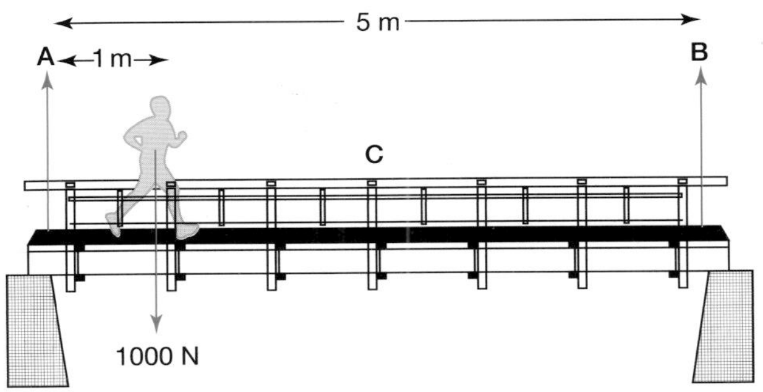

Figure 11.6

There are three easy places where you can use common sense to calculate the forces:

- When he is at A, the upward force at A is 1000 N and the upward force at B is zero.
- When he is at B, the upward force at B is 1000 N and the upward force at A is zero.
- When he is at C, the centre of the bridge, the upward force at each of A and B is 500 N.

When the man is 1 m from A you can use your knowledge of moments to see that A will support more of his weight than B. The exact force can be calculated using the principle of moments.

The turning moment of the man's weight about A is balanced by the turning moment of the upward force at B.

$$1000\ \text{N} \times 1\ \text{m} = (\text{force at B}) \times 5\ \text{m}$$
$$\text{force at B} = 200\ \text{N}$$

The upward force at A is 800 N, as together the two forces A and B must add up to 1000 N.

STUDY QUESTIONS

1 a) Explain why cranes use counterbalances.
b) In Figure 11.3 the dock workers want to load a heavier weight into the boat. Which way should they move the counterbalance to balance the crane? Explain your answer by referring to turning moments.
c) A load of 1200 N is now lifted into the boat. Calculate where the counterbalance should be placed now.

2 This question refers to the window cleaner in Figure 11.4.
a) Calculate the turning moment due to the weight of the bucket about the window cleaner's shoulder.
b) Calculate the turning moment due to the weight of the ladder about his shoulder.
c) State the size of the turning moment his arm must produce to balance these turning moments.
d) Calculate the size of the force, *F*, which the window cleaner must exert.
e) A friend suggests that his ladder would be easier to carry if he put the bucket at the other end. Use your knowledge of moments to discuss why this is a good idea.
f) Can you make a further suggestion, which would make the ladder easier for the man to carry?

3 The diagram shows a pair of cutters.

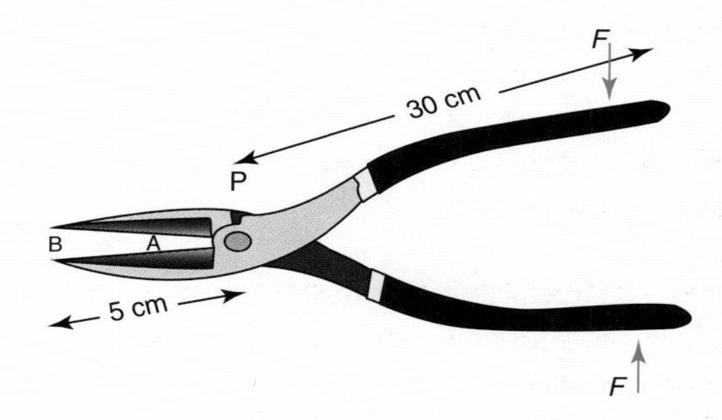

a) Explain why they have long handles.
b) You want to cut some thick wire. Use your knowledge of moments to explain whether it is better to put the wire at A or B.
c) Wire that needs a force of 210 N to cut it is placed at B, 5 cm from the pivot. How big a force will you need to exert at the ends of the handle to cut the wire?

4 This question refers to the man running across the bridge in Figure 11.6.
a) Calculate the size of the upward forces at A and B when he is 1 m away from B.
b) Using the information in the text and your answer to part **a)**, sketch a graph of the way the upward force at A changes as the man runs from A to B.
c) Calculate the sizes of the forces at A and B when he is 3 m away from A and 2 m away from B.

Summary

I am confident that:

✓ I can recall these facts about forces and motion

- Average speed = distance/time
- The unit of speed is m/s.
- Acceleration = change of velocity/time
- The unit of acceleration is m/s^2.
- The gradient of a distance–time graph is the speed.
- The area under a velocity–time graph is the distance travelled.
- The gradient of a veolocity–time graph is the acceleration.
- A force is a push or a pull.
- The unit of force is the newton, N.
- Force is a vector quantity.
- Air resistance increases with speed.
- Newton's first law: if balanced forces act on an object, it remains at rest or moves with constant speed in a straight line.
- Newton's second law: resultant force = mass × acceleration
- Newton's third law: to every force there is an equal and opposite force.

✓ I can recall these facts about stopping distances

- Thinking distances are affected by the driver's reaction time.
- The reaction time can be slower if the driver is sleepy, has taken drugs or alcohol or is distracted.
- Braking distances depend on the speed and mass of the car.
- Braking distances are increased: if the tread on the tyres is worn; if the brakes are worn; if the road is covered with water, ice or mud; if you are going downhill.

✓ I understand and can describe these facts about stretching

- The extension varies with the applied force for springs, metal wires and rubber bands.
- Hooke's law states that the extension of a spring or wire is proportional to the applied force. This law is associated with the initial linear region of a force–extension graph.
- Elastic materials return to their original shape when a force is removed.

✓ I can recall these facts about momentum

- momentum = mass × velocity
- The units of momentum are kg m/s.
- Momentum is a vector quantity.
- force = change of momentum/time
- In collisions, the forces acting are smaller when the time is longer. This is important when designing crumple zones in cars.
- Momentum is conserved in collisions and explosions.

✓ I can recall and use these facts about turning forces

- moment = force × perpendicular distance from the pivot
- The unit of moment is Nm
- The weight of a body acts through its centre of gravity.
- The principle of moments states that in equilibrium the turning moments balance on a body.
- The upward forces on a light beam, supported at its ends, vary with the position of a heavy object placed on the beam.

Sample answers and expert's comments

1 The graph shows how the velocities of two cars, A and B, change after they accelerate from rest.

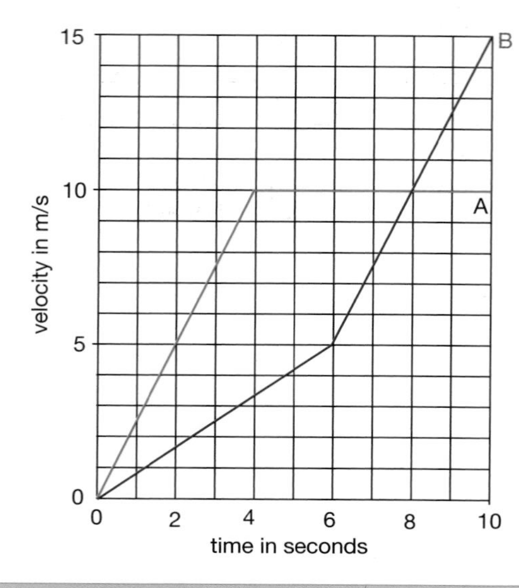

a) Use the graph to estimate the speed of car B, in m/s, after 3 seconds. (1)

b) Which car accelerates faster during the first 4 seconds? Explain your answer. (2)

c) Calculate how far car A travels, in m, between 4 and 10 seconds. (2)

d) State the equation which links acceleration, change of speed and time. Use the graph to calculate the acceleration of car B between 6 and 10 seconds. State the unit of acceleration. (4)

(Total for question = 9 marks)

Student response Total 7/9	Expert comments and tips for success
a) 2.5 m/s ✔ 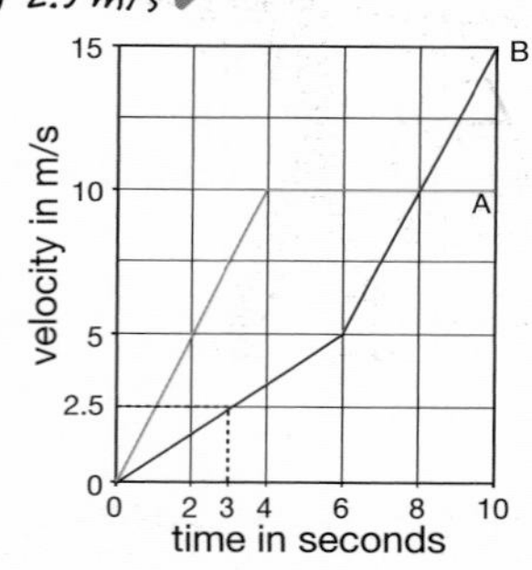	The answer is correct. Working is shown on the graph.
b) A accelerates faster. The gradient is steeper. ✔	This gets 1 mark for understanding that gradient can be used to calculate the acceleration. However, the answer needs to say that A increases in speed more in the same time.
c) The distance travelled is the area under the graph. ✔ $d = 10 \times 6 = 60$ m ✔	This is correct. The area under the graph can also be used to calculate the distance when the speed is changing.
d) $a = \frac{v-u}{t}$ ✔ $a = \frac{15}{10}$ O $= 1.5$ m/s² ✔✔	The equation is correct. You need to remember this equation, and be able to use it. A mark is lost because the change of speed was worked out over the whole 10 seconds, not the last 4 seconds. But the calculation is correct and the unit is given. In the last 4 seconds the speed increased from 5 m/s to 15 m/s. So the acceleration is $a = \frac{10}{4} = 2.5$ m/s²

2 A student sets up the apparatus shown below to investigate the principle of moments.

She hangs a 4 N weight from the 50 cm mark on the ruler.

She uses a force meter to hold the ruler in a horizontal position.

The force meter reads from 0 N to 20 N.

a) Explain how the student checks that the ruler is horizontal. (2)

b) i) State the equation which links moment, force and distance from the pivot. (1)

ii) Calculate the moment of the 4 N weight. State the unit. (3)

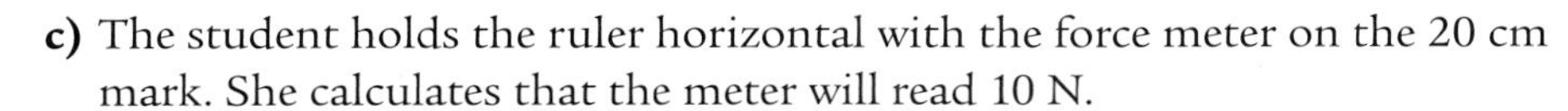

c) The student holds the ruler horizontal with the force meter on the 20 cm mark. She calculates that the meter will read 10 N.

i) Show how she reached the answer of 10 N. (3)

ii) The actual reading is 12 N. Explain why the correct reading should be greater than 10 N. (2)

d) The picture shows two farmers using a pole to carry some vegetables.

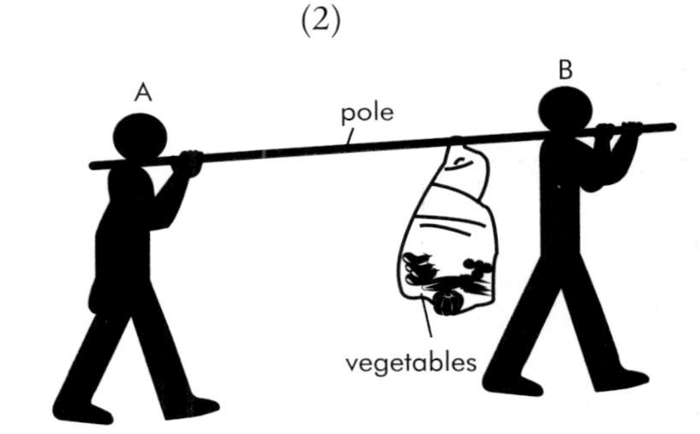

The two farmers feel different forces from the pole. Use ideas about moments to explain why farmer B feels the larger force. (3)

(Total for question = 14 marks)

Student response Total 9/14	Expert comments and tips for success
a) *You can see that it is horizontal by eye* O	You measure that it is horizontal by making sure each end of the ruler is the same height above the bench – using a ruler
b) i) *Moment = force × perpendicular distance* ✔ **ii)** *Moment = 4 × 0.5* ✔ *= 2 N/m* ✔ O	Correct, you must include the word perpendicular The calculation is correct, but it is a common mistake on the unit. The correct unit is N m: the units N and m are multiplied. You could also use N cm when the answer is 200 N cm.
c) i) *10 × 20 = 4 × 50* ✔ ✔ *The moments balance when the force metre reads 10 N* ✔ **ii)** *The ruler has a weight too.* ✔	This gets the marks, but it could be explained better. The anticlockwise moment of the force meter about the pivot (F × 20 N cm) balances the clockwise moment of the weight (200 N cm). Therefore F × 20 = 200 F = 10 N This is a common mistake, to give half the answer. The ruler has a weight which therefore exerts a turning moment about the pivot.
d) *B experiences a larger force because he is close to the vegetables.* ✔ *The two turning moments balance.* ✔	This is a difficult question and this is a good attempt. You need to add that the turning moment on B = a large force × a small distance and the turning moment on A = a small force × a large distance.

3 The diagram shows a spacecraft which is ready to take off from the Moon's surface.

The mass of the spacecraft is 30 000 kg.

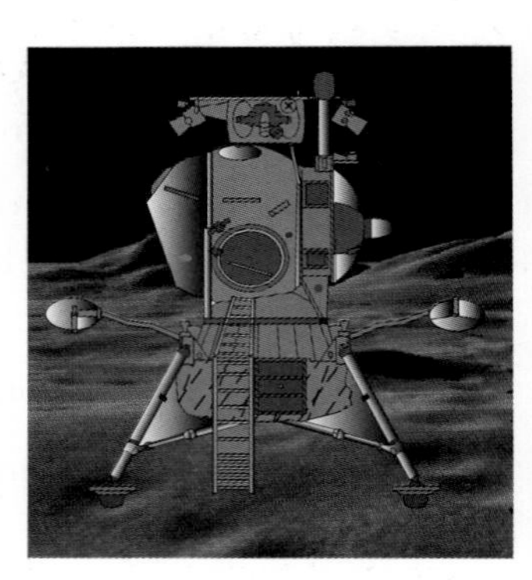

a) The weight of the spacecraft of the Moon is 48 000 N. Calculate the Moon's gravitational field strength. State the unit of gravitational field strength. (3)

b) When the spacecraft takes off, the upward force on it is 63 000 N. Calculate the resultant force acting on the spacecraft. (1)

c) Write an equation which links resultant force, mass and acceleration. Calculate the acceleration of the spacecraft, in m/s², when it takes off. (3)

d) Write an equation which links change of speed, acceleration and time. Calculate the speed of the spacecraft, in m/s, after 60 seconds. (3)

(Total for question = 10 marks)

Student response Total 9/10	Expert comments and tips for success
a) $mg = W$ $30\,000 \times g = 48\,000$ ✔ $g = 48\,000/30\,000$ $= 1.6$ ✔O	The equation has been used correctly and the value of g is correct. The unit is missing, so this answer loses a mark.
b) 15 000 N ✔	The unbalanced force is the difference between the weight and the forward force, i.e. 63 000 − 48 000 = 15 000 N
c) $F = ma$ ✔ $15\,000 = 30\,000 \times a$ $a = 15\,000/30\,000$ ✔ $= 0.5$ m/s² ✔	The equation is correct. Correct substitution gains a mark. Correct answer gains a mark.
d) a = change of speed/time ✔ 0.5 = change of speed/60 ✔ change of speed = 60 × 0.5 = 30 m/s ✔	A correct equation, the right substitution and accurate calculation with the right units gains full marks. It is always a good idea to substitute the numbers before you do the calculation. You will earn a mark for the correct substitution, even if you then make a mistake later.

Exam-style questions

1 The unit of acceleration is:

a) m/s **b)** m/s^2 **c)** ms^2 **d)** (m/s)^2

2 The unit of a turning moment is:

a) Nm **b)** N/m **c)** m/N **d)** N/m^2

3 In diagram A the skydiver is falling at a constant speed. Which of the following is the value of the drag force X?

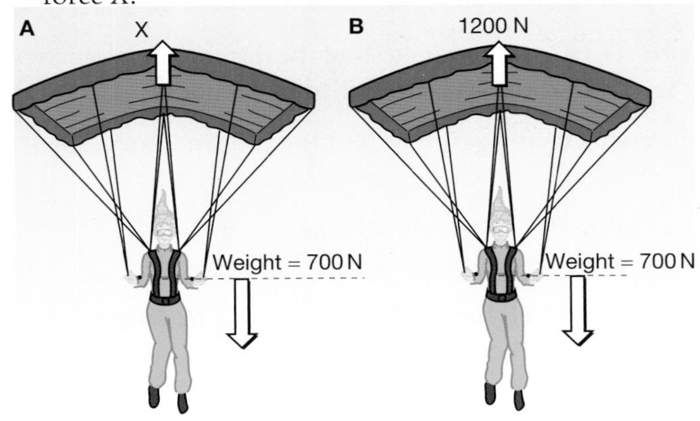

a) Just less than 700 N

b) Slightly more than 700 N

c) 700 N

d) It is impossible to predict because we do not know her speed

4 In diagram B the skydiver has just opened her parachute. Which of the following correctly describes her motion in the next few seconds?

a) She flies upwards

b) She decelerates until the drag force is 700 N and she then falls at a constant speed.

c) She slows down until she stops falling.

d) She slows down until the drag force is just a little more than 700 N

5 A cyclist increases his speed from 5 m/s to 7 m/s in 5 seconds. This acceleration is:

a) 2 m/s **b)** 2 m/s^2 **c)** 0.4 m/s **d)** 0.4 m/s^2

6 The cyclist and his bicycle in question 5 have a mass of 50 kg. The resistive force, *R*, acting on him is:

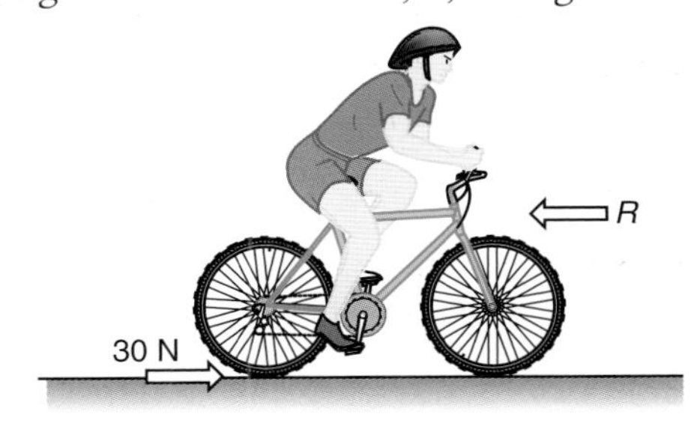

a) 50 N **b)** 70 N **c)** 10 N **d)** 20 N

7 The unit of momentum is:

a) kg/s **b)** kg m/s^2 **c)** kg m **d)** kg m/s

8 A car of mass 800 kg travels at a constant speed. It travels 80 m in 4 seconds. The momentum of the car is:

a) 6400 kg m **b)** 3200 kg s **c)** 16 000 kg m/s

d) 160 0000 N m/s

9 The moment of the bucket of water about P is:

a) 360 Nm **b)** 360 W **c)** 1440 Nm **d)** 160 N/m

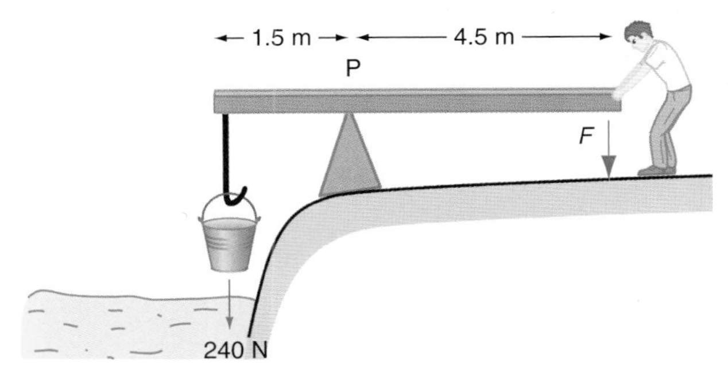

10 The force *F* the man must exert to lift the bucket is:

a) 240 N **b)** 80 N **c)** 720 N **d)** 60 N

11 A triathlon race has three parts: swimming, riding a bicycle and running.

a) The diagram shows the force responsible for the forward movement of the athlete in the swimming race.

Copy the diagram and show two other forces acting on the athlete. [2]

b) The table shows the distance of each part of a triathlon race and the time an athlete takes for each part.

Part of race	Distance in m	Time in s
swimming	1500	1200
riding a bicycle	40 000	3600
running	10 000	2000

i) Calculate the athlete's average speed during the swim. [1]

ii) Calculate the athlete's average speed for the whole race. [2]

iii) The graph shows how the distance varied with time for the running part of the race.

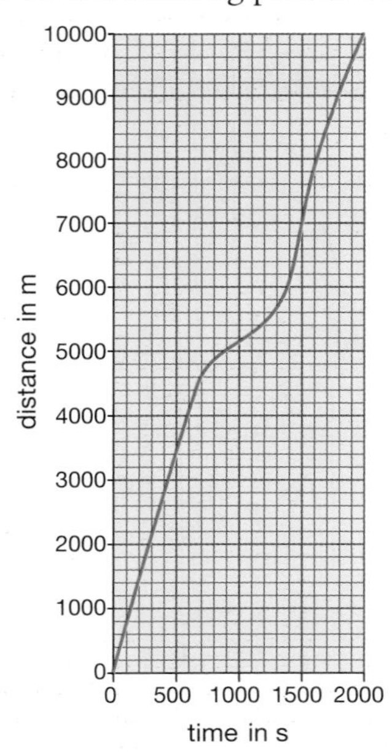

Describe how the athlete's speed changed during this part of the race. [2]

12 The table gives information about a journey made by a cyclist.

Time / hours	Distance in km
0	0
1	15
2	30
3	45
4	60
5	75
6	90

a) Plot a graph using the data in the table. [3]

b) i) Use your graph to find the distance in kilometres that the cyclist travelled in 4.5 hours. [1]

ii) Use the graph to find the time in hours taken by the cyclist to travel 35 kilometres. [1]

c) State the equation that connects average speed, distance moved and time taken. [1]

13 A train travels between two stations. The velocity–time graph shows the train's motion.

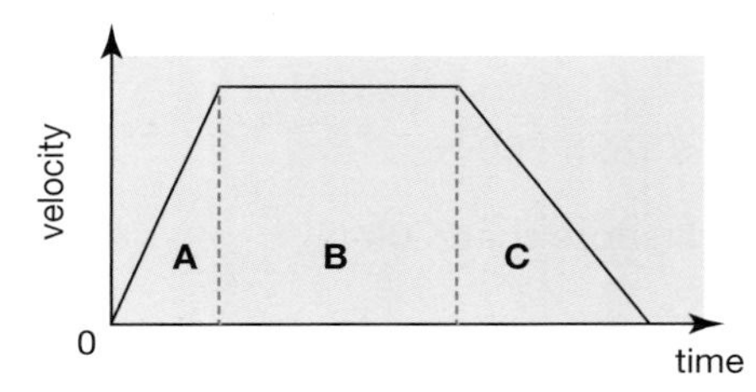

a) Give a reason why you know that the train is decelerating in part C. [1]

b) State the features of the graph that represent the distance travelled between the two stations. [1]

c) A second train travels between the two stations at a constant velocity and does not stop. It takes the same time as the first train. Sketch a copy of the graph and then draw a line showing the motion of the second train. [2]

14 a) The diagram shows a lorry. It is travelling in a straight line and it is accelerating. The total forward force on the lorry is F and the total backward force is B.

i) Which is larger, force F or force B? Explain your answer. [2]

ii) State an equation that connects acceleration, mass and resultant force. [1]

iii) A resultant force of 15 000 N acts on the lorry. The mass of the lorry is 12 500 kg. Calculate the lorry's acceleration and give the unit. [3]

b) The thinking distance is the distance a vehicle travels in the driver's reaction time. The braking distance is the distance a vehicle travels when the brakes are applied.

i) State one factor that increases the thinking distance. [1]

ii) State one factor that increases the braking distance. [1]

15 The graph shows the minimum stopping distances, in metres, for a car travelling at different speeds on a dry road.

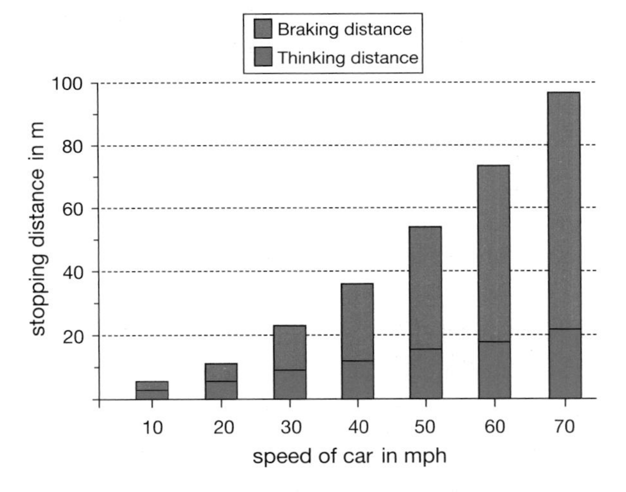

a) Write an equation to show the link between stopping distance, thinking distance and braking distance. [1]

b) Describe the patterns shown in the graph. [2]

c) Use the graph to estimate the stopping distance for a car travelling at 35 miles per hour. [1]

d) To find the minimum stopping distance, several different cars were tested. Suggest how the data from the different cars should be used to give the values in the graph. [1]

e) The tests were carried out on a dry road. If the road is icy, explain what change if any there would be to:

i) the thinking distance [2]

ii) the braking distance. [2]

16 The graph shows how the velocity of an aircraft changes as it accelerates along a runway.

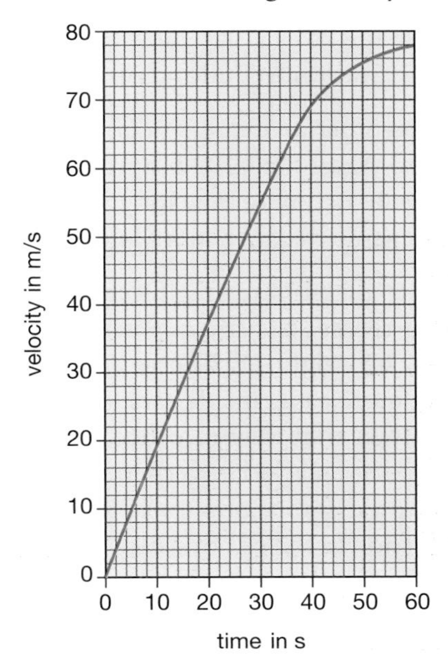

a) Use the graph to find the average acceleration of the aircraft. [3]

b) Explain why the acceleration is not constant, even though the engines produce a constant force. [3]

17 The following graph is a velocity–time graph for the Maglev train, which travels from Longyang Station, on the outskirts of Shanghai, to Pudong International Airport.

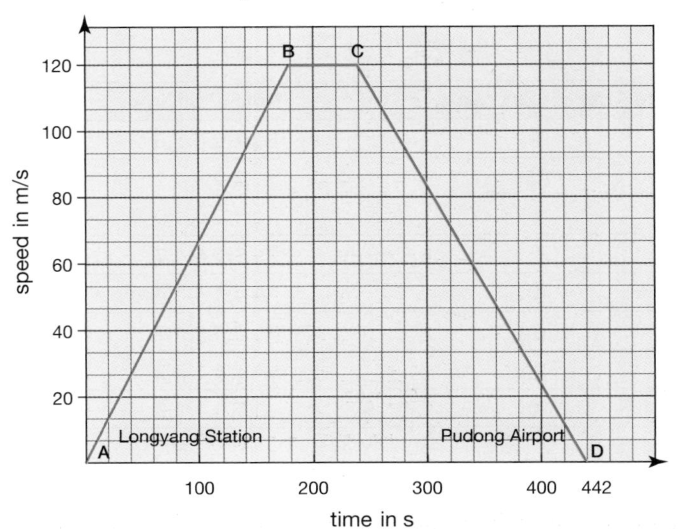

a) Use the graph to calculate the train's acceleration over the region AB. Give the unit of acceleration. [4]

b) How far does the train travel at its maximum speed? [3]

c) Use the graph to calculate how far it is from Longyang Station to Pudong Airport. [3]

d) The Trans-Siberian express takes 6 days and 4 hours to travel the 9289 km from Moscow to Vladivostok. Estimate how long it would take to cover the same distance in a Maglev train. [3]

18 Gina plans to investigate drag forces on falling objects. She decides to use a cupcake case as a parachute. The weight of the case can be increased by stacking more cases inside each other. The table shows the times that were recorded for the different numbers of cases stacked together, when they were timed falling a distance of 4 m. Where a number of 1.5 cases is shown, she cut one case in half and put it inside another case.

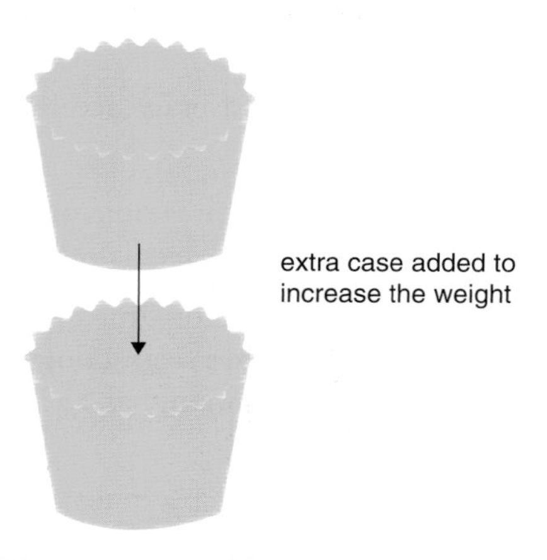

Number of cake cases	Time of fall in s
1	2.7, 2.6, 2.6
1.5	2.2, 2.3, 2.2
2	2.0, 2.0, 1,9
3	1.5, 1.6, 1.7
4	1.4, 1.4, 1.4
6	1.3, 1.3, 1.2
8	1.1, 1.1, 1.2
10	1.1, 1.1, 1.0

a) Why did Gina record three times for each number of cake cases? [1]

b) Paul said that Gina should have used a more precise stopwatch. Gina replied that her readings were accurate enough, as her reaction time introduced an error anyway. Evaluate this discussion. [2]

c) Copy the table and add two further columns to show:

i) the average time of fall [2]

ii) the average speed of fall. [2]

d) Plot a graph of the number of cases on the y-axis against the speed of fall on the x-axis. Draw a smooth line through the points and comment on any anomalous results. [5]

e) Use the graph to predict the speed of fall for a weight of:

i) 2.5 cases [1]

ii) 7 cases. [1]

f) Explain what is meant by terminal velocity. [2]

g) Draw a diagram to show the forces that act on a cupcake case, falling at its terminal velocity. [2]

h) Assuming that the cases reach their terminal velocity quickly after release, comment on how the drag force on a falling cupcake case depends on its speed. [2]

i) Paul did an experiment using a cake case with twice the area of Gina's cases. He discovered that one case fell 4 m in a time of 2.63 s. He concluded that 'this means that the area has no effect on the drag force'. Do you agree? Give a reason for your answer. [2]

19 A group of students uses a special track. The track is about 2 metres long and is horizontal. Two gliders, P and Q, can move along the track.

The surface of the track and the inside surface of the gliders are almost frictionless. The diagram shows that the gliders can move though two light gates, A and B.

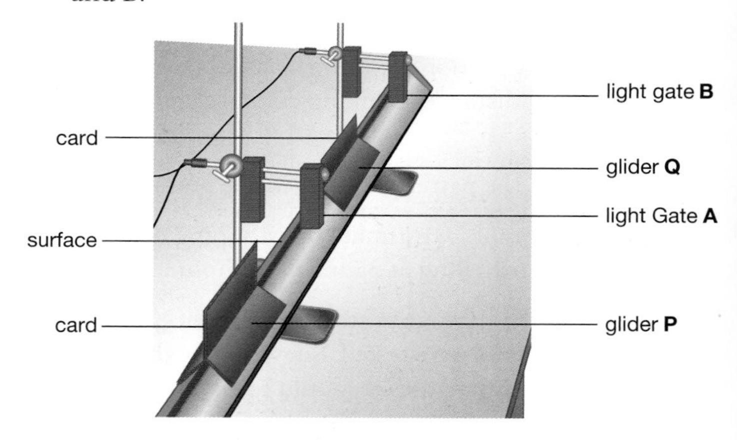

The mass of glider P is 2.4 kg. This glider is moving toward Q at a constant velocity of 0.6 m/s. Glider Q is stationary.

The diagram below shows a side view. Each glider has a card and a magnet attached. Light gate A records the time for which the card is in front of the light gate.

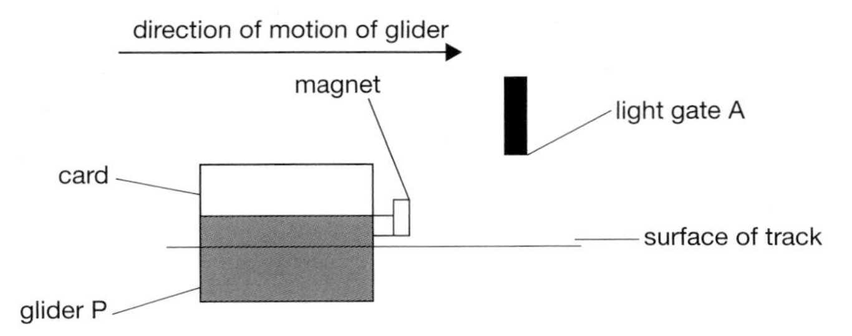

a) i) Apart from the time recorded by the light gate A, state the other measurement that would needed to calculate the velocity of glider P. [1]

ii) Why does the surface of the track need to be frictionless and horizontal? [1]

b) Momentum is a vector quantity.

i) State what is meant by a vector quantity. [1]

ii) Calculate the momentum of glider P. [2]

iii) State the momentum of glider Q. [1]

c) Glider P collides with glider Q and they move off together at a speed of 0.4 m/s.

i) State the combined momentum of P and Q after the collision. [1]

ii) Calculate the mass of glider Q. [2]

iii) Calculate the change in momentum of glider P during the collision. [2]

iv) The time taken for the collision was 0.05 s. Calculate the force that acted on glider P during the collision. [3]

v) State the size of the force acting on glider Q during the collision. [1]

20 The diagram shows a crane being used to lift a container into a boat.

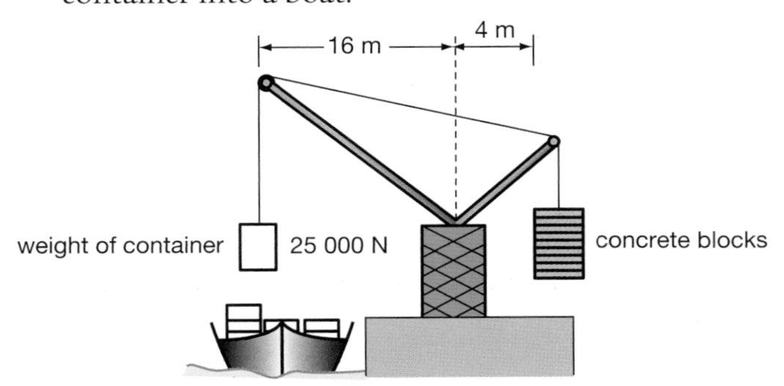

a) Calculate the turning moment of the container, about the pivot P. [3]

b) Explain the purpose of the concrete blocks. [1]

c) Calculate the weight of concrete blocks that would balance the turning moment of the container. [3]

21 The diagram shows a man climbing a step ladder.

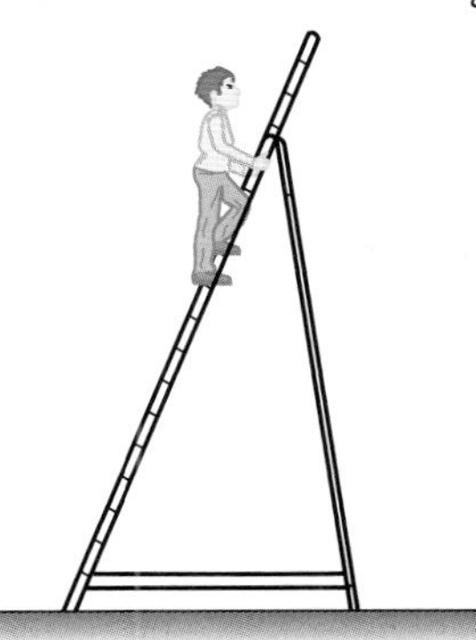

Use the term 'centre of gravity' to explain why the man is unlikely to topple over in this position. [3]

22 The diagram shows a windsurfer in action.

a) Explain why the windsurfer leans out on her sailboard. [2]

b) The weight of the windsurfer is 750 N. Calculate the turning moment caused by this weight about the mast. [3]

c) Explain why the windsurfer leans out further when the wind blows with greater force. [2]

EXTEND AND CHALLENGE

1 **a)** A rocket has a mass of 3 000 000 kg at take-off. The thrust from the engines is 33 MN. Calculate the initial acceleration.

b) The rocket burns 14 000 kg of fuel per second. Calculate the mass of the rocket 2 minutes after lift-off.

c) Calculate the acceleration of the rocket 2 minutes after lift-off, assuming that the engines produce the same thrust.

2 This question is about the take-off of from the Moon a lunar landing craft. Use the data provided:
mass of craft 20 000 kg; take-off thrust from engines on craft 52 000 N; Moon's gravitational field strength 1.6 N/kg.

a) Calculate the craft's weight on the Moon.

b) At take-off, determine the resultant force on the craft.

c) Calculate the acceleration of the craft at take-off.

d) Calculate the craft's speed 30 seconds after take-off.

e) Calculate how far above the Moon's surface the spacecraft will be after 30 seconds.

f) Give two reasons why the craft will actually have travelled slightly further than you estimate.

g) Explain why this spacecraft could not take off from the surface of the Earth.

3 The graph shows the size of the forces required to stretch samples of steel and spider's silk. Both samples are of the same thickness and length.

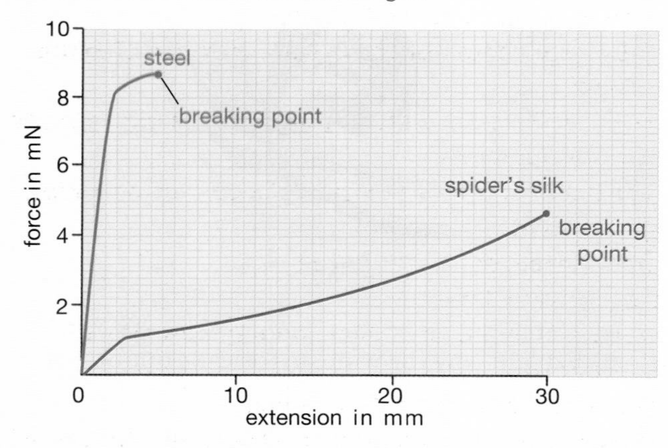

a) Explain why it was decided to use samples of silk and steel that were exactly the same length and thickness.

b) A student observed that neither material obeys Hooke's law. Evaluate this remark.

Both samples break as shown on the graph.

c) Determine the force that breaks the sample of steel.

d) Determine extension of the spider's silk when it breaks.

e) Although the sample of steel is stronger, the spider's silk is less likely to break when catching a fly. Use your understanding of forces and momentum to explain this.

4 You are in outer space, and you are worried that your crew are looking thin. How are you going to work out their mass in a weightless environment? Describe an experiment to calculate an astronaut's mass, using a chair, a large spring balance, a stop clock and a tape measure.

5 Greg test drives his new car. He accelerates rapidly until he reaches a speed of 20 m/s. The mass of the car (with him in it) is 750 kg. The graph shows how the speed changes.

a) Calculate his acceleration over the first 4 seconds.

b) Work out the force acting on the car to accelerate it.

c) Greg invites three friends (total mass 250 kg) into his car to demonstrate its rapid acceleration. Copy the graph, and show how it now accelerates over the first 4 seconds.

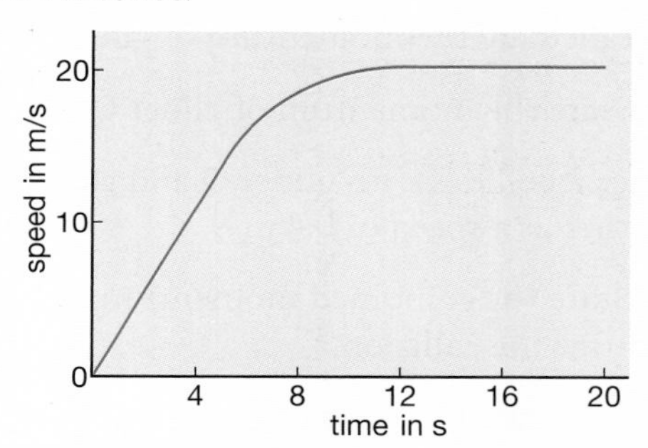

6 The diagram below shows an alpha particle colliding head on with a stationary proton in a cloud chamber. Before the collision the alpha particle has a speed of 10^7 m/s. After the collision its speed is 6×10^6 m/s. The mass of the alpha particle is 4 times that of the proton. Calculate the speed of the proton after the collision.

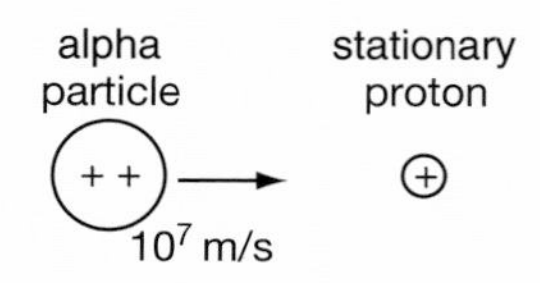

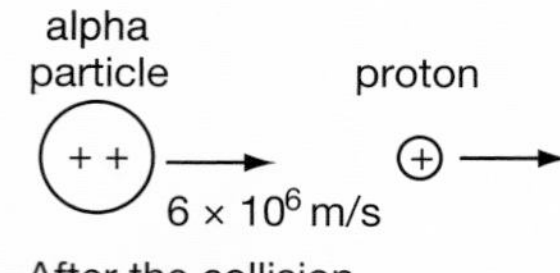

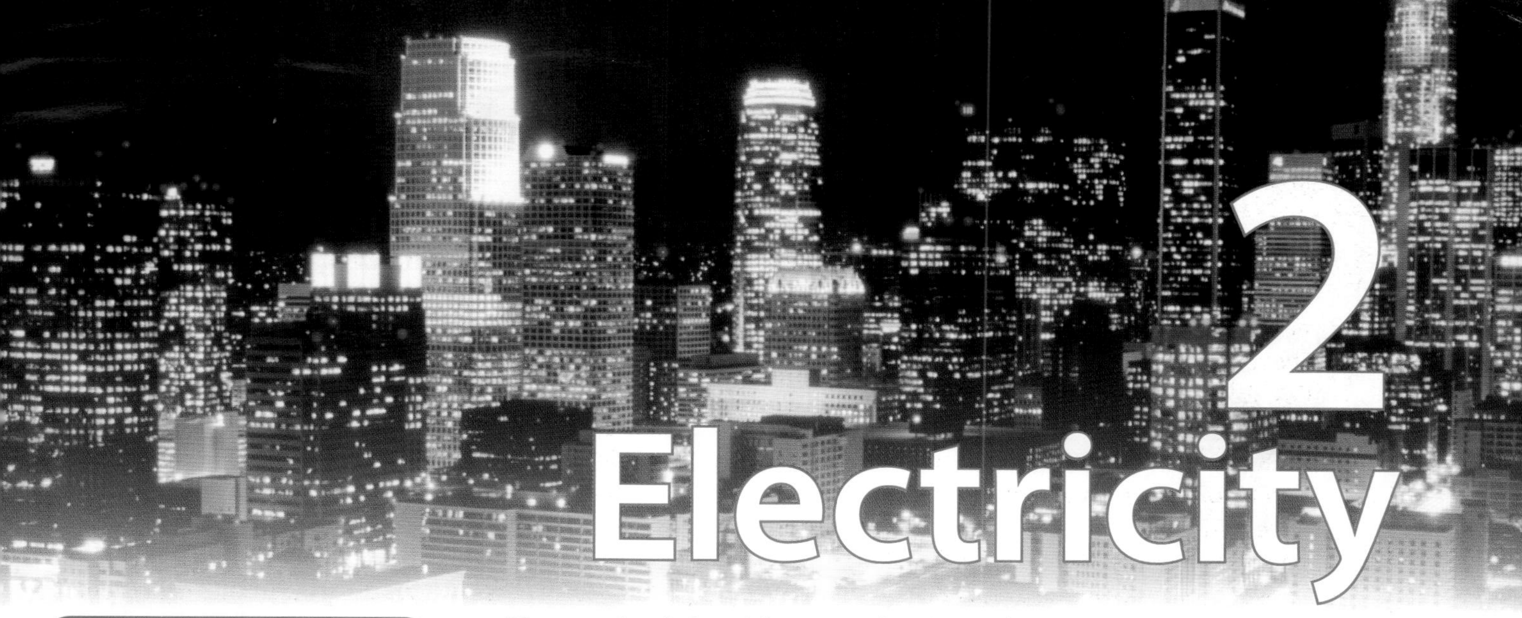

2 Electricity

DISCUSS • CALCULATE

1. How would your life have been different had you lived in the same country as you do now, in 1880?
2. How much does it cost to provide electricity to your home each year? How much does it cost to provide electricity to your school each year?
3. What impact does electricity generation have on our environment?
4. What hazards are associated with the use of electricity?

The supply of electricity to our homes and city centres is an essential part of twenty-first century life. Lights in the streets, our homes and offices allow us to be active 24 hours a day. However, electricity comes at a price and care must be taken with how much we use it.

By the end of this section you should:

- know and be able to use electrical units
- understand the use of insulation, earthing, fuses and circuit breakers
- understand why a current in a resistor results in the transfer of energy
- know that power = current × voltage
- know that energy transferred = current × voltage × time
- know the difference between alternating and direct currents
- be able to explain the use and benefits of parallel and series circuits
- understand how the current in a series circuit depends on the applied voltage and the nature of components in the circuit
- be able to describe and investigate how current varies with voltage in wires, resistors, filament lamps and diodes
- be able to describe the qualitative effect on current of changing the resistance in the circuit
- describe the variation of light-dependent resistors with light intensity and thermistors with temperature
- know that lamps and LEDs show the presence of a current in a circuit
- know that voltage = current × resistance
- know that current is the rate of flow of charge
- know that charge = current × time
- know that electric current in solid metallic conductors is a flow of negatively charged electrons
- understand that current is conserved at a junction
- know that the voltage across two components in parallel is the same
- be able to calculate the currents, voltages and resistances of two resistive components connected in a series circuit
- know that voltage is the energy transferred per unit charge
- know that energy transferred = charge × voltage
- be able to identify common materials that are insulators and conductors
- be able to explain the charging of materials by the loss or gain of electrons
- know that there are attractive forces between unlike charges and repulsive forces between like charges
- be able to explain the potential dangers of electrostatic charges
- be able to explain some uses of electrostatic charges

2.1 Introducing electricity

Figure 1.1

Figure 1.2

Figure 1.1 shows how important electricity is in our lives. Electrical devices provide entertainment in the family room and convenience in the kitchen. How many electrical devices do you use every day?

Electrical transfer of energy

Below are two statements about a torch cell connected to its lamp. Which one is correct?

Statement A: The cell does electrical work to transfer energy from one store to another. Energy is transferred from the chemical store in the cell, into the thermal store of the lamp. Energy is then transferred to the thermal store of the surroundings, by heating and radiation.

Statement B: The cell in an electrical circuit is like a pump in a central heating system that circulates water. Electricity in the wires is pushed round by the cell; water is pushed through the pipes by the pump.

The answer is that both ideas are helpful and correct. Statement A provides a good description of electrical transfer of energy from one store to another. (You meet this idea again in Section 4.) Statement B uses a model (or an analogy) of water flow, which is designed to help the student visualise how electrical circuits work.

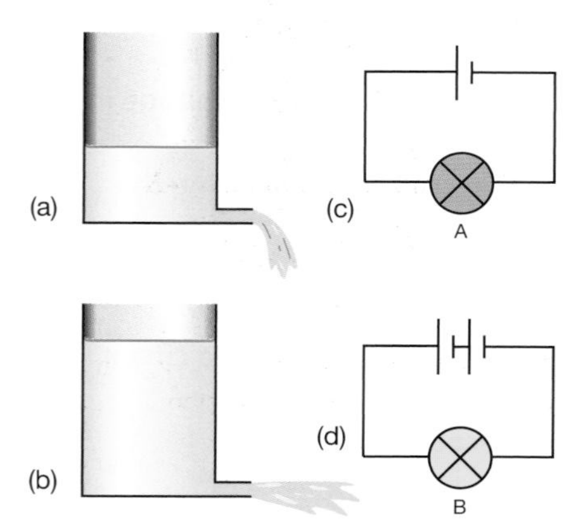

Figure 1.3

Cells and brightness

Figure 1.3(a) shows a tank of water with an outlet pipe at the bottom; water flows out at a slow rate. When the depth of water in the tank is increased, as in Figure 1.3(b), the pressure of water at the bottom of the tank is increased and water flows out at a faster rate.

The cell in an electrical circuit provides an 'electrical pressure' that pushes electricity through the lamp. In Figure 1.3(c) one cell pushes electricity through the lamp at a slow rate and the lamp is dull; energy is being transferred at a slow rate. In Figure 1.3(d), two cells push the electricity round the circuit more quickly and the lamp is brighter; electrical energy is now being transferred more quickly.

Voltage and current

The words voltage and current are important in describing electrical circuits.

- When describing a stream or river we use the word 'current' to describe the flow of water. When the current is large, this means lots of water flows past us each second.
- In an electrical circuit we use the same word. Current describes the flow of electricity. When the current is large a lot of electricity flows through the circuit.

The unit of current is the **ampere** (A), though this is often shortened to amp.

Figure 1.4 Two men are required to direct this fire hose due to the high pressure of the water.

- Figure 1.4 shows two fire fighters practising to use their hose. The pressure in the hose is high and water flows quickly.
- The voltage of a cell is like an 'electrical pressure.' A high voltage drives a large current around the circuit. High voltages transfer large amounts of energy and can be dangerous.
- Electric shocks delivered from a high voltage source can be very painful, and can even cause death. This is why we are warned about high voltages.

The unit of voltage is the **volt** (V).

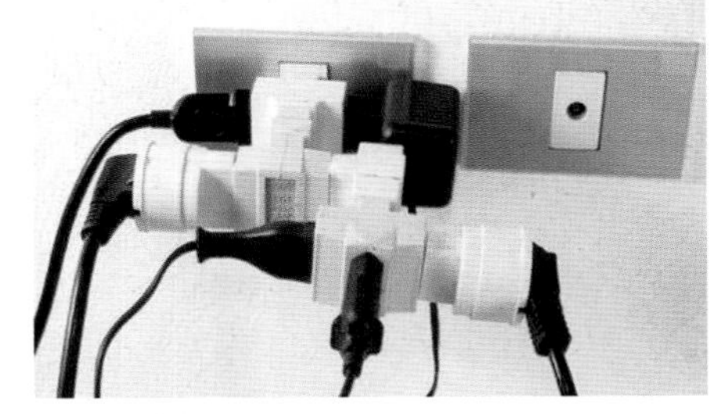

Figure 1.5 This socket has many electrical devices plugged into it. Together they might draw too much current – a large current can melt wires.

Electrical hazards

Electricity provided for the home is called the **mains supply**. The voltage of the mains supply differs from country to country but is usually either about 230 V or 120 V. Either voltage is large enough to be dangerous.

Due to the dangers of electricity, we must be on the lookout for hazards. Some of these are listed below:

- Look out for damaged plugs or bare wires.
- Remember that an electric fires and toasters expose you to high voltages. Do not try to remove toast from the toaster with a metal knife or fork.
- Do not overload a socket by connecting many plugs. A large current might melt a wire.
- Be careful outside. Do not cross railway lines. Keep clear of overhead cables.

Figure 1.6 Be careful near overhead cables. You could get a shock by flying a kite or carrying a fishing rod too close to them.

Figure 1.7 Keep clear of anything that displays this hazard warning sign.

STUDY QUESTIONS

1 a) Choose two electrical hazards mentioned in the chapter and explain what you would do to remove the danger.
b) Identify two electrical hazards that have not been mentioned in the chapter. Explain why we need to be careful of these hazards.

2 Describe the energy transfers that occur when a cell is used to light an electric lamp.

2.2 Using mains electricity

Figure 2.1 Electricity can be used to keep us warm.

In the home we use electricity for heating, lighting, cooking and powering many other devices, which we use for leisure or tasks.

Alternating and direct current

In Figure 2.2(a) and (b) you can see two lamps that are lit by two different electrical supplies. In Figure 2.2(a) the lamp is lit by a 6 V battery. A battery provides a direct current (**d.c.**) – this means that there is a constant current in one direction. In electrical diagrams a cell is represented by two vertical lines, one long and the other shorter. Here the longer line is on the left. This is the positive side or 'terminal' of the cell. We use the convention that a current flows from the positive terminal to the negative. The current flow is marked with an arrow.

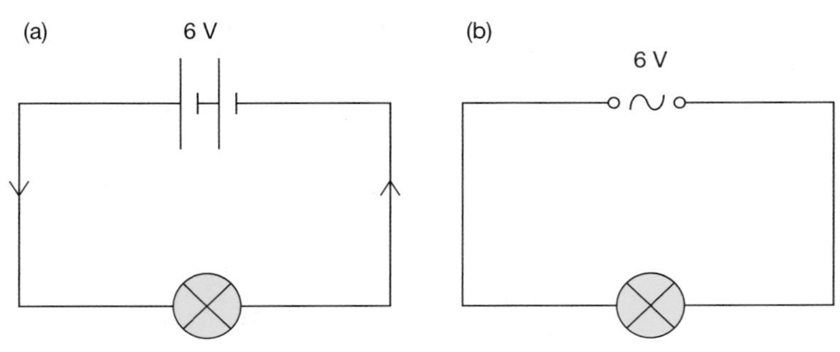

Figure 2.2 (a) In a direct current circuit the current is constant in one direction round the circuit.
(b) In an alternating circuit the current constantly changes direction, from one direction to the other.

d.c. supply
6 V
0
0.02
voltage in V
a.c. supply
time in s

Figure 2.3 Direct and alternating supply voltages.

In Figure 2.2(b) the lamp is lit by an alternating supply (**a.c.**). In an alternating supply, the voltage switches direction many times each second and the current changes direction too.

Figure 2.3 shows graphs of how the a.c. and d.c. supplies change with time. The d.c. supply remains constant at 6 V; the a.c. supply changes from positive to negative. The peak of the a.c. supply rises above 6 V to make up for the time when the voltage is close to zero.

Mains supply

All countries use a.c. supplies for their mains electricity. Most countries use a mains voltage of about 120 V or 230 V. By using a standardised voltage, electrical devices can be used in different countries.

We also describe an alternating current by its **frequency**. The frequency of most mains supplies is either 50 cycles per second or 60 cycles per second. Figure 2.3 shows one cycle of an a.c. supply. So if the frequency of a supply is 50 cycles per second, the current completes each cycle in 1/50 of a second.

The unit of frequency is the **hertz**. A frequency of 50 cycles per second is written as 50 hertz or 50 Hz.

Sometimes direct currents fluctuate. But provided the current is always in the same direction, it is still called a direct current.

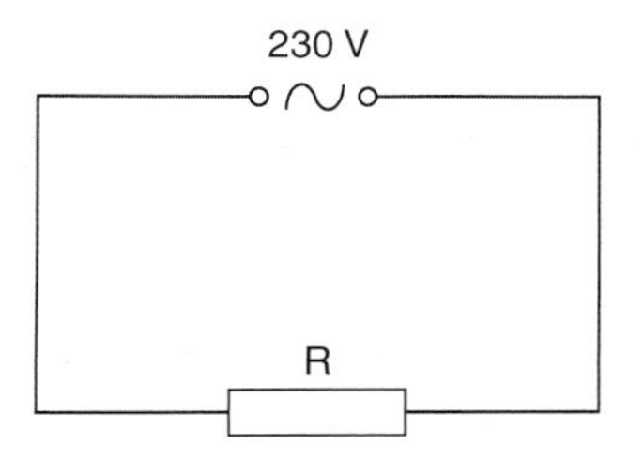

Figure 2.4 The electrical work done by the power supply transfers thermal energy to the surroundings.

Electrical heating

Figure 2.4 shows a 230 V mains supply. It provides current to a heater, which can also be called a **resistor**. This is shown in the diagram, marked 'R'. The symbol R is used, because we say that the heater offers **resistance** to the current.

When there is a current through a resistor, there is an electrical transfer of energy into the resistor's thermal store. This results in an increase in the temperature of the resistor. Some resistors glow red hot.

Electricity at home

We use electricity in our homes for several things, e.g. lighting, cooking, heating, computers and TVs. Each electrical device is designed to take a different current, depending on its use. Table 1 shows the size of current through some electrical appliances attached to a 230 V mains supply.

Table 1 Some electrical appliances and the currents drawn by them

Appliance	Current in A
lamp	0.1
computer monitor	0.5
hairdryer	3.0
toaster	4.0
convector heater	8.5
kettle	12.0

Fuses

Fuses are used with electrical devices for safety. Their main purpose is to prevent a fire.

A fuse contains a thin wire that heats up and melts if there is too much current. If this happens we say that the fuse has 'blown'. The rating of a fuse is the maximum current that can pass through it without melting the fuse wire.

You need to choose the correct fuse to put into a plug or fuse board, so that it blows if there is a fault. In the UK it is common to use 3 A or 13 A fuses. Look at the appliances in Table 1. You need to use a 3 A fuse for the lamp and the computer monitor, and a 13 A fuse for the other appliances. A lamp will work with a 13 A fuse but is it safer to use a 3 A fuse. A lamp works on a low current such as 0.1 A or 0.2 A. So, if 3 A go through a lamp circuit it means there is a fault – a 3 A fuse will blow but a 13 A fuse will not.

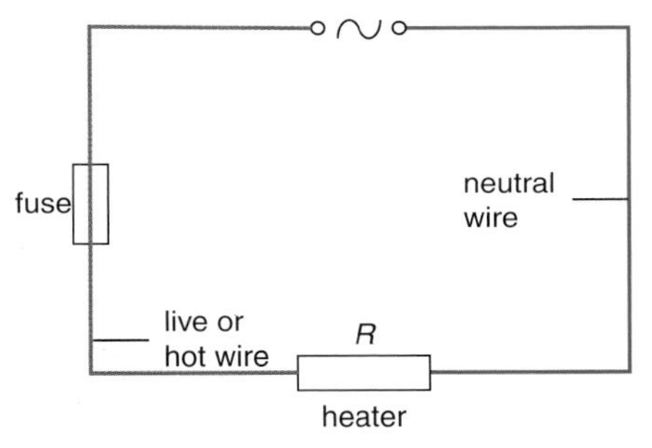

Figure 2.5 A fuse in a circuit.

Figure 2.5 shows the symbol for a fuse. In most countries mains electricity is supplied by a **live** wire (or **hot** wire in the USA) together with a **neutral** wire. The live or hot wire has a large voltage, which supplies the energy. This is the dangerous wire. The fuse is put into the live wire so that it is disconnected when a fuse blows.

Circuit breakers

A circuit breaker is an electromagnetic switch that opens or 'trips' when a current is bigger than a certain value. The advantages of circuit breakers over fuses are that they are sensitive and also they can quickly be reset once a fault has been put right. Circuit breakers are often found in modern houses instead of fuses in the fuse boxboard.

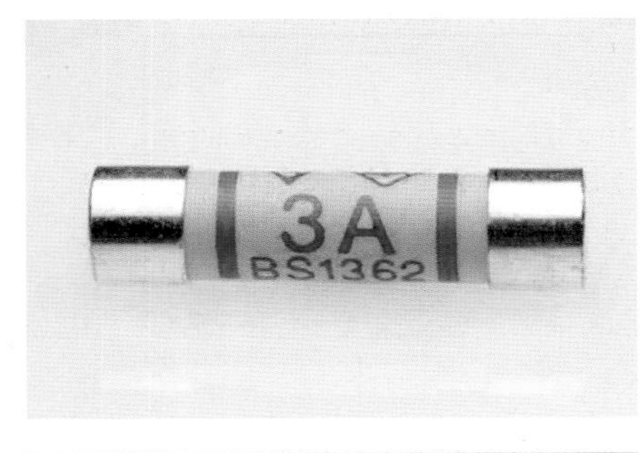

Figure 2.6 Different fuses.

Earthing

Many countries use three-pin plugs for the electrical supply. The third wire is called the earth wire, which as its name suggests is attached to the earth. Figure 2.7 shows a diagram of a correctly wired three-pin plug in use in the UK. The live wire is brown, the neutral wire is blue and the earth wire is green and yellow.

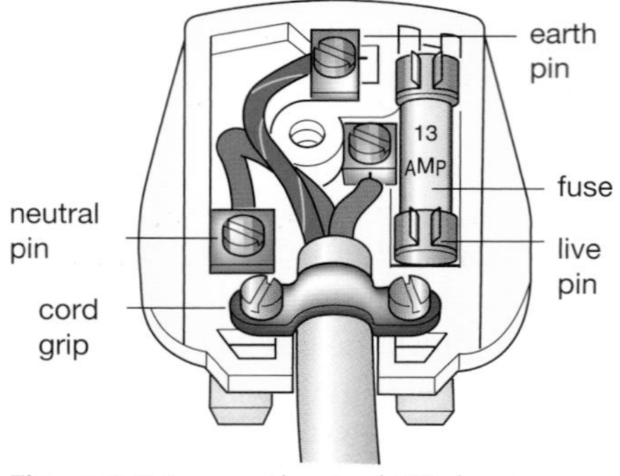

Figure 2.7 A correctly wired UK plug.

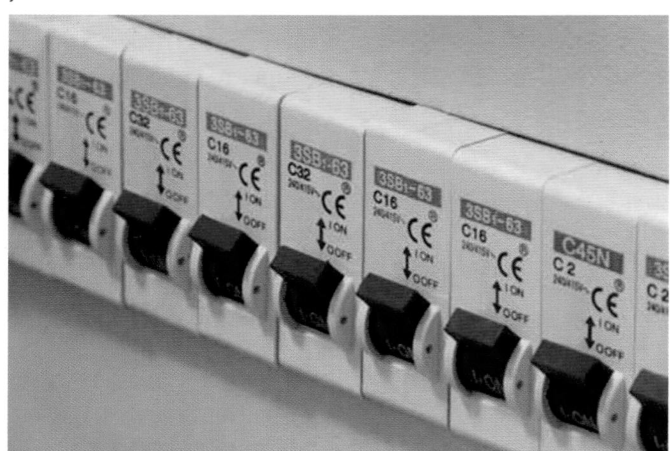

Figure 2.8 Circuit breakers in a modern fuse box.

TIP

The earth wire and fuse together protect the person using an electrical device.

Figure 2.9 explains why the earth wire is an important safety feature. Figure 2.9 (a) shows the circuit diagram for an electric heater, which is enclosed in a metal case. The current is goes from the live wire to the neutral and the fire is working safely. The earth wire is attached to the metal case. When the heater is working correctly the earth wire does not carry a current.

Figure 2.9 (b) shows what happens when a fault develops. Now a wire is touching the metal case. Without the earth wire, someone could touch the metal case and get a shock. With the earth wire in place current goes through the earth wire and not the person touching the case. A large current going to earth can also blow the fuse, which disconnects the dangerous live wire.

TIP

Metals are conductors that allow a current to pass.
Plastics are insulators which do not allow currents to pass.

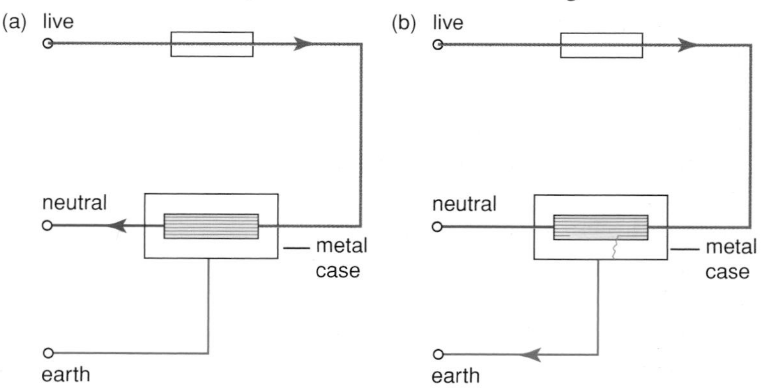

Figure 2.9 (a) The heater works safely. **(b)** A broken wire touches the metal case and the current goes to earth; if the current is large, the fuse blows, disconnecting the live wire

Double insulation

Many modern electrical appliances are made out of plastic, which does not conduct electricity. Plastic is an **insulator**. Appliances that are made of plastic can safely be connected to the mains using only two wires. The earth wire is not needed as you cannot get a shock from the plastic.

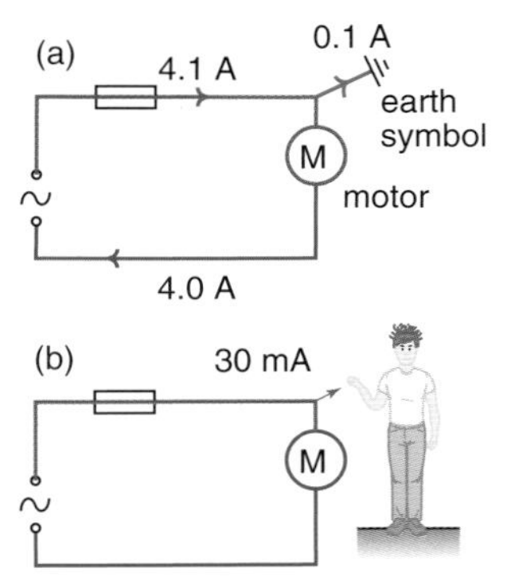

Figure 2.10 (a) An electric lawnmower will continue to work even though a damaged live wire allows a small current to goes to earth.
(b) A small current (30 mA) goes through a person's body. This current is too small to blow a fuse.

TIP

Make sure you understand these terms about electricity in the home:
- fuse
- earthing
- circuit breaker
- double insulation

This method of protection is called **double insulation** and it is shown by the sign ⧈. When you see this sign on an appliance you know you do not need an earth wire.

Residual-current circuit breaker (RCCB)

A residual-current circuit breaker (RCCB) is a device that disconnects a circuit when it detects that the current is not the same in the live and neutral wires. When the currents are different there is a small current leaking to earth, which could be through the body of a person who accidentally touches the live wire.

Figure 2.10 shows two examples of where there could be an imbalance of current. A gardener is mowing his lawn with an electric mower. The mower is double insulated, so the gardener is protected if the live wire becomes detached inside the mower.

It is possible that the wire to the mower becomes damaged – it could be cut by the mower itself. In Figure 2.10 (a), a damaged live wire allows a small current to earth. This extra current is not large enough to melt the fuse, so the mower keeps working.

In Figure 2.10 (b), the gardener has turned off the mower and picks up the wire. A residual current passes through him. This current is small – only 30 mA, but it could deliver a lethal shock.

An RCCB, which detects an imbalance of currents between the live and neutral wires, will disconnect the supply before an injury occurs. RCCBs are widely used in different countries.

STUDY QUESTIONS

1 **a)** Explain the purpose of a fuse in a mains circuit.
  b) State the advantages circuit breakers have over fuses.

2 Explain carefully each of the following:
  a) why the fuse is put into the live or hot wire of a mains circuit
  b) why the rating of a fuse should be slightly higher than the normal working current through an appliance
  c) why appliances with metal cases need to be earthed, but appliances with plastic cases do not need to be earthed.

3 Name the colours of the live, neutral and earth wires in your country.

4 Explain the difference between an a.c. and a d.c. electrical supply.

5 State and explain the energy transfers that occur in an electrical circuit when a battery is connected to a resistor.

6 The diagram shows the trace on an oscilloscope that is being used to measure the voltage of an electrical supply.

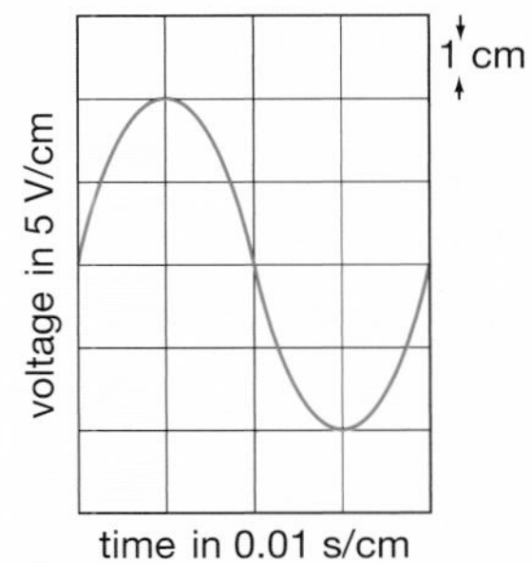

  a) State the type of supply.
  b) Determine the peak (maximum) voltage of the supply.
  c) Calculate the frequency of the supply.

7 Do some research to find out how an RCCB works.

2.3 Electrical power

Figure 3.1 This car runs on electricity instead of petrol. Here its battery is being charged up from the mains.

TIP

Remember these letters for electrical quantities:
- power, P
- voltage, V
- current, I

How do electric cars shape up against petrol-driven cars? Find out: how long it takes to charge up an electric car; how much it costs to drive 100 km; what the average range of an electric car is. By 2040, only electric cars will be for sale in the UK.

Calculating the power

When you use an electrical appliance at home you are interested in how much energy it provides you with in each second. This determines how bright a light is, or how effective a heater is at keeping you warm. The energy supplied per second is called the **power**.

$$\text{power} = \text{energy per second}$$

Power is measured in **joules per second** (J/s) or **watts** (W). Sometimes power is supplied in large quantities. An electric fire can do 1000 J of electrical work per second (which is the same as 1000 W), transferring energy to the thermal store of the surroundings; 1000 W is called a **kilowatt** (1 kW). If you want to talk about the output of a power station you will need to use **megawatts** (MW); 1 MW is a million watts. Some power stations produce 2000 MW of electrical power.

You can work out the electrical power, P, used in a circuit, if you know the voltage, V, of the supply and the current, I, using the formula:

$$\text{power} = \text{current} \times \text{voltage}$$
$$P = I \times V$$

Tables 1 and 2 give examples of electrical appliances with their power, operating voltage and the current supplied.

Table 1 Power used by low-voltage applications

Appliance	Power in W	Operating voltage in V	Current in A
torch LED	0.12	3	0.04
torch lamp	0.9	3	0.3
pocket calculator	0.0003	3	0.0001
car window heater	90	12	7.5
electric van	3600	72	50

Table 2 Power used by household appliances with a mains voltage of 230 V

Appliance	Power in W	Operating voltage in V	Current in A
lamp	23	230	0.1
TV	69	230	0.3
fridge	115	230	0.5
hairdryer	690	230	3.0
microwave	1380	230	6.0
electric shower	1955	230	8.5

Choosing a fuse

Countries with a 230 V mains supply commonly use 3 A, 5 A or 13 A fuses. It is safest to use a fuse which is just slightly above the maximum current that the appliance uses.

Example. You buy a kettle that is marked 230 V 2.4 kW. Which fuse should you use in the plug?

You can calculate the current using the formula:

$$P = I \times V$$
$$2400 = I \times 230$$
$$I = \frac{2400}{230}$$
$$2400 = 10.4 \text{ A}$$

You must use a 13 A fuse to protect the kettle because the current will blow a smaller fuse.

Paying for energy

When you pay an electricity bill you are paying the electricity company for the energy you have used. You can calculate how much energy has been used with the formula:

energy = power × time

or

energy = current × voltage × time

TIP

Remember the formula
$E = I \times V \times t$

Example. Calculate the energy used when a 1 kW heater is on for 1 hour.

$$\text{energy used} = \text{power} \times \text{time}$$
$$= 1000 \text{ W} \times 3600 \text{ s}$$
$$= 3\,600\,000 \text{ J}$$

STUDY QUESTIONS

1 Calculate the power supplied to each of the following appliances when they are attached to the correct operating voltage.
 a) a 12 V, 5 A lamp
 b) a 230 V, 2 A heater
 c) a 115 V, 4 A toaster

2 a) Calculate the current flowing in each of the following electrical appliances when they are supplied with the correct operating voltage.
  i) a 230 V, 11.5 W lamp
  ii) a 115 V, 2 kW shower
  iii) a 230 V, 800 W microwave oven
  iv) a 115 V, 460 W TV set
 b) You have available 3 A, 5 A, 13 A and 20 A fuses. Which would you choose for each of the appliances in part a) of the question?

3 a) Make a list of some household electrical appliances that run off the mains supply, for example television, heater, fridge.
 b) Look on the back of the appliance to find out the recommended voltage supply for each, and the power used.
 c) Calculate the current used by each device and recommend a safe fuse to use.

4 A householder goes on holiday for two weeks. To deter burglars she leaves four lights that come on for 6 hours a day. Each bulb is rated at 23 W.
 Calculate the energy used over the two weeks in:
  i) joules
  ii) MJ.

5 The battery for an electric car is charged from a 230 V supply using a current of 65 A. The charging time to charge the battery fully is 3.5 hours. Calculate the electrical energy used to charge the battery in joules.

2.4 Electric circuits

Figure 4.1 A continuous circuit is needed to light this lamp.

A continuous circuit

Figure 4.2(a) shows how some components are connected together to make an electrical circuit. This is called a **circuit diagram**. You can see that the current passes from the positive terminal of the cell, through an ammeter and resistor, before returning to the cell. The current passes through all the circuit components, one after another. These components are in **series**.

The ammeter is a device that measures the size of the current – 1 A in this example. In Figure 4.2(a) the switch, S, is closed. This means there is a complete, continuous circuit for the current. When the switch is open (Figure 4.2(b)), there is a break in the circuit and there is no current.

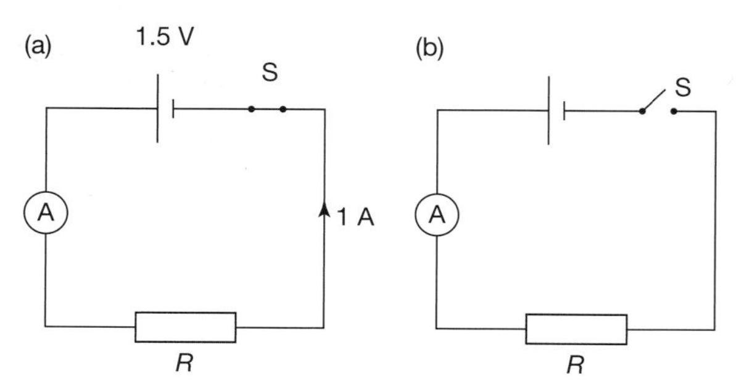

Figure 4.2 Circuit diagrams with the components in series.

TIPS

- Doubling the voltage doubles the current in a series circuit.
- Increasing the resistance reduces the current.

Controlling the current

You can change the size of the current in a circuit by changing the voltage of the cell or battery, or by changing the components in the circuit. Figure 4.3 shows the idea.

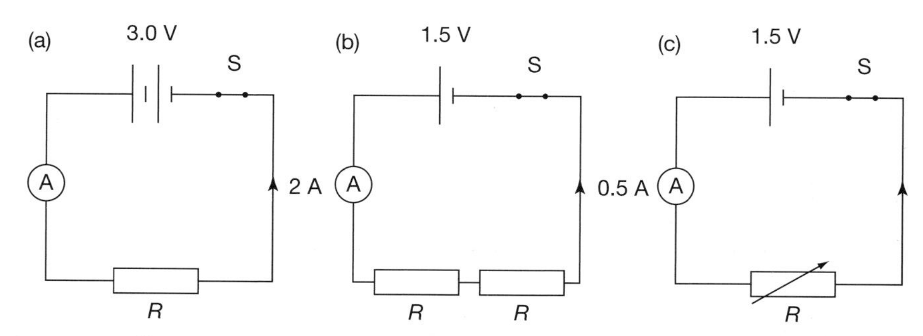

Figure 4.3 Different series circuits.

In Figure 4.3(a), an extra cell has been added to the circuit in Figure 4.2. Now the current is larger.

In a series circuit increasing the voltage increases the current.

In Figure 4.3(b), the voltage of the cell is 1.5 V but now the resistance has been increased by adding a second resistor to the circuit. This makes the current smaller.

PRACTICAL

You should be able to set up a series circuit and use an ammeter to measure the current.
Note: Avoid connecting the ammeter across the cell or battery without there being a resistor in series - the ammeter may be damaged if you do.

In a series circuit increasing the resistance makes the current smaller.

In Figure 4.3(c) a variable resistor has replaced the fixed resistor in the circuit. You can control the size of the current by adjusting the resistor.

Figure 4.4 shows how the variable resistor works. If you use terminals B and C you have a fixed resistor. But if you use terminals A and C you have a variable resistor. The current passes through the thick metal bar at A, then along the resistance wire to C via the sliding contact. The metal bar has low resistance. So the resistance gets lower as the slider moves away from A.

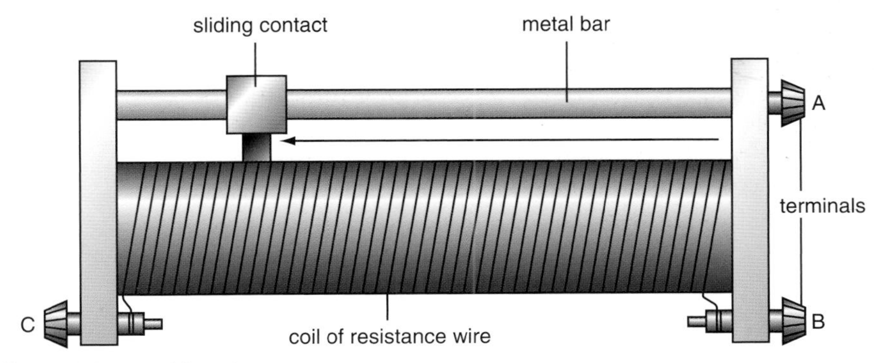

Figure 4.4 A variable resistor.

STUDY QUESTIONS

1 a) Draw a circuit diagram to show a cell connected in series to an ammeter, a diode and a fixed resistor.
b) Draw a circuit diagram to show how you can control the brightness of a lamp.

2 In the circuit diagrams, lamp A is connected to a single cell. The lamp lights with normal brightness.

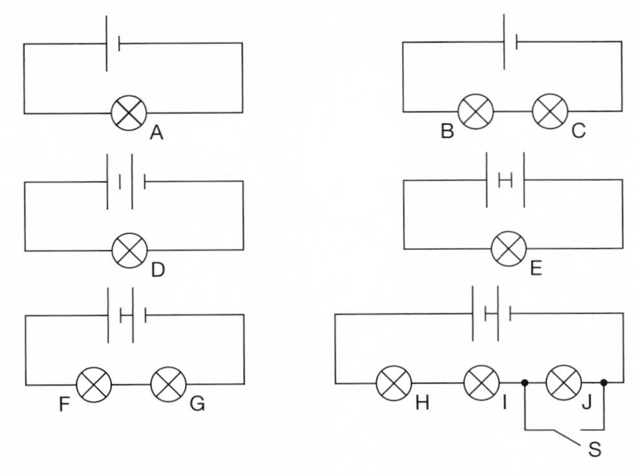

a) Use the words normal, bright, dim or off to describe the lamps B, C, D, E, F and G.
b) Describe the brightness of the lamps H, I and J when the switch S is:
i) open
ii) closed.

3 a) Which of the lamps in the diagram, K or L, will light?

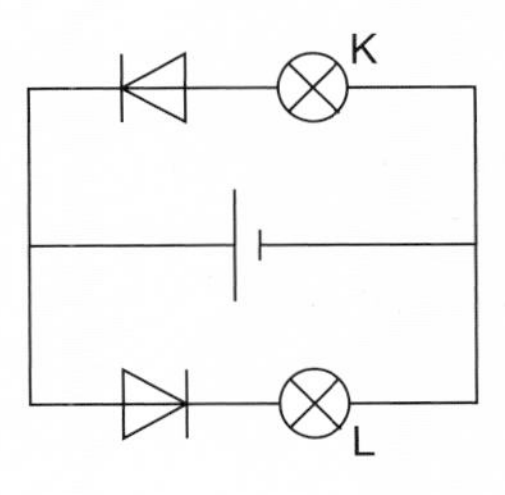

b) What will happen to the lamps if the cell is turned around?
c) Draw a new circuit diagram with two diodes and two lamps, so that both lamps light.

4

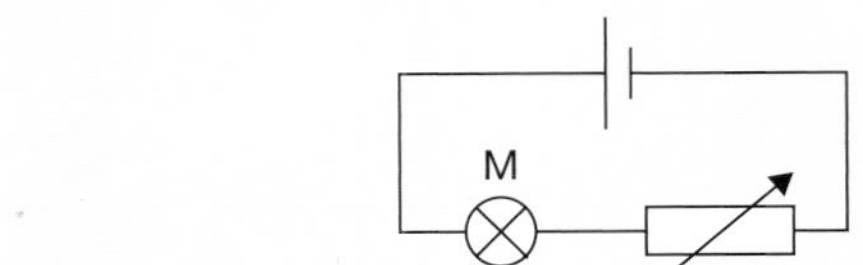

Describe what happens to the brightness of lamp M as the resistance of the variable resistor is decreased.

Make sure you know your circuit symbols.

Circuit symbols

Table 1 shows the circuit symbols for the electrical components that you will meet in the next few chapters. A brief explanation of the function of each component is given next to each symbol.

Table 1 Circuit symbols

Description	Symbol
Conductors crossing with no connection	
Junction of conductors	
Open switch	
Cell	
Battery of cells	
Power supply (DC)	+ –
Power supply (AC)	
Transformer	
Ammeter	A
Voltmeter	V
Fixed resistor	
Variable resistor	
Heater	

Description	Symbol
Thermistor	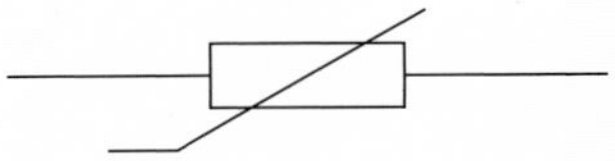
Light-dependent resistor (LDR)	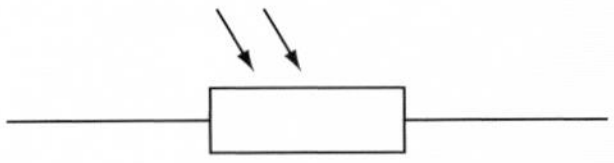
Diode	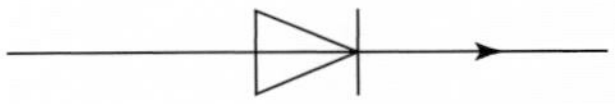
Light-emitting diode (LED)	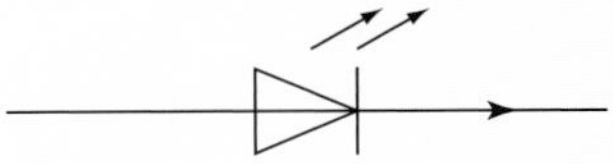
Lamp	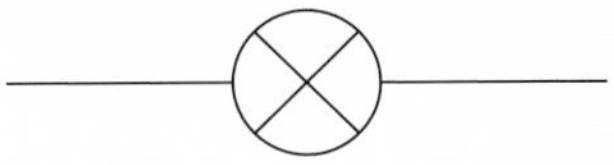
Loudspeaker	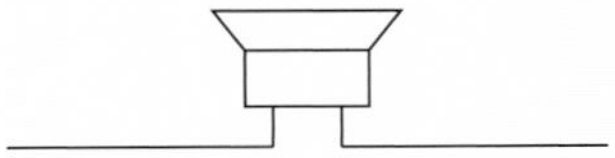
Microphone	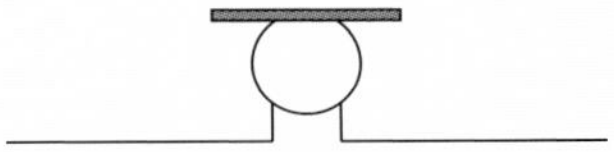
Electric bell	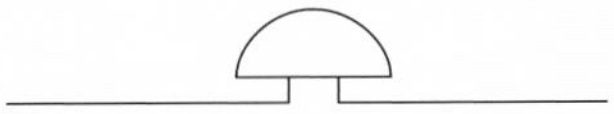
Earth or ground	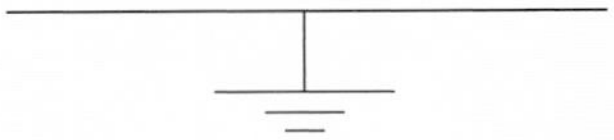
Motor	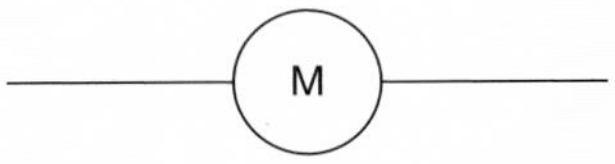 M
Generator	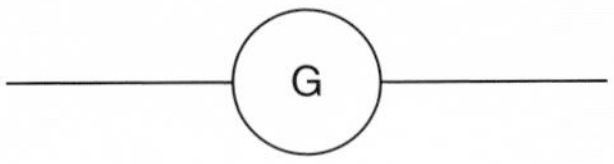 G
Fuse/circuit breaker	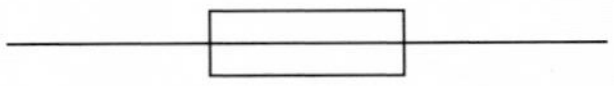

2.5 Calculating the resistance

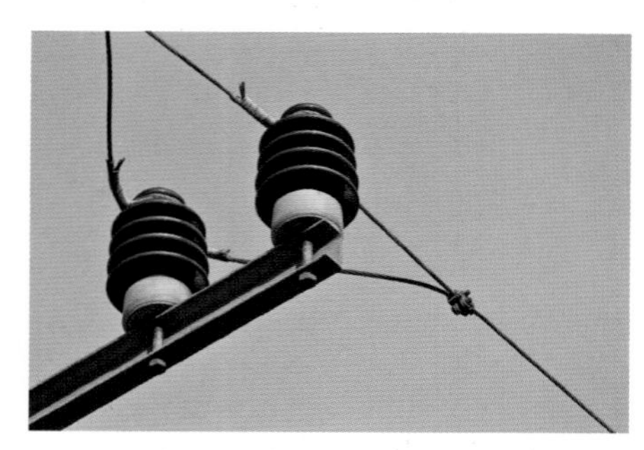
Figure 5.1

These ceramic discs have a very high resistance so that the power line is insulated from the metal pylon. Each ceramic disc is connected to the next with a metal core to provide strength.

When the ceramic insulators are wet, a small leakage current can flow down the outside of the ceramic discs. Discuss how the design of the discs reduces the size of this current.

What changes would you make to this insulator to make it suitable for a higher voltage power line?

Ammeters and voltmeters

Figure 5.2 shows you how to set up a circuit using an ammeter and a voltmeter.

- The ammeter is set up in **series** with the resistor, *R*. The same current passes through the ammeter and the resistor. The ammeter measures the current in amperes (A), or milliamperes (mA). 1 mA is 0.001 A.
- The voltmeter is placed in **parallel** with the resistor, *R*. The voltmeter measures the **voltage** across (between the ends of) the resistor. The unit of voltage is the volt (V). The voltmeter only allows a very small current to pass through it, so it does not affect the current passing around the circuit.

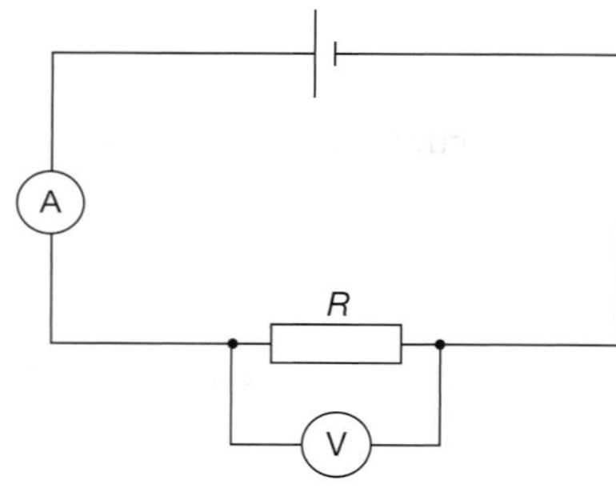

Figure 5.2 Meters in a simple circuit.

Resistance

The circuit in Figure 5.2 can be used to calculate the resistance in the circuit. The resistance of a resistor or electrical component is defined using this equation:

$$\text{resistance} = \frac{\text{voltage}}{\text{current}}$$

This equation is usually written as:

$$R = \frac{V}{I}$$

where *R* is the resistance, *V* is the voltage across the resistor and *I* is the current flowing through the resistor.

The unit of resistance is the **ohm** (Ω). Resistances are also measured in kΩ, where 1 kΩ is 1000 Ω.

TIP

You will often see the equation on the right written in this form:
voltage = current x resistance
or $V = I \times R$

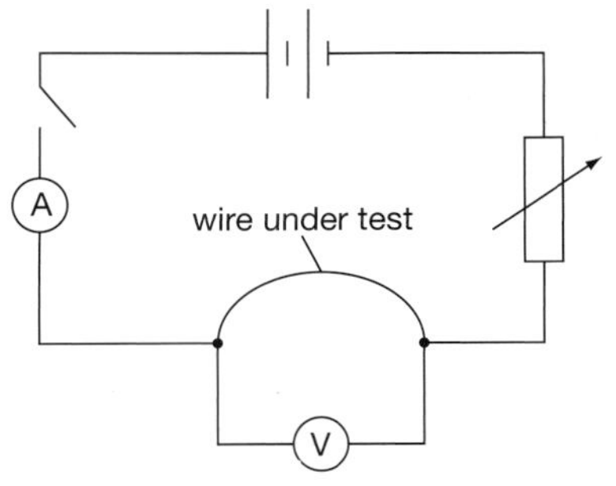

Figure 5.3 Investigating the resistance of a metal wire.

Ohm's law

The circuit diagram in Figure 5.3 shows how you can investigate the resistance of a wire. A variable resistor is used to vary the current flowing. You can take pairs of readings of current and voltage. Table 1 shows some typical results.

To avoid risk of damaging the meters, before switching on check that the ammeter is in series and the voltmeter is connected in parallel with the resistor. Also check that they are able to read the expected current and voltage.

Table 1

Voltage in V	0.5	1.0	1.5	2.0	2.5	3.0
Current in A	0.02	0.04	0.06	0.08	0.1	0.12

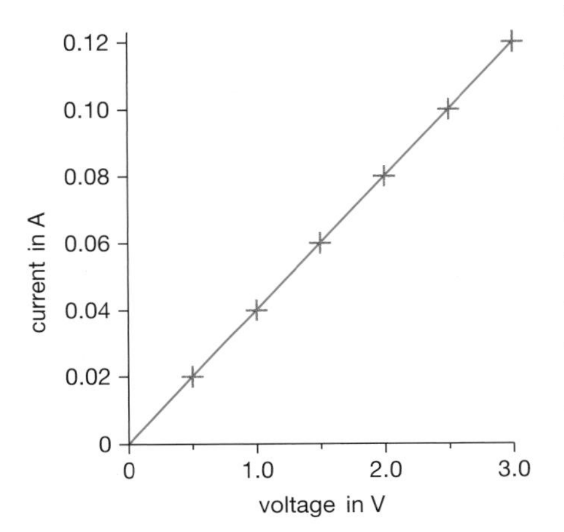

Figure 5.4 A current–voltage graph.

You can then plot a graph to show how the current flowing through the resistor depends on the voltage across it. Figure 5.4 shows the points plotted. This is called a **current–voltage graph**. A straight line has been drawn through the points, which passes through the origin. This shows that the current is *directly proportional* to the voltage. If the direction of the voltage is reversed, the graph has the same shape. The resistance is the same when the current direction is reversed.

When a resistor or wire behaves in this way, it is said to obey **Ohm's law**. This states that:

For some resistors, at constant temperature, the current through the resistor is proportional to the voltage across it.

The resistor in this case is said to be **ohmic**.

The resistance can be calculated using any voltage and the corresponding current:

$$R = \frac{V}{I} = \frac{2.0}{0.08} = 25\,\Omega$$

Check that the resistance is the same using another pair of voltage and current readings.

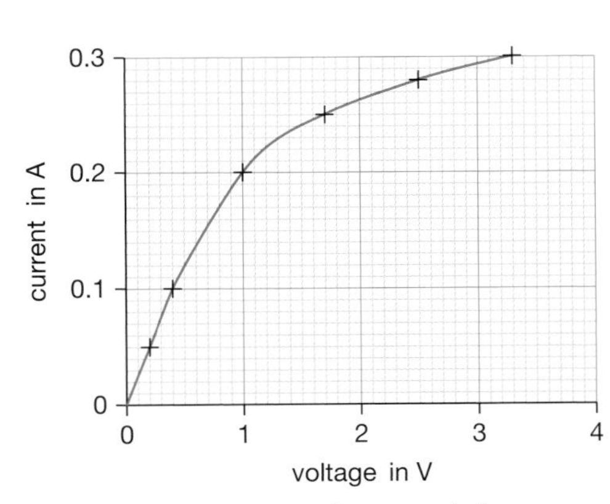

Figure 5.5 A current–voltage graph for a filament lamp.

The filament lamp

The circuit in Figure 5.3, with the test wire replaced by a filament lamp, can be used to investigate how the current through a filament lamp changes with the applied voltage. The results of the investigation are plotted in a graph, which is shown in Figure 5.5. The line is not straight – it curves away from the current axis (y-axis). The current is not proportional to the applied voltage. So the filament lamp does not obey Ohm's law. We describe the lamp as a **non-ohmic** resistor.

As the current increases, the resistance gets larger. The temperature of the filament increases when the current increases. So we can conclude that the resistance of the filament increases as the temperature increases.

PRACTICAL

Make sure you can describe an experiment to investigate how the current for a filament lamp or diode changes as the applied voltage changes.

The diode

A similar investigation can be carried out for a diode. Figure 5.6 shows a current–voltage graph for a diode.

A diode is a component that allows current to pass through it in only one direction. For a 'forward' voltage, current starts to pass when the voltage reaches about 0.7 V. When the voltage is 'reversed', there is no current at all.

A light-emitting diode (LED) lights up when there is a current through it. This is useful because it allows an LED to be used as an indicator – to show us when there is a current.

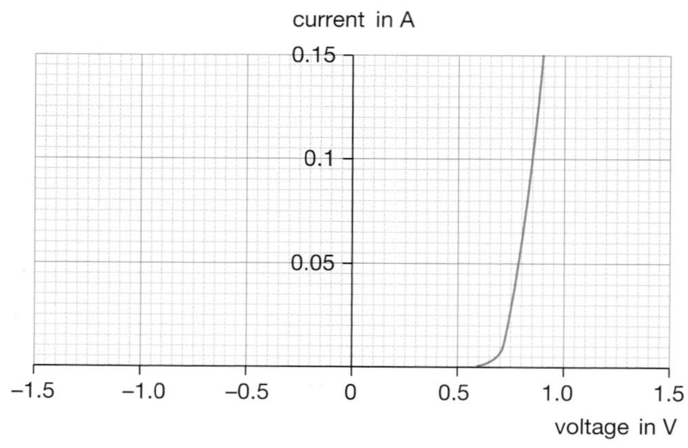

Figure 5.6 A current–voltage graph for a diode; the current only flows for a positive or forwards voltage.

Changing resistance

Some resistors change their resistance as they react to their surroundings. The resistance of a **thermistor** decreases as the temperature increases. You can control its temperature by putting it into a beaker of warm or cold water, as in Figure 5.7.

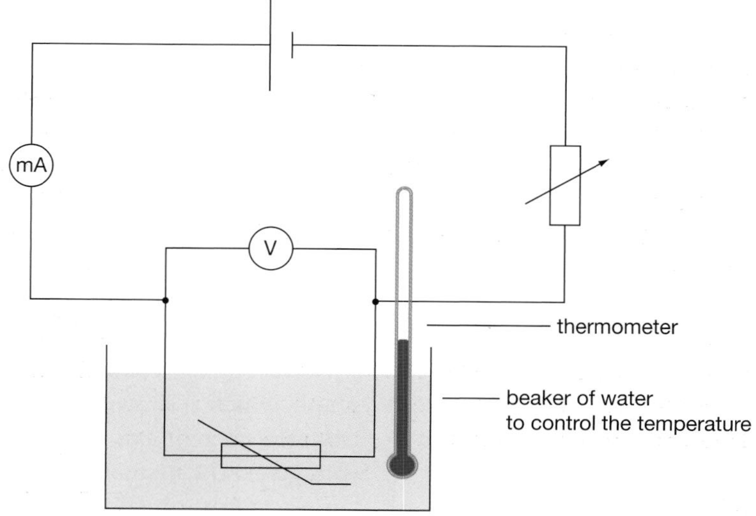

Figure 5.7 Investigating how the resistance of a thermistor changes with temperature.

Figure 5.8 shows how one type of thermistor's resistance changes as the temperature changes.

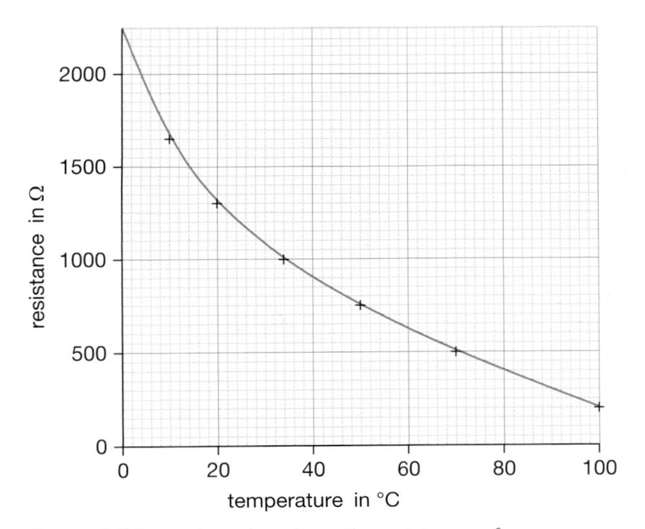

Figure 5.8 A graph to show how the resistance of a thermistor changes with temperature.

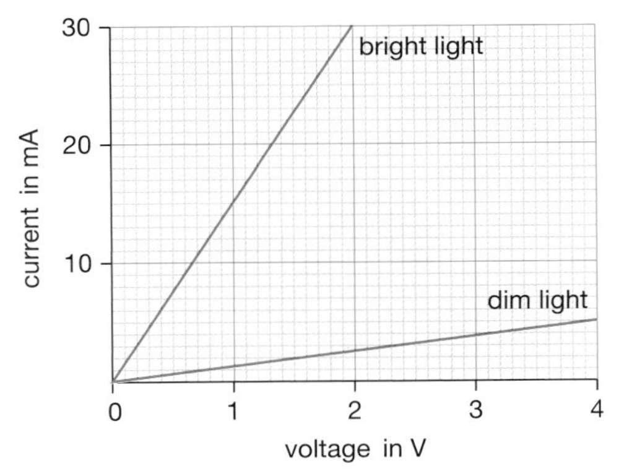

Figure 5.9 A current–voltage graph for a light-dependent resistor in bright and dim light.

The resistance of a **light-dependent resistor** (LDR) changes as the light intensity changes. In the dark the resistance is high but in bright light the resistance of an LDR is low. This is shown in Figure 5.9. There is a higher current through the resistor in bright light because the resistance is lower.

Adding resistance

When you add resistors in series you increase the resistance. Here is a simple example: you place a 6 Ω, an 8 Ω and a 12 Ω resistance in series. State the total resistance. The answer is 26 Ω. You just add up the resistors, as each one limits the current.

STUDY QUESTIONS

1 Explain what is meant by an ohmic resistor.

2 This question refers to Figure 5.3. Explain what happens to the current through the ammeter when you increase the resistance of the variable resistor.

3 Figure 5.5 shows the current–voltage graph for a filament lamp.

a) Calculate the resistance of the filament when the applied voltage is:

i) 1 V

ii) 3 V.

b) Explain what causes this resistance to change.

4 Figure 5.9 shows two current–voltage graphs for an LDR in dim and bright light. Calculate the resistance of the LDR in:

a) bright light

b) dim light.

5 a) Draw a circuit diagram to show a cell, a 1 kΩ resistor, and an LED used to show that there is a current flowing through the resistor.

b) Draw a circuit diagram to show how you would investigate the effect of light intensity on an LDR.

6 The table shows the values of current, voltage and resistance for five resistors, A, B, C, D and E. Use the equation $R = V/I$ to fill in the missing values.

Resistor	Current in A	Voltage in V	Resistance in Ω
A	0.4	12.0	
B		36.0	216
C	0.002		5000
D	10.0	230	
E		115	460

2.6 Current, charge and voltage

Figure 6.1 This motorbike has been electroplated with chromium, to make it attractive.

Electroplating with metals is a common industrial process. How does electroplating work?

Give two more examples of the use of electroplating.

Figure 6.2 shows a diagram of an electrical circuit. In this circuit a cell provides a **voltage** of 1.5 V, and a **current** of 0.1 A flows round the circuit. You have already met the terms voltage and current, but in this chapter they are defined and explained in greater detail.

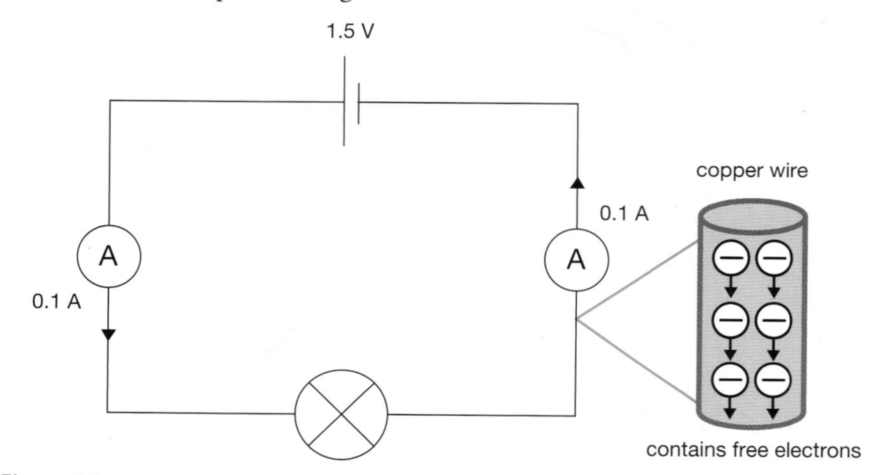

Figure 6.2.

Current and charge

The current is a measure of the rate at which electrical charge flows round the circuit. In a metal the current is carried by electrons which are free to move. The electrons are repelled from the negative terminal of the cell and attracted towards the positive terminal.

The direction of the electric current is defined as the direction in which positive charge would flow – from the positive terminal of the battery to the negative terminal. Current was defined in this way before the electron was discovered, at a time when people did not understand how a wire carried a current. So in Figure 6.2 the direction of current is shown from positive to negative.

The amount of charge flowing round in the circuit is measured in coulombs, C. 1 coulomb of charge is equivalent to the charge on 6 billion billion electrons.

The unit of current is the ampere, A. This unit is often abbreviated to amp. Small currents can be measured in milliamps or mA.

$$1 \text{ mA} = 0.001 \text{ A}$$

The current at all points of the circuit is the same. So the two ammeters on either side of the lamp read the same current – in this case 0.1 A.

Current and the flow of charge are linked by the equation.

$$Q = I\,t$$

$$\text{charge flow} = \text{current} \times \text{time}$$

charge flow, Q, is measured in coulombs, C

current, I, is measured in amperes, A

time, t, is measured in seconds, s

Example. In Figure 6.2 the current of 0.1 A flows for 30 minutes. Calculate how much charge flows round the circuit.

$$Q = I\,t$$

$$= 0.1\text{ A} \times 1800\text{ s}$$

$$= 180\text{ C}$$

TIP

You will not be examined on ionic solutions, but it is included here to help you understand that conventional current is in the direction of positive charge flow.

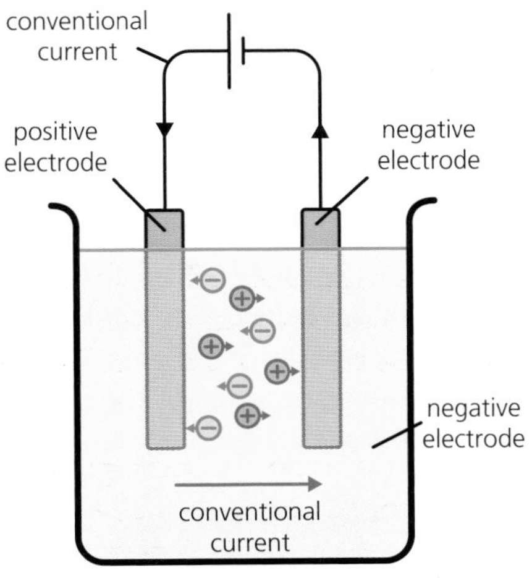

Figure 6.3

Ionic solutions

Figure 6.3 shows an ionic solution which contains both positive and negative ions. A cell is making a current flow through the solution. In the solution the current is carried by positive ions moving to the right and also by negative ions moving to the left. In the connecting wires the current is carried only by electrons. The conventional current flows around the circuit as shown by the arrows.

Defining the volt

You have already met the idea that voltage is a measure of energy. Voltage is defined as the energy transferred for each unit of charge that passes through a component such as a resistor. This allows us to define the volt:

A volt is one joule per coulomb.

$$\text{voltage} = \frac{\text{energy}}{\text{charge}}$$

This equation can also be written in the form:

$$\text{energy transferred} = \text{charge} \times \text{voltage}$$

$$E = Q \times V$$

energy, E, is in joules, J

charge, Q, is in coulombs, C

voltage, V, is in volts, V

So in figure 6.2 the cell transfers 1.5 joules of energy to each coulomb of charge that flows through the lamp.

A worked example will help you understand this idea.

TIP

When calculating the current, remember to turn time in minutes into time in seconds.

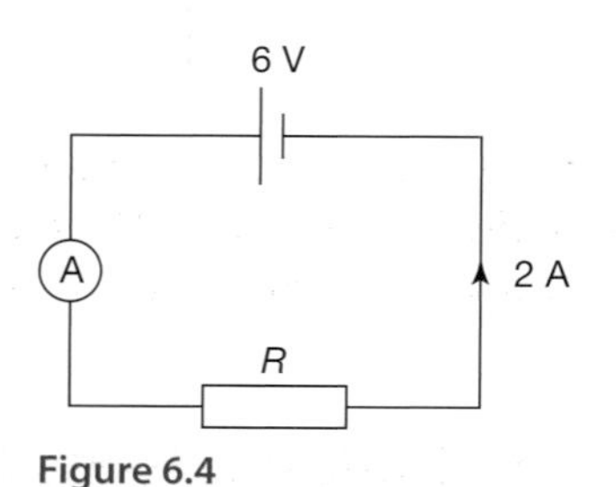

Figure 6.4

Example. In Figure 6.4 there is a current of 2 A through a resistor. The battery has a voltage of 6 V. Calculate the energy transferred to the resistor in one minute.

First, we must calculate the charge that flows around the circuit in 1 minute.

$$\begin{aligned} \text{charge} &= \text{current} \times \text{time} \\ &= 2\text{ A} \times 60\text{ s} \\ &= 120\text{ C} \end{aligned}$$

Then the energy is calculated using the equation:

$$\begin{aligned} \text{energy} &= \text{charge} \times \text{voltage} \\ &= 120\text{ C} \times 6\text{ V} \\ &= 720\text{ J} \end{aligned}$$

STUDY QUESTIONS

1 State the unit of each or the following quantities:
  a) charge
  b) current
  c) voltage.

2 a) Write an equation that links current, charge and time.
  b) Write an equation that links energy, charge and voltage.

3 a) Calculate the charge that flows round a circuit when a current of 0.1 mA passes for 2 minutes.
  b) Calculate the current when it takes 30 s for a charge of 6 C to flow.
  c) A charge of 7200 C is required to electroplate a metal object. The current in the process is 0.5 A. Calculate the time taken to complete the electroplating.

4 a) A charge of 0.8 C flows through a resistor that has a voltage of 9 V across it. Calculate the energy transferred to the resistor.
  b) A current of 3 A passes through a resistor for 5 minutes. During this time 6300 J of energy are transferred to the resistor. Calculate the voltage across the resistor.
  c) A resistor has a voltage of 20 V across it. A current is switched on and 4000 J of energy are transferred to the resistor. Calculate the charge that flows through the resistor.

5 A student decides to copper plate a spoon by placing it in a solution of copper sulfate and passing a current through the solution. The process takes 80 minutes.

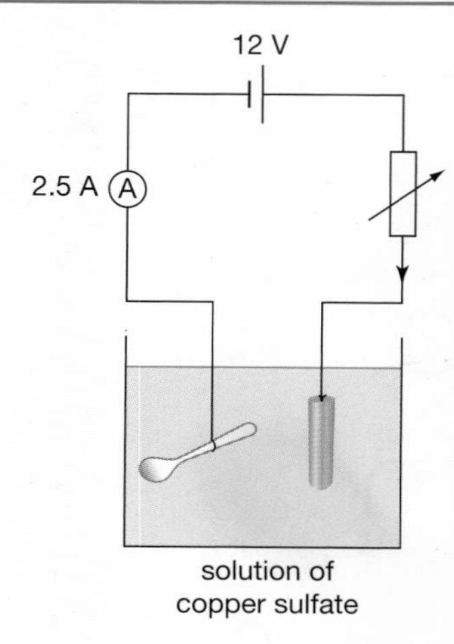

  a) Use the information in the diagram to calculate how much charge is passed by the battery during this time.
  b) Calculate how much energy is delivered by the battery during the process of electroplating the spoon.
  c) Draw a diagram to show the movement of ions through the copper sulfate solution during the copper plating process.

6 A camera flash bulb is lit for 0.02 s. During this time 0.1 C of charge flows through the bulb. Calculate the average current flowing through the bulb while it is lit up.

2.7 Current and voltage rules

Some decorative lights are wired in series and some are wired in parallel. Which method is suitable for each of the examples in Figure 7.1? What are the advantages of using a parallel circuit and what are advantages of using a series circuit?

Figure 7.1 **(a)** These small decorative lights are used to light up lots of trees in the street.

(b) These bright decorative lights spread light around the tree and its surroundings.

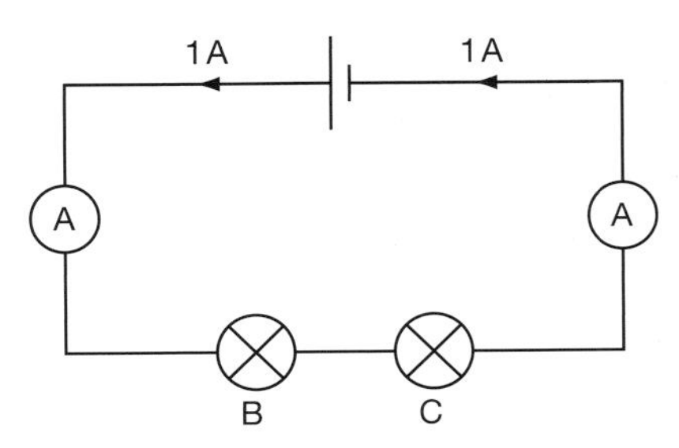

Figure 7.2 A series circuit.

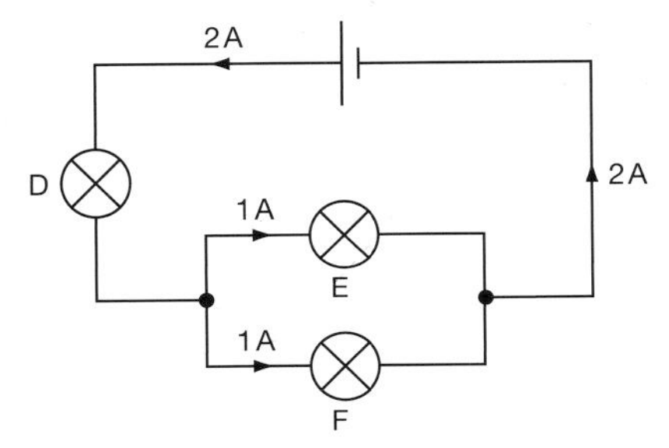

Figure 7.3 The current is divided between two lamps E and F in a parallel circuit.

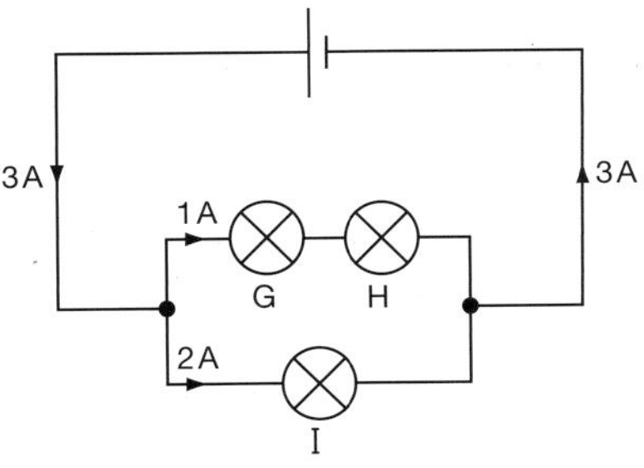

Figure 7.4 A parallel circuit with unequal branches.

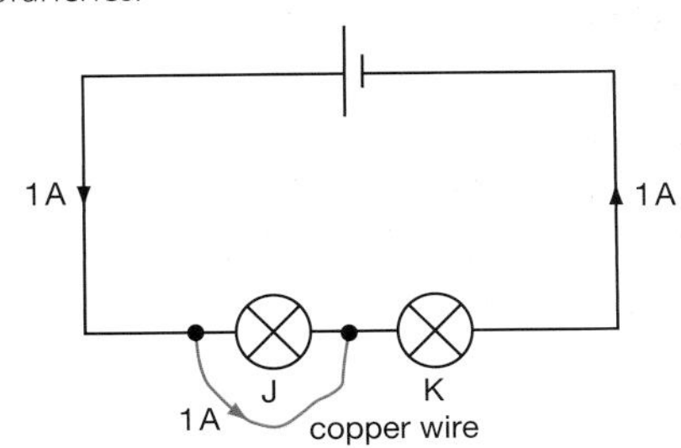

Figure 7.5 A short circuit.

Current paths

Figure 7.2 reminds you of the rules about a **series** circuit. In this case the current in the circuit is 1 A. It is the same in each part of the circuit. So the current is 1 A in each of the lamps B and C, and 1 A through both ammeters and the cell. A circuit cannot lose current – in every circuit the current leaving the cell is the same as the current returning to the cell.

Figure 7.3 shows a circuit with branches in it. In this case, 2 A leaves the cell and passes through lamp D. Then, the current splits equally through the identical lamps E and F, and a current of 2 A returns to the negative terminal of the cell. When the current splits between two lamps in this way, we say they are in **parallel**.

The current does not always split equally. Figure 7.4 shows another parallel circuit with three identical lamps. More current flows through lamp I than through lamps G and H. Lamp I provides a lower resistance path. The sum of the currents going in the two parallel paths is 3 A – the same as the current flowing out of and back into the cell.

Figure 7.5 shows an example of a **short circuit**. A copper wire placed across lamp J provides a very low resistance path, so virtually no current flows through the lamp; it goes through the wire. Lamp J does not light up.

Current rules

The current is the same in all parts of a series circuit.

The same current flows into a junction as comes out of the junction.

Adding the voltage

Figure 7.6 shows a car headlamp connected to a 12 V car battery. The battery does electrical work to transfer energy from its chemical store into the thermal store of the lamp and the surroundings. The voltage across the lamp is the same as the battery voltage.

TIP

Remember: voltage is a measure of energy provided by a cell.

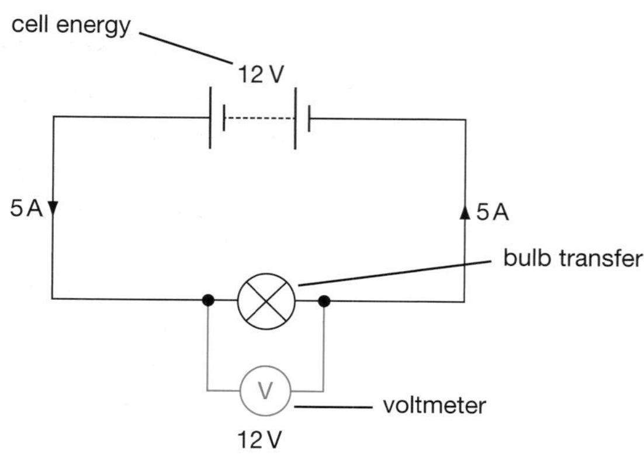

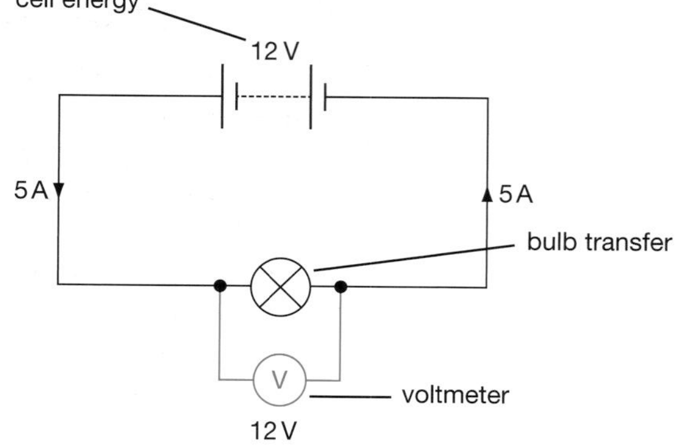

Figure 7.6

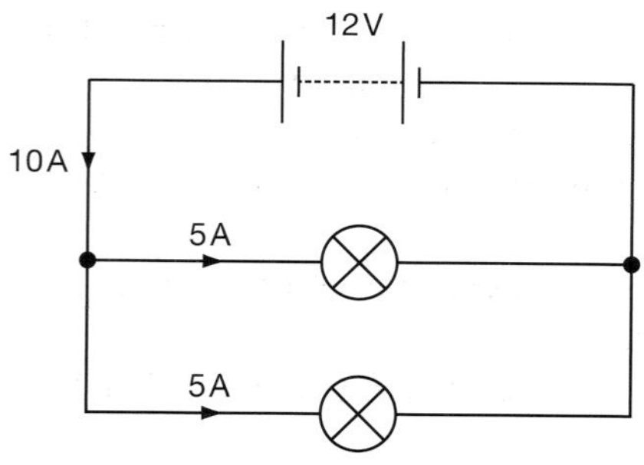

Figure 7.7 A circuit to supply two headlamps from a car battery.

In Figure 7.7 two headlamps are connected in parallel to the car battery; a current of 5 A passes through each. The voltage across each lamp is 12 V because both lamps are connected directly to the battery.

Figure 7.8 can help you understand why the voltage is the same across each headlamp. The voltage is like an 'electrical pressure'. The pressure on the water in each pipe from the water tank in Figure 7.8 is the same. The water tank empties more quickly when water flows through two pipes rather than one. The same idea works for the headlamps in a car. The battery transfers energy more quickly when two lamps are on rather than one.

In Figure 7.9 two identical lamps are connected in series to the car battery. In this example each lamp has a voltage of 6 V across it. As the charge flows through the lamps, it transfers half its energy to each lamp.

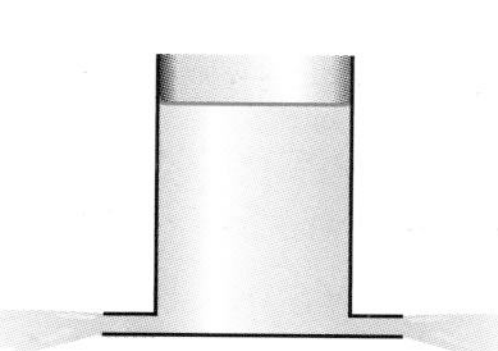

Figure 7.8 The pressure on the water in each pipe is the same. These pipes are parallel, like two headlamps connected in parallel.

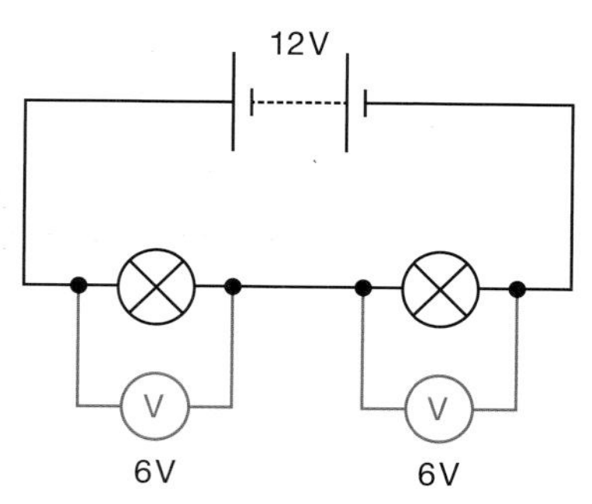

Figure 7.9 Each voltmeter reads 6 V.

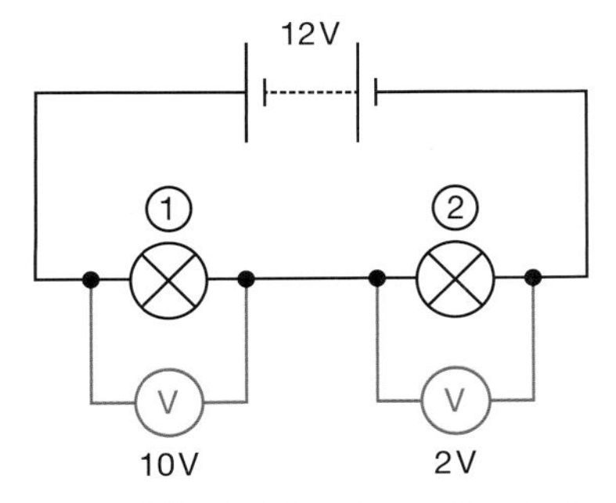

Figure 7.10 The brighter lamp (1) gets the larger voltage.

The voltages do not always split equally. In Figure 7.10, lamp 1 has a larger resistance than lamp 2. There is a larger voltage across lamp 1 than across lamp 2. The charge flowing transfers more of its energy to lamp 1, so it is the brighter lamp.

Voltage rules

The voltage across electrical components in parallel is the same.

The sum of the voltages across electrical components in series adds up to the battery voltage.

Uses of parallel circuits

- Lamps in a car are placed in parallel across a 12 V battery. Each lamp is designed to work on 12 V. If the lamps are placed in series they get less than their operating voltage.
- All domestic appliances are designed to work at mains voltage (about 230 V or 120 V in most countries). By putting them in parallel, domestic appliances receive the full mains voltage and can be independently controlled.
- Voltmeters are placed in parallel with appliances to measure the voltage across them.

Uses of series circuits

- Decorative lights are an example of a string of small lamps placed in series. The lamps are designed to work on low voltages so they share the mains voltage. They can all be turned off with the same switch.
- A switch is placed in series with a light to turn it on and off.
- An ammeter is placed in series with a resistor to measure the current through it.
- A resistor is placed in series with a component such as a diode or thermistor to protect it from overheating.
- A fuse is placed in series with a domestic appliance to protect it. The fuse melts if the current is too large.

Cells and batteries

A battery consists of two or more electrical cells. Figure 7.11 shows three ways in which two cells may be connected.

In series:

a) the voltages add up to produce a larger voltage, provided the cells 'face' the same way;
b) if the cells face in opposite directions the resultant voltage is zero.

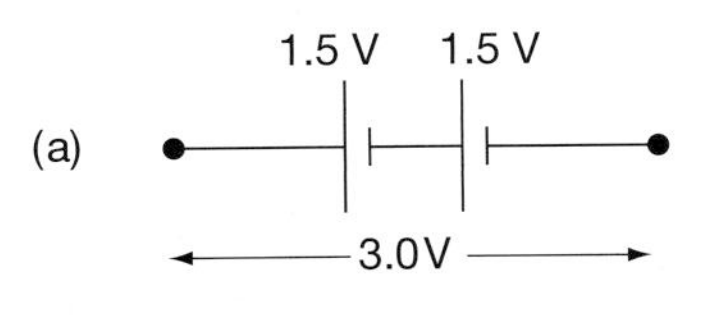

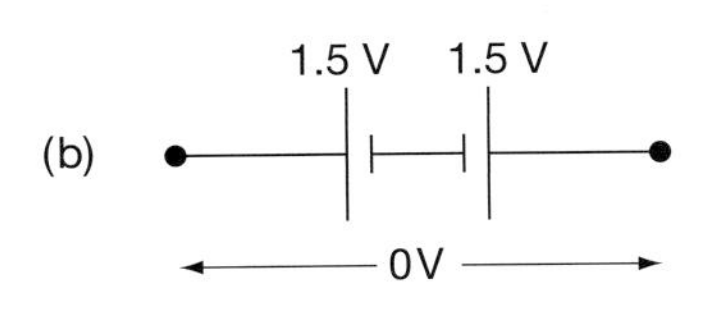

Figure 7.11 Combining cells.

STUDY QUESTIONS

1 State which current in the diagram below is the largest, *p*, *q* or *r*. Which is the smallest?

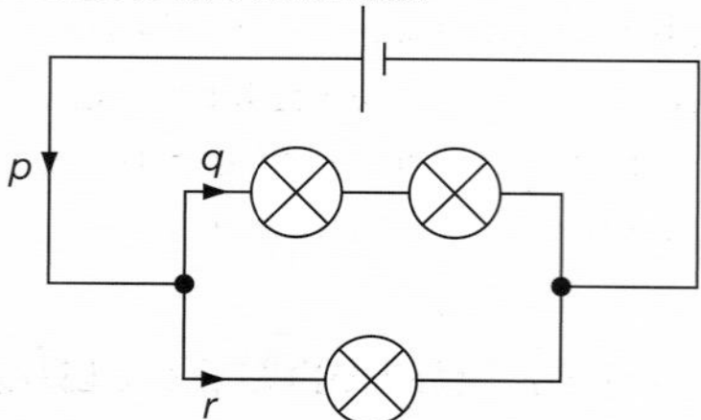

2 Copy the diagrams below and mark in the missing values of current.

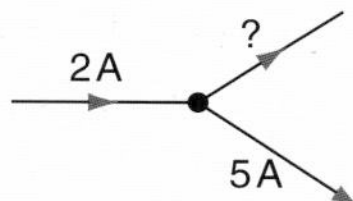

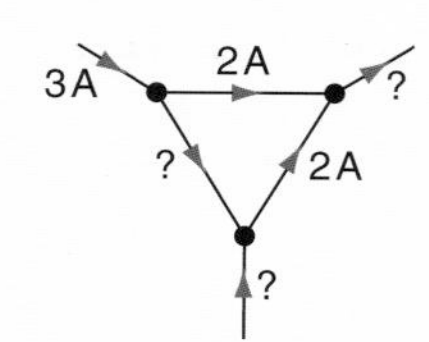

3 In the circuit shown in the diagram, each cell provides a voltage of 1.5 V.

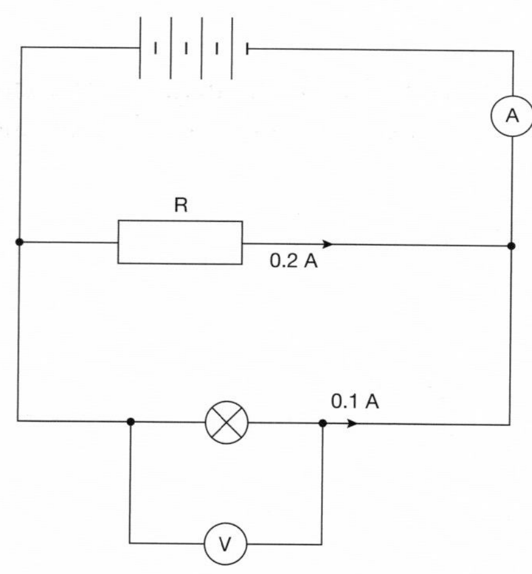

a) State the total voltage provided by the battery.
b) State the reading on the voltmeter.
c) Calculate the current through the ammeter.
d) Calculate the resistance of the lamp.

4 A set of decorative lights has 40 identical lamps connected in series.

Each lamp is designed to take a current of 0.1 A. The set of lights plugs directly into the 230 V mains electricity supply.

a) Write down an equation that links voltage, current and resistance.
b) Calculate the voltage across one of the lamps.
c) Calculate the resistance of one lamp.
d) Calculate the resistance of the 40 lamps in series.
e) Calculate the power transferred by the 40-lamp set.

5 The circuit diagram shows a 16 Ω resistor in series with a component X.

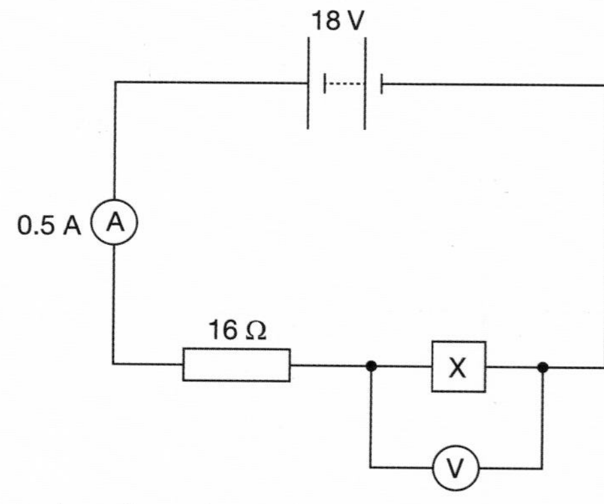

a) Calculate the voltage across the 16 Ω resistor.
b) State the voltage across the component X.
c) Now calculate the resistance of X.

6 The diagram shows six elements of a car rear-window heater, which are connected in series to a 12 V car battery. The resistance of each element is 2.5 Ω.

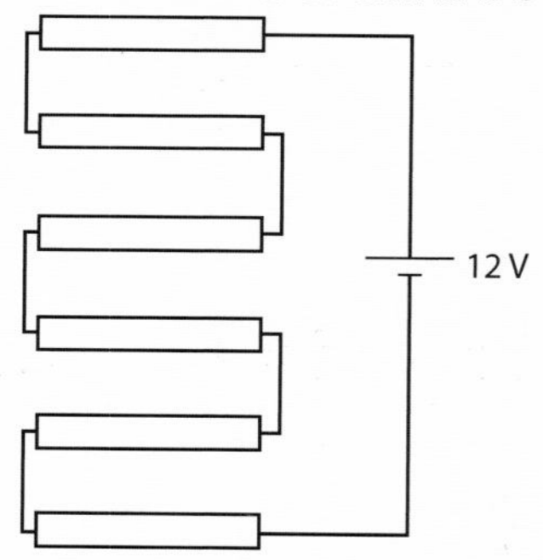

a) Calculate the total resistance of the heater.
b) Calculate the current flowing through the heater.
c) Use the formula: power = voltage × current to calculate the power generated by the heater.
d) A student suggests that the power could be increased by using elements with a greater resistance. A second student suggests that using less resistance would be better because the current would be larger. Who is correct?
e) The elements of some car rear-window heaters are connected in parallel. Suggest an advantage of this different arrangement.

2.8 Circuit calculations

In the previous chapters you have learnt the basic rules about electrical circuits. The best way to make sure that you have understood them is to practise by solving problems. In this chapter, some examples are given, but most of the work is to be done by you.

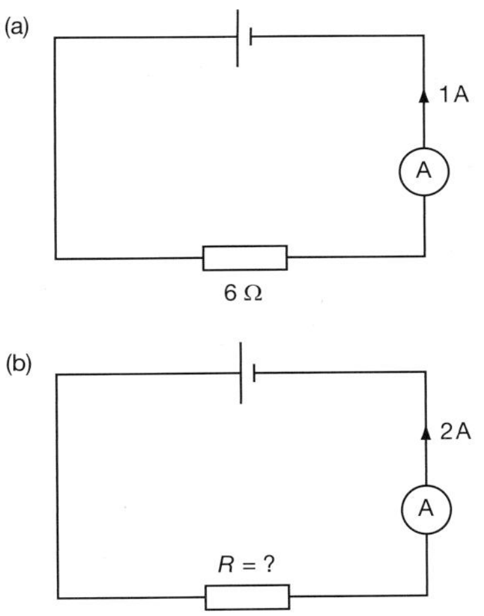

Resistance, voltage and current calculations

You can compare the value of two resistors using an ammeter and a cell; the larger resistor lets less current through. In Figure 8.1(a) the cell pushes a current of 1 A through a 6 Ω resistor; in Figure 8.1(b) the same cell pushes a current of 2 A through a second resistor whose value is not known. Because the current is more than 1 A, the second resistor must have a resistance less than 6 Ω. Can you explain why it is a 3 Ω resistor?

Figure 8.1

The value of a resistor may be calculated using an ammeter and a voltmeter; for example, in Figure 8.2,

$$R = \frac{V}{I} = \frac{10\text{ V}}{2\text{ A}} = 5\ \Omega$$

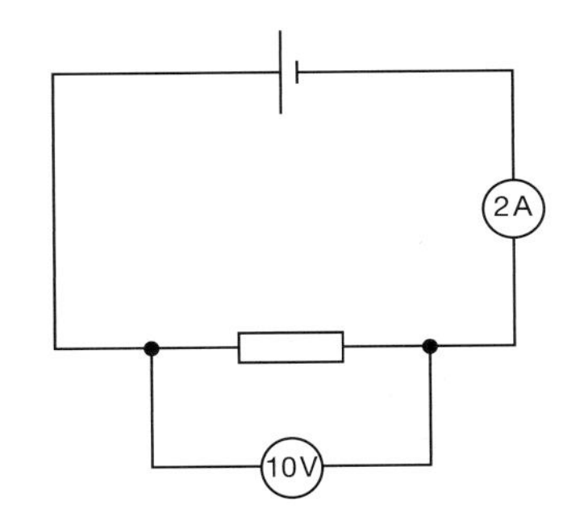

Figure 8.2

More complicated problems may be solved with the help of the current and voltage rules in Chapter 2.6. For example, you might be asked to calculate the cell voltage in Figure 8.3.

The current through the 12 Ω resistor is:

$$I = \frac{V}{R} = \frac{6\text{ V}}{12\ \Omega} = 0.5\text{ A}$$

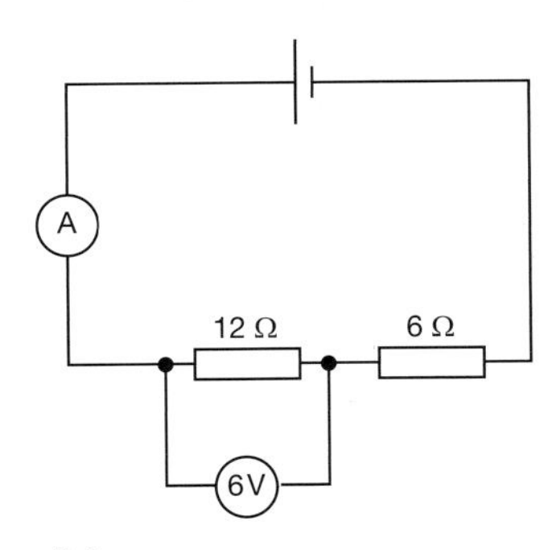

Figure 8.3

The voltage across the 6 Ω resistor is:

$$V = I \times R = 0.5\text{ A} \times 6\ \Omega = 3\text{ V}$$

The cell voltage is the sum of the voltages across the two resistors, which is 9 V.

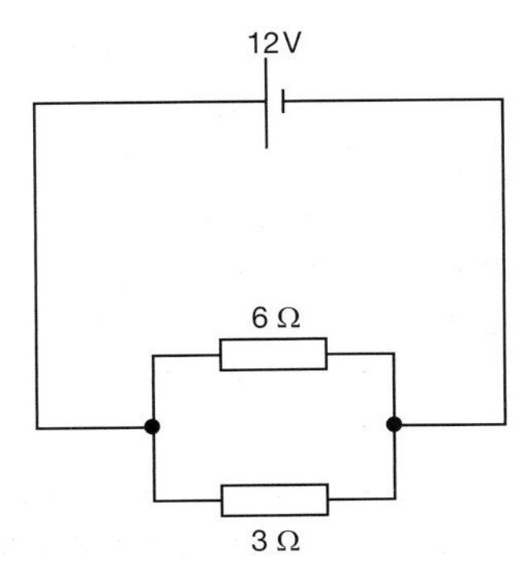

Figure 8.4

Power calculations

Similar care needs to be taken to calculate the power developed in a resistor. For example how much power is transferred by the 3 Ω resistor in Figure 8.4?

Each resistor has 12 V across it. So the current through the 3 Ω resistor is:

$$I = \frac{12\,\text{V}}{3\,\Omega}$$
$$= 4\,\text{A}$$

So the power transferred is:

$$P = 12\,\text{V} \times 4\,\text{A}$$
$$= 48\,\text{W}$$

STUDY QUESTIONS

1 Determine which of R_1 and R_2 has the larger resistance.

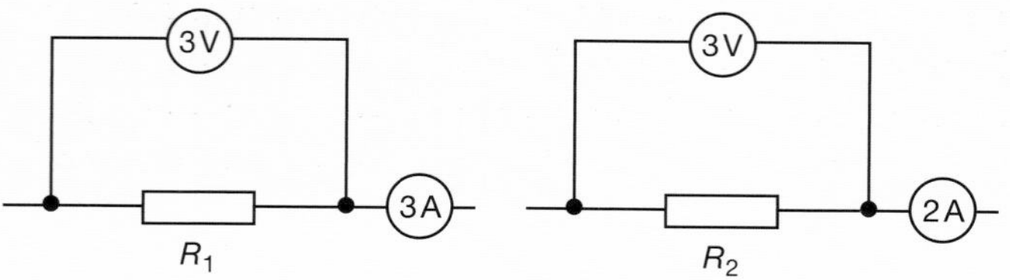

Figure 8.5 Both voltmeters read the same.

2 Explain why you can deduce that R_3 has a larger resistance than R_4.

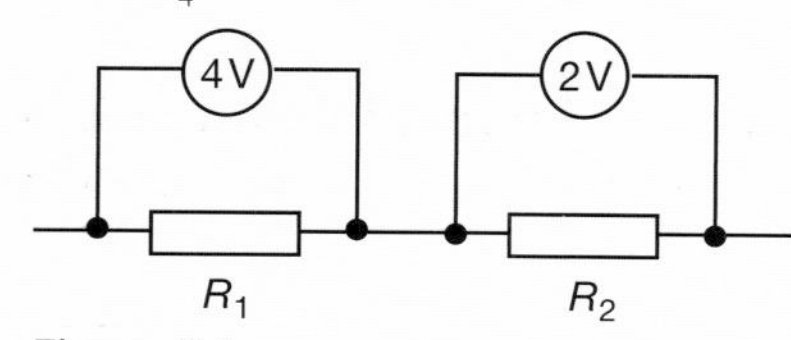

Figure 8.6

3 **a)** Calculate the current in Figure 8.7
b) Calculate the power dissipated by the 24 Ω resistor.
c) State the voltage across R.
d) Calculate the resistance R.

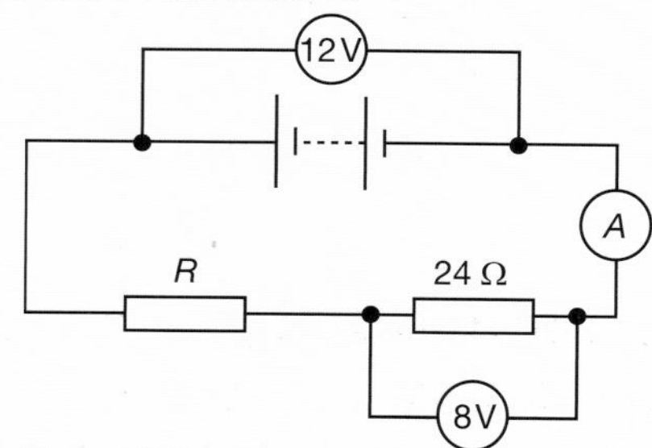

Figure 8.7

4 **a)** Calculate the current in Figure 8.8.
b) Use your answer to part **a)** to calculate the voltage V_1.
c) Determine the voltage V_2.

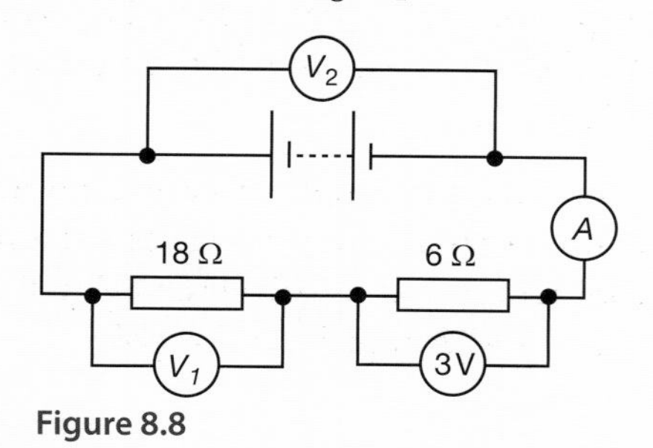

Figure 8.8

5 Calculate the resistance R in Figure 8.9.

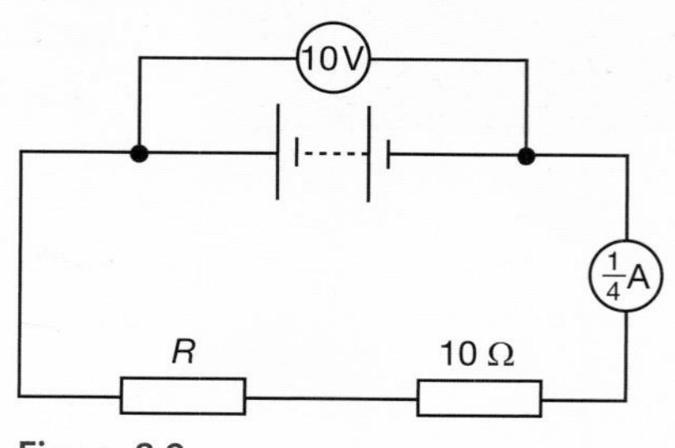

Figure 8.9

6 **a)** Calculate the voltage V in Figure 8.10.
b) State the ammeter readings A_1 and A_2.

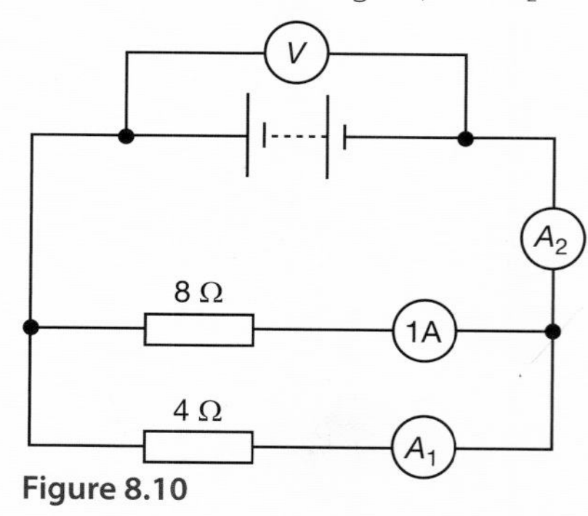

Figure 8.10

7 **a)** Determine the ammeter readings A_1 and A_2 in Figure 8.11.
b) i) Calculate the voltmeter reading V_2.
ii) State the voltmeter reading V_1.
c) Calculate the power delivered by the battery in Figure 8.11.

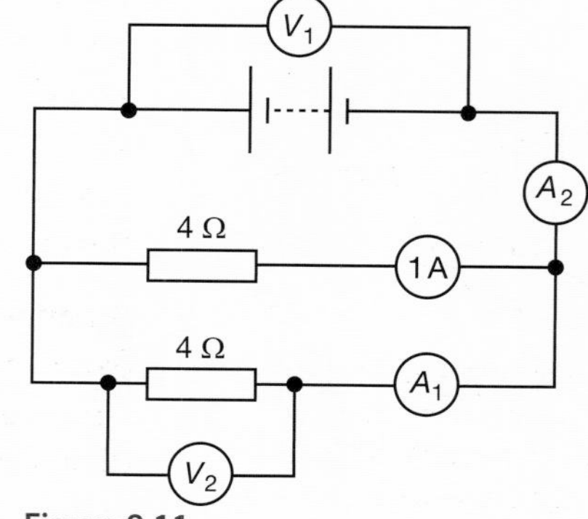

Figure 8.11

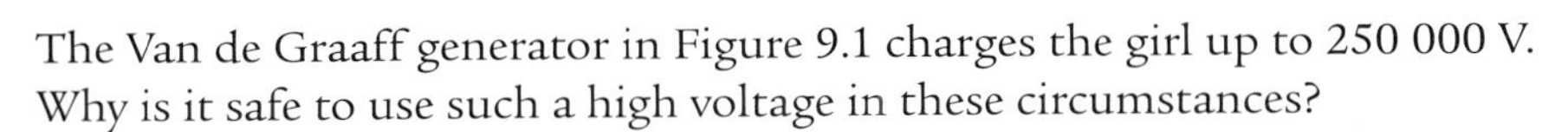

2.9 Electrostatics

Figure 9.1 Using a Van de Graaff generator to demonstrate that like charges repel each other.

The Van de Graaff generator in Figure 9.1 charges the girl up to 250 000 V. Why is it safe to use such a high voltage in these circumstances?

Electrical charge

Electricity was discovered a long time ago when the effects of rubbing materials together were noticed. If you pull off a shirt containing synthetic fibres you may hear the shirt crackle and in a dark room you may also see some sparks. A well-known trick is to rub a balloon and stick it on to the ceiling. You may have felt an electrical shock after walking across a carpet. In these examples, you, or the balloon, have become charged as a result of **friction** (rubbing).

TIP

Like charges repel; unlike charges attract.

There are two types of electrical charge, positive and negative. A **positive** charge is produced on a Perspex ruler when it is rubbed with a woollen duster. You can put **negative** charge onto a plastic comb by combing it through your hair.

Some simple experiments show us that like charges repel each other, and unlike charges attract each other. These experiments are shown in Figure 9.2.

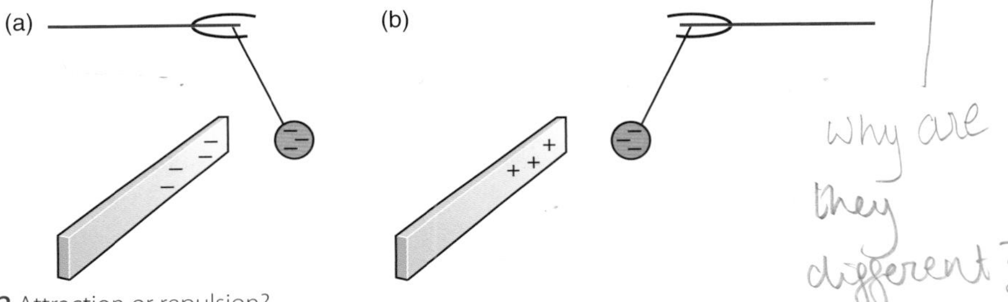

Figure 9.2 Attraction or repulsion?
(a) Like charges repel.
(b) Unlike charges attract.

Where do charges come from?

There are three types of small particle inside atoms. There is a very small centre of the atom called the **nucleus**. Inside the nucleus there are **protons** and **neutrons**. Protons have a positive charge but neutrons have no charge. The electrons carry a negative charge and they move around the nucleus. The size of the charge on an electron and on a proton is the same. Inside the atom there are as many electrons as protons. This means that the positive charge of the protons is balanced by the negative charge of the electrons. So the atom is neutral or uncharged (Figure 9.3).

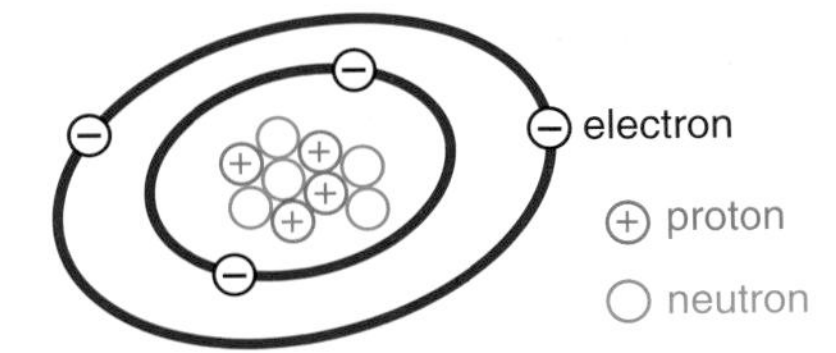

Figure 9.3 This beryllium atom is neutral. Four negatively charged electrons balance four positively charged protons. (The diagram is not drawn to scale – see page 219.)

When you rub a Perspex ruler with a duster, some electrons are removed from the atoms in the ruler and are transferred to the duster. As a result the ruler has fewer electrons than protons and so it is positively charged. But the duster has more electrons than protons and is negatively charged (Figure 9.4).

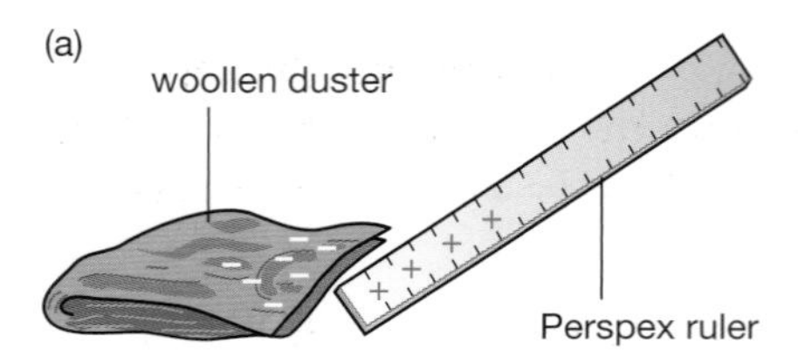

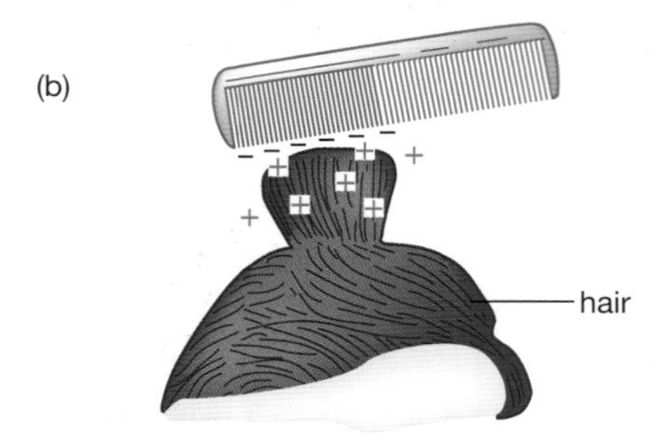

Figure 9.4 Charging by friction.

Picking up litter

You can use a plastic comb to pick up small pieces of paper. Figure 9.5 shows the idea. The comb is negatively charged after combing your hair (Figure 9.4(b)). When it is placed close to the paper, electrons in the paper are pushed to the bottom or repelled. The top of the paper becomes positively charged and the bottom negatively charged. The negative charges on the comb attract the top of the paper upwards. The comb's charges repel the bottom, but the positive charges at the top of the paper are closer to the comb. Therefore, the upwards force is bigger than the repulsive downwards force and the piece of paper is picked up.

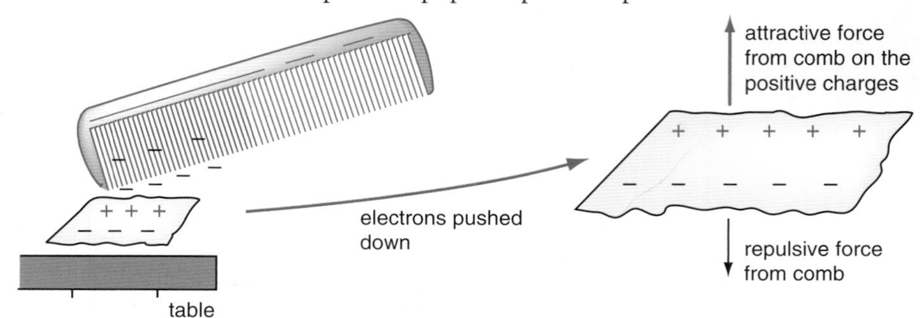

Figure 9.5 Your comb can pick up small pieces of paper. The attractive force on the positive charges is larger than the repulsive force on the negative charges.

PRACTICAL

A gold leaf electroscope can be used to investigate the nature of the charge on a plastic rod. The electroscope can be charged safely using an approved 5 kV laboratory power supply.
Note:
The gold leaf electroscope is not a required part of the specification. However, it can be used easily to show that there are two types of charge.

The gold leaf electroscope

The construction of a gold leaf electroscope is shown in Figure 9.6. A metal rod and cap are insulated from a metal box. A thin piece of gold leaf is attached to the end of the rod. When the cap is negatively charged, the electrons flow in the metal and the charge is spread out over the electroscope including the gold leaf. Like charges repel, so the leaf moves away from the rod next to it.

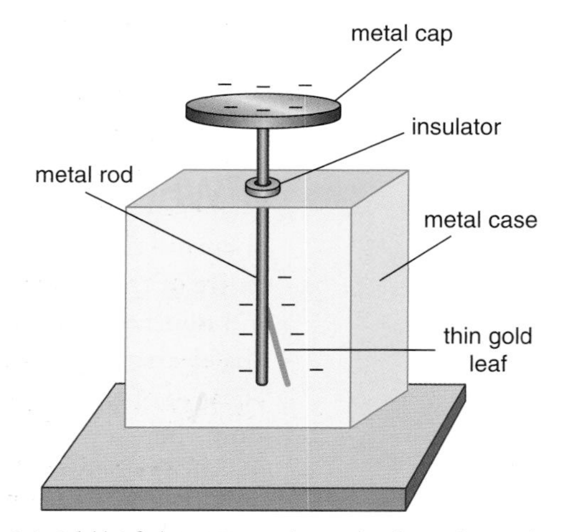

Figure 9.6 A gold leaf electroscope is used to investigate charges.

Figure 9.7 shows how a gold leaf electroscope is used to investigate the charges on two plastic rods. In Figure 9.7 (a) the electroscope has been charged positively by connecting it to the positive terminal of a high-voltage supply. It has lost electrons and has an overall positive charge.

In Figure 9.7 (b) a negatively charged rod is held near the cap. This repels some of the negatively charged electrons from the cap towards the leaf. This neutralises some of the positive charge near the gold leaf, so there is less overall positive charge there. The gold leaf falls.

In Figure 9.7 (c) a positively charged rod is held near the cap. This attracts some electrons away from the gold leaf towards the cap, so the overall positive charge near the gold leaf increases. The gold leaf rises further.

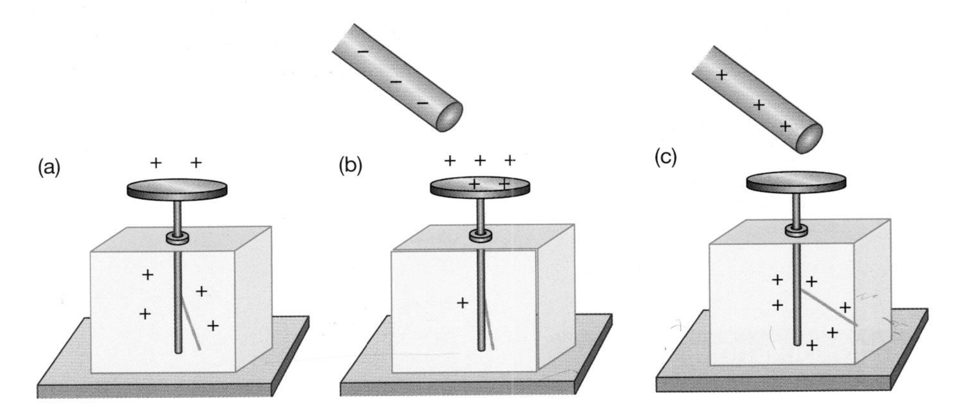

Figure 9.7 (a) Electroscope with positive charge. **(b)** A negatively charged rod repels electrons towards the leaf. Positive charge is neutralised and the leaf falls. **(c)** A positively charged rod attracts electrons from the leaf to the cap. This neutralises the cap and the leaf rises further as there is more positive charge near it.

Investigating charges

REQUIRED PRACTICAL

The purpose of this experiment is to investigate how insulating materials can be charged through friction.

The purpose of this experiment is to investigate how insulating materials can be charged through friction. You will need a variety of plastic rods and dusters, and a gold leaf electroscope. The class also need access to a high voltage power supply.

1 Charge your electroscope using the high voltage supply. Note whether your electroscope is positively or negatively charged.
2 Rub an insulator with a duster and hold the insulator close to the electroscope.
3 From the movement of the electroscope leaf, deduce whether
   a) the insulator is charged,
   b) and then the sign of the charge.
4 Record your results, noting the duster, the insulator and the sign of the charge on the duster and on the insulator.
5 Repeat the process for other pairs of insulators and dusters.

Which materials conduct electricity?

Metals are good conductors of electricity. Most metal atoms have one or two loosely held electrons, which are free to move. When a cell is placed across a metal, the electrons inside it are attracted to the positive terminal of the cell and repelled from the negative terminal. So, electrons move around a circuit, carrying the electric current (Figure 9.8).

Currents can also be carried by positive or negative ions. When an atom gains an electron it is called a **negative ion**. When an atom loses an electron it becomes a **positive ion**. You conduct electricity because your body is full of ions. This is why you can get an electric shock.

Table 1 Examples of conductors and insulators

Conductors	Insulators
Good:	rubber
metals, e.g. copper, silver, aluminium	plastics, e.g. polythene, PVC, Perspex
Moderate:	china
carbon	air
silicon	
Poor:	
water	
humans	

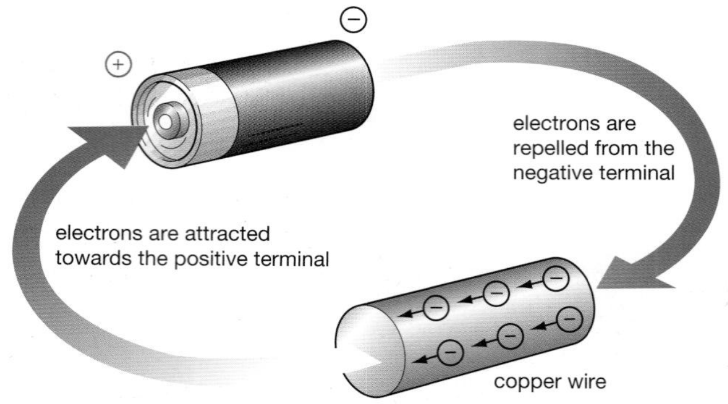

Figure 9.8 In a metal wire current is a flow of electrons.

Materials such as plastics and rubber do not have electrons that are free to move around. These materials do not conduct electricity. They are called **insulators**.

STUDY QUESTIONS

1 a) Two balloons are rubbed with the same woollen cloth so that they become electrically charged. They are then suspended close to each other as shown in Figure 9.9. Explain what you can deduce about the charges on the two balloons.

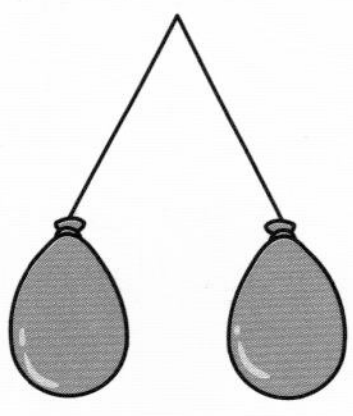
Figure 9.9

b) The cloth is then held close to one of the suspended balloons as shown in Figure 9.10.

i) Explain why you can deduce that the balloon and cloth have opposite charges.

ii) Explain why it is not possible to produce the same sign of charge on the balloon and cloth by rubbing them together.

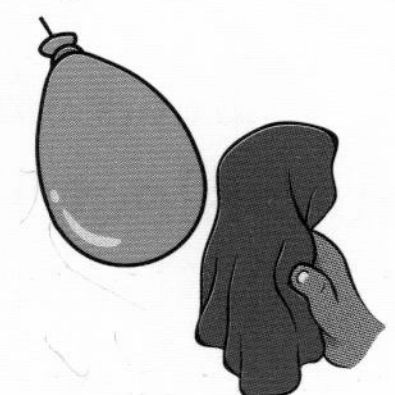
Figure 9.10

2 a) A plastic rod is rubbed and it becomes positively charged. Explain, in terms of electron movement, what has happened to some of the atoms in the rod.

b) A plastic rod is rubbed and it becomes negatively charged. Explain, in terms of electron movement, what has happened to some of the atoms in the rod.

c) Explain why it is not possible to charge a metal rod that is earthed.

3 Design an experiment to investigate whether a plastic rod has a positive or a negative charge. You can use a gold leaf electroscope and a high-voltage power supply to help you. Describe how you will use the apparatus, and how the result of the investigation will allow you to come to a conclusion.

4 You are given a number of materials that you have not seen before. Design an experiment to investigate which of the materials conduct electricity. You should explain clearly which electrical components you will use, drawing a circuit diagram to show how you will use these components. You should also explain how you will use your results to draw conclusions.

2.10 Electrostatics at work

Figure 10.1 A lightning strike on the Eiffel Tower, photographed in 1919.

Large thunderclouds have strong convection currents inside them. Ice crystals are carried up and down by these currents and they become charged as a result of friction. The base of the cloud gains a large negative charge; the electric forces are so large that air becomes ionised. A lightning flash occurs when electrons on the base of the cloud flow through a conducting pathway in the air, to the ground or a tall object. A flash of lightning can transfer a billion joules of electrostatic energy to thermal energy in the surroundings.

Spark hazards

A sailor who works on board an oil tanker has to wear shoes that conduct electricity. Metals are **conductors** of electricity. Materials like plastic and rubber do not conduct electricity; they are **insulators**. Wearing shoes with rubber soles on board a tanker could be very dangerous.

When the sailor moves around the ship in rubber shoes, charges can build up on him as he works. When he touches the ship, there will be a small spark as electrons flow to neutralise his body. Figure 10.2 explains this process. Such a spark could ignite oil fumes and cause an explosion. Some very large explosions have destroyed tankers in the past. So sailors wear shoes with soles that conduct. Now any charge on him flows away and he cannot make a spark.

Sparks are most likely to ignite the oil when it is being unloaded. To avoid this, the surface of the oil is covered with a 'blanket' of nitrogen. This gas does not burn, so a spark will not cause an explosion.

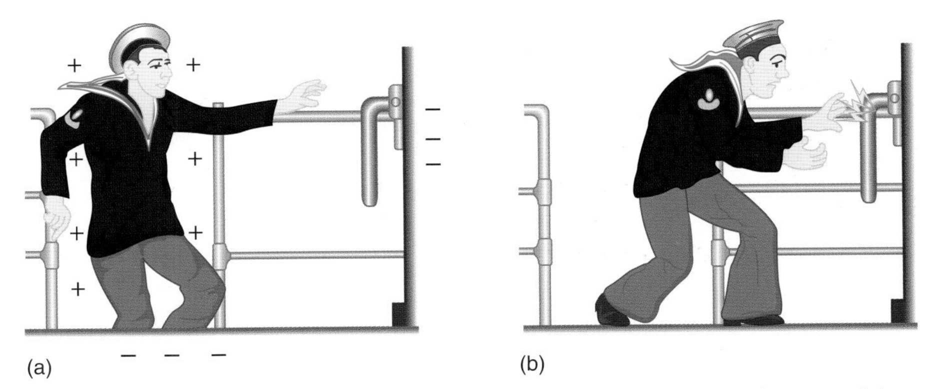

Figure 10.2 (a) Jack is positively charged. His charge attracts electrons on to the parts of the ship near to him.
(b) Jack touches the ship with his hand. Electrons flow onto him to neutralise his positive charge. The spark could cause an explosion.

Electrostatic precipitation

Many power stations burn coal to produce electrical energy. When coal is burnt a lot of soot is produced. It is important to remove this soot before it gets into the atmosphere. One way of doing this is to use an **electrostatic**

precipitator. Inside the precipitator there are wires that carry a large negative charge. As the soot passes close to these wires the soot particles become negatively charged. These particles are repelled away from the negative wires and are attracted to positively charged plates. The soot sticks to the plates, and can be removed later. Some large precipitators in power stations remove 30 or 40 tonnes of soot per hour.

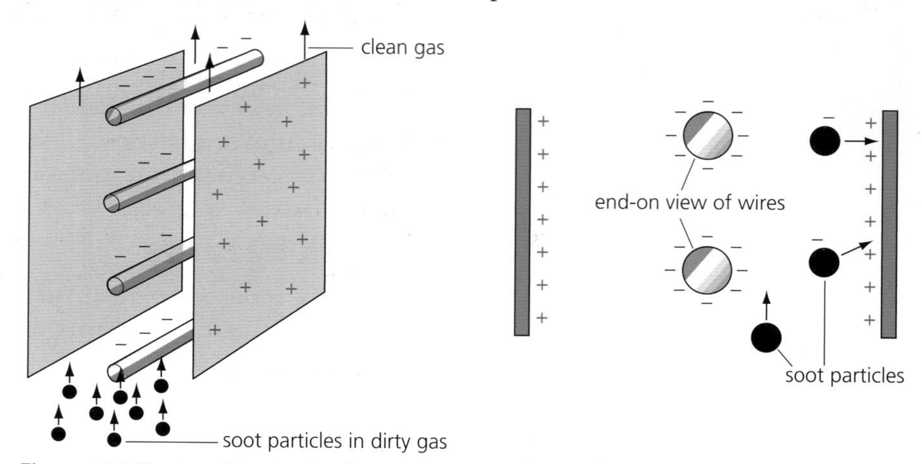

Figure 10.3 Soot particles in dirty fumes are removed in an electrostatic precipitator.

Photocopying

Most offices use photocopying machines. The key to photocopying is a plate that is affected by light. When the plate is in the dark its surface is positively charged. When the plate is in the light it is uncharged.

An image of the document to be copied is projected on to the plate (Figure 10.4(a)). The dark parts of the plate become charged. Now the plate is covered with a dark powder, called **toner**. The particles in the toner have been negatively charged (Figure 10.4(b)), so the toner sticks to the dark parts of the plate, leaving a dark image.

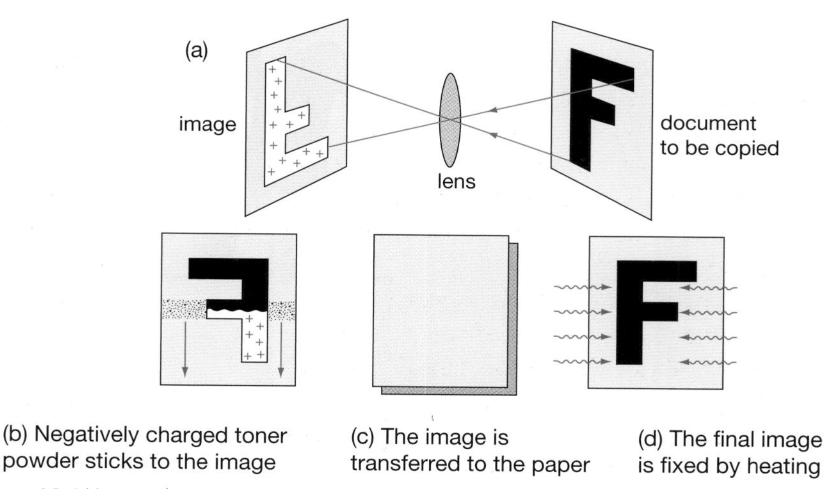

Figure 10.4 How a photocopier works.

Next, a piece of paper is pressed on to the plate. This paper is positively charged, so the toner is attracted to it (Figure 10.4(c)).

Finally, the paper is heated. The toner melts and sticks to the paper, making the photocopy of the document (Figure 10.4(d)).

Electrostatic paint spraying

Electrostatic paint spraying is used to ensure that cars and bicycles get an even coat of paint. Figure 10.5 shows the idea.

- The spray nozzle is charged negatively.
- Paint droplets leaving the nozzle become negatively charged. They repel each other and make a fine mist.
- The droplets are attracted to the positively charged car. The droplets get into places that are difficult to reach.

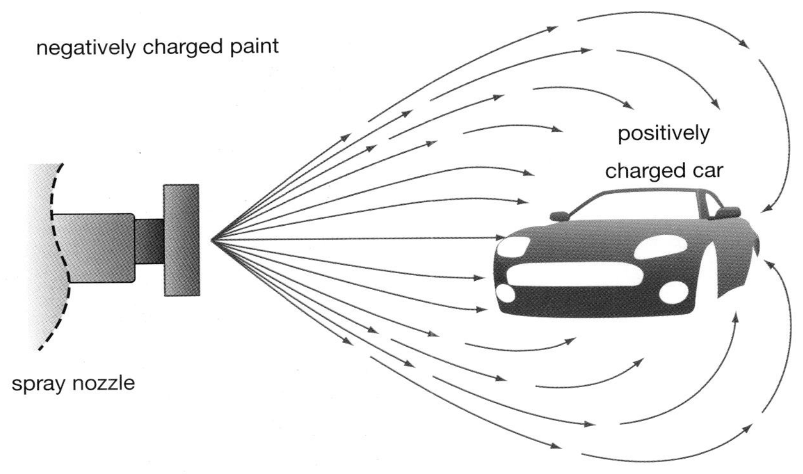

Figure 10.5 Using electrostatic forces to spray-paint a car.

STUDY QUESTIONS

1 When an aircraft is being refuelled it is fitted with a bonding line to electrically connect it to earth. Explain why this is an important safety precaution.

2 **a)** A photocopier needs toner powder. Explain why does the powder needs to be charged.

b) When you get your photocopy out of the copier it is usually warm. Why?

c) The lens in Figure 10.4(a) produces an image of the document, which is upsidedown and back-to-front. Explain why the final image is the right way round.

3 **a)** When you polish a window using a dry cloth on a dry day the window soon becomes dusty. Why does this not happen on wet days?

b) Cling film is a thin plastic material that is used for wrapping up food. When you peel the film off the roll it sticks to itself. Suggest why this happens.

4 **a)** Electrostatic charges can be dangerous. Name one example of a dangerous situation. Explain what we do to reduce the danger.

b) Name a process in which electrostatics is put to good use in industry. Give an account of how the process works.

5 A lightning flash delivers a charge of 5 C. The voltage between the cloud and Earth is 100 million volts.

a) The flash lasts 0.002 s. Calculate the average current during this time.

b) Calculate the electrical energy transferred by the flash.

c) Explain what transfers of energy take place.

Summary

I am confident that:

✓ **I can recall these facts about mains electricity:**

- Electric shocks delivered from a high voltage source can be very painful, and can even cause death.
- A direct current (d.c.) flows in one direction at a constant rate.
- An alternating current (a.c.) switches direction many times each second.
- All countries use a.c. supplies for their mains electricity.
- Current in a resistor results in an electrical transfer of energy and a rise in temperature.
- The wire in a fuse heats up and melts if there is too much current.
- In a circuit breaker the electromagnetic switch opens when the current is too big.
- An earth wire in a plug prevents electrical shocks.
- Insulators are materials that do not conduct electricity.
- Appliances made of an insulating material do not need an earth wire. This is called double insulation.

✓ **I can recall these equations about electrical power and energy:**

- power = current × voltage ($P = I \times V$)
- energy = current × voltage × time ($E = I \times V \times t$)

✓ **I can recall these facts about series and parallel circuits:**

- In a series circuit the same current passes through all the components, one after another.
- In a series circuit increasing the voltage increases the current.
- In a series circuit increasing the resistance makes the current smaller.
- An ammeter is placed in series with a resistor to measure the current through it.
- A domestic appliance is placed in series with a fuse to protect it. The fuse melts if the current is too large.
- In a parallel circuit the current is split between two (or more) components.
- Components in parallel have the same voltage across them.
- Car headlights and domestic appliances are connected in parallel so they receive the full voltage.
- Voltmeters are placed in parallel with appliances to measure the voltage across them

✓ **I can recall these facts about resistance:**

- resistance = voltage / current ($R = V / I$)
- voltage = current × resistance ($V = I \times R$)
- For some resistors and wires, at constant temperature, the current through the resistor is proportional to the voltage across it; these are Ohmic resistors.
- As the current across a filament lamp increases, the resistance gets larger.
- The resistance of a thermistor decreases as the temperature increases.
- The resistance of a light-dependent resistor (LDR) changes as the light intensity changes

✓ **I can recall these facts about current, charge and voltage:**

- Current is a measure of the rate at which electrical charge moves round the circuit.
- charge = current × time
- An amp is equivalent to a flow of charge of 1 coulomb per second.
- A volt is one joule per coulomb.

✓ **I can recall these facts about electrostatics:**

- Insulators can be charged by friction.
- When electrons are removed from a material, it becomes positively charged.
- When electrons are added to a material, it becomes negatively charged.
- Like charges repel and unlike charges attract.
- Metals and ionic materials are good conductors.
- Rubber and plastics are good insulators.
- Electrostatics has applications in industry, such as paint spraying and electrostatic precipitation.

Sample answers and expert's comments

1 The diagram shows a 12 V battery in series with an ammeter, a variable resistor and a lamp.

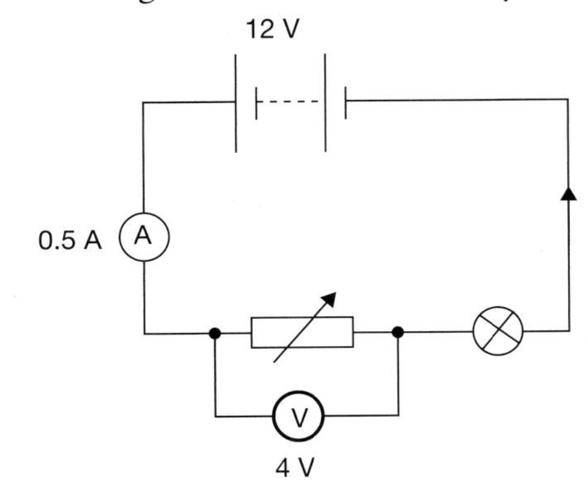

a) Explain how the circuit can be adjusted, using the same components, to make the lamp brighter. (2)

b) i) State the equation which links voltage, current and resistance. (1)

ii) Calculate the resistance of the lamp. State the unit of resistance. (3)

c) i) State the equation which links charge, current and time. (1)

ii) Calculate the charge which flows through the bulb in 5 minutes. State the unit of charge. (2)

(Total for question = 9 marks)

Student response Total 4/9	**Expert comments and tips for success**
a) change the resistance ○	This does not earn a mark, as the answer is too vague. The resistance of the variable resistor must be decreased; this will increase the current, which will brighten the lamp.
b) i) voltage = current × resistance ✓	The formula has been correctly written.
ii) R = $\frac{12}{0.5}$ ○ ✓ = 34 ○	The current of 0.5 A has been identified. However, the voltage across the lamp is 12 – 4 = 8 V. The voltage provided by the battery is equal to the voltages across the resistor and lamp. Always remember to give the unit. Here the unit of resistance is Ω. If the unit is not given in the question, you might gain a mark for writing it.
c) i) charge = current × time ✓	The correct formula has been used. You need to remember this formula.
ii) charge = 0.5 × 5 ○ = 2.5 coulombs ✓	The time should be in seconds, not minutes, as 1 amp = 1 coulomb/second. This is therefore the wrong substitution, which gives an incorrect answer. A mark is given for the correct unit $Q = I \times t$ = 0.5 × 300 = 150 C

2 A student uses this circuit to investigate how the current in a diode changes with the voltage across it. The graph shows the results of this investigation.

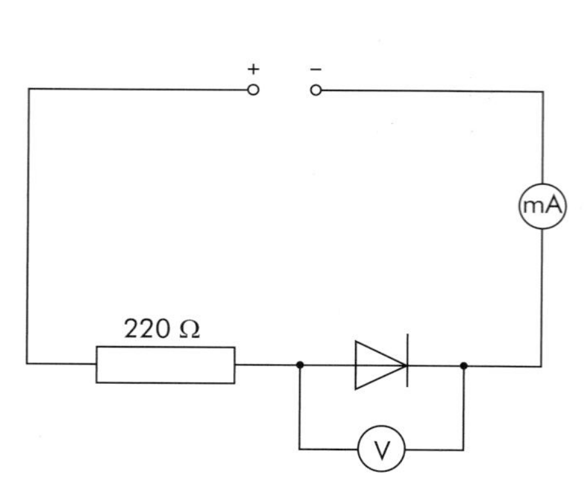

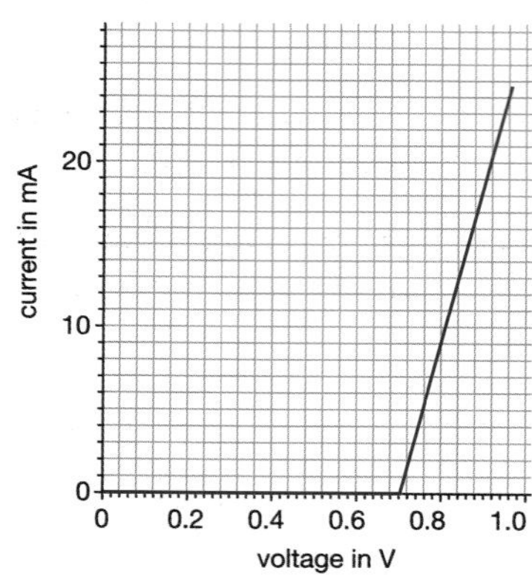

a) Describe how the current changes with the applied voltage. (2)

b) Use the graph to calculate the voltage across the diode when the current is 10 mA. (1)

c) Calculate the voltage across the resistor when the current is 10 mA. (3)

d) Use your answers to b) and c) to calculate the battery voltage. Explain your answer. (1)

(Total for question = 7 marks)

Student response Total 5/7	Expert comments and tips for success
a) The current is zero then it goes up in a straight line. ✓	This answer gains a mark for describing the broad outline. To gain the second mark the answer needs to state that the increase starts at about 0.7 V.
b) 0.8 ✓	The value is correct, but the student has forgotten the unit. This time no mark is lost in the mark scheme, but you should always include the unit for a quantity.
c) $V = I \times R$ ✓ $= 10 \times 220$ ○ $= 2200$ V ✓	Correct equation. The current is 10 mA (0.01 A), so the substitution is incorrect. But the next calculation has been done correctly and the unit is right
d) The battery voltage is 2200.8 V, because it has to provide both voltages. ✓	Although the answer to part c) is wrong, the answer correctly explains that the battery must provide the voltage across the resistor and the voltage across the diode

3 Using electricity can be dangerous.

a) Suggest two safety precautions you should take when putting a plug into a mains socket. (2)

b) Mains electricity provides an alternating current (a.c.). A battery provides a direct current (d.c.). Describe the difference between a.c. and d.c. (2)

c) The photograph shows two mains plugs. Mains plug A has a connection for an earth wire. Mains plug B does not have an earth connection.

A

B

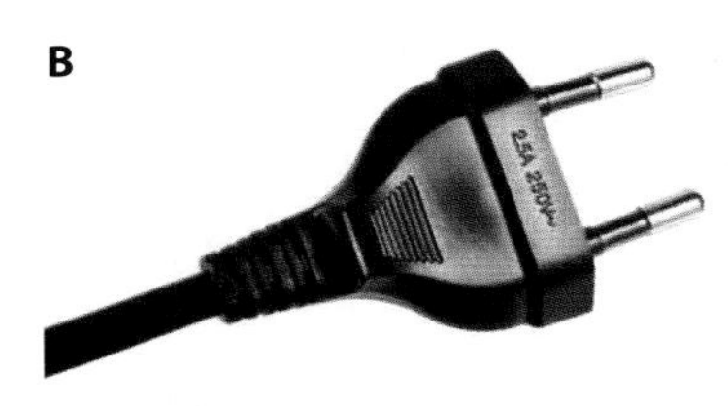

i) Describe how the earth wire can act together with a fuse as a safety device. (3)

ii) Explain why mains plug B can be safe to use even though it has no earth connection. (2)

(Total for question = 9 marks)

Student response Total 3/9	Expert comments and tips for success
a) *Take care when plugging something in.* ○	This offers no specific advice e.g. keep your fingers away from the prongs as you plug it in. Better still, turn the switch off at the mains.
b) *d.c. goes one way, a.c. goes both ways* ✓	This earns one mark, but is too vague for both marks. You need to say something along these lines: In a d.c. circuit the current is always one way ✓ and has a constant value – this is direct current. In an a.c. (alternating current) circuit the current goes one way and then switches to the other direction. ✓ The switching can take place many times each second.
c) i) *The Earth wire stops you getting a shock when the fuse blows.* ✓	Again this is a difficult idea to explain and you need to explain carefully. If a live wire touches the metal case of an appliance (e.g. toaster or fire), the case becomes live and you can get a shock. ✓ But if the case is earthed, a large current flows and blows the fuse (which is in the live wire). ✓ Now you cannot get a shock as the live is disconnected. ✓
ii) *Plug B is safe if it is used for something with a plastic case.* ✓	This is correct, but the sentence must be finished off … because plastic is an insulator so you cannot get a shock if you touch it.

Exam-style questions

1 The unit of the volt can be expressed as:

A J × s **B** J/C **C** J × C **D** J/s [1]

2 The unit of the amp is equivalent to:

A C × s **B** C **C** C/s **D** s/c [1]

The diagrams A to D show current-voltage graphs for 4 electrical components.

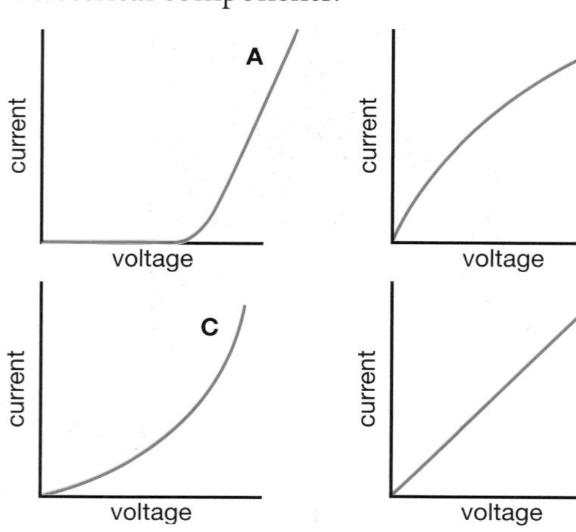

3 Which is a lamp? [1]

4 Which is an Ohmic resistor? [1]

5 Which is a diode? [1]

6 The resistance of R is decreased. Which of the following is true about the brightness of lamps X and Y? [1]

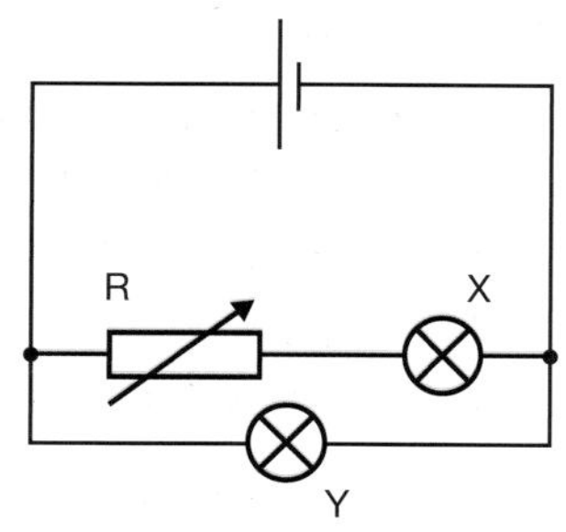

	X	Y
A	Duller	Same
B	Brighter	Brighter
C	Brighter	Same
D	Same	Same

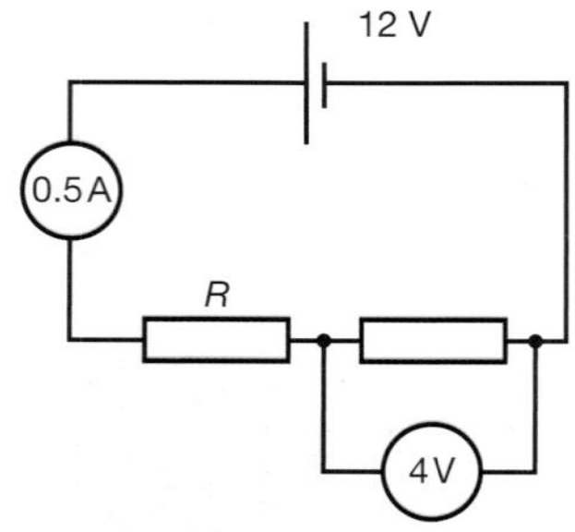

7 The value of the resistance R is:

A 24 Ω **B** 16 Ω **C** 13 Ω **D** 8 Ω [1]

8 Which of the following correctly shows the reading on ammeters A_1 and A_2?

	A_1	A_2
A	0.6 A	0.6 A
B	0.3 A	0.3 A
C	0.4 A	0.2 A
D	0.2 A	0.4 A

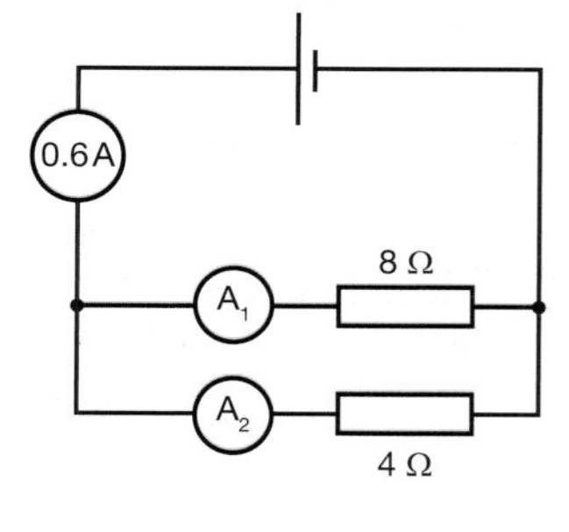

[1]

9 Which of the following correctly shows what happens to the readings on the voltmeter and ammeter when the light intensity increases?

	V	A
A	increases	increases
B	increases	decreases
C	decreases	increases
D	decreases	decreases

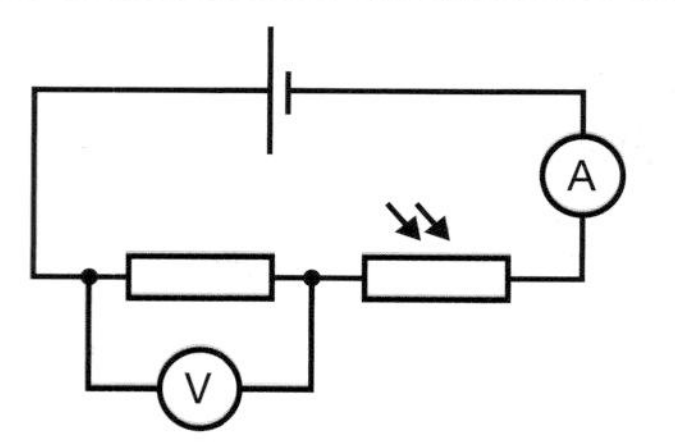

[1]

10 A kettle heating element has a resistance of 20 Ω. It is connected to a 220 V supply. Which of the following correctly shows the electrical transfer of energy by the kettle in 2 minutes?

A 4.4 kJ **B** 4.8 kJ **C** 290 kJ **D** 528 kJ [1]

11 The diagram shows an electric mains plug with a fuse and an earth wire.

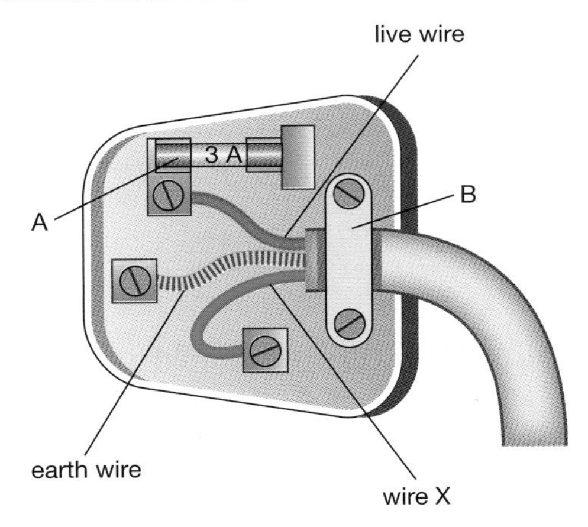

a) Name the third wire X, and the parts labelled A and B. [3]

b) Why is part A put in the live wire? [2]

c) The diagram below shows a plug that has been wired dangerously. Explain how you would change the wiring to make it safe. [4]

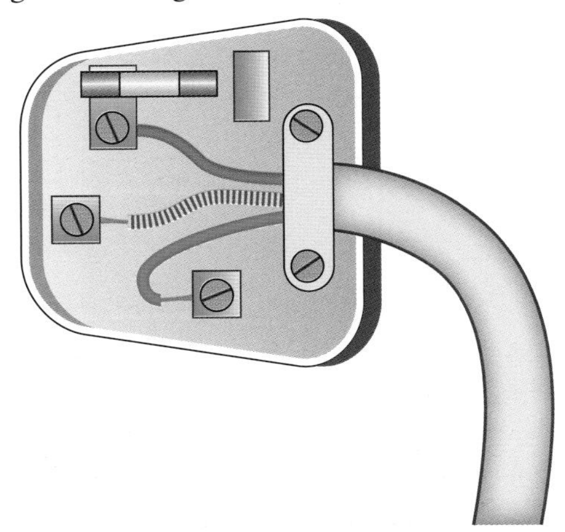

12 An electric iron has been wired without an earth connection. After years of use the live wire becomes loose and touches the metal iron.

a) A man touches the iron and receives an electric shock. Sketch a diagram to show the path of the current through him. [1]

b) Explain how an earth wire would have made the iron safe for him to touch. [2]

c) The mains voltage is 230 V. The man's resistance is 46 kΩ.

Calculate the current through the man. [3]

13 A four-way adaptor is plugged into a mains supply of 230 V. The adaptor has a 13 A fuse. Four appliances may be powered from the adaptor in parallel. The table shows some appliances that might be plugged in to the adaptor and the current that each draws from the mains.

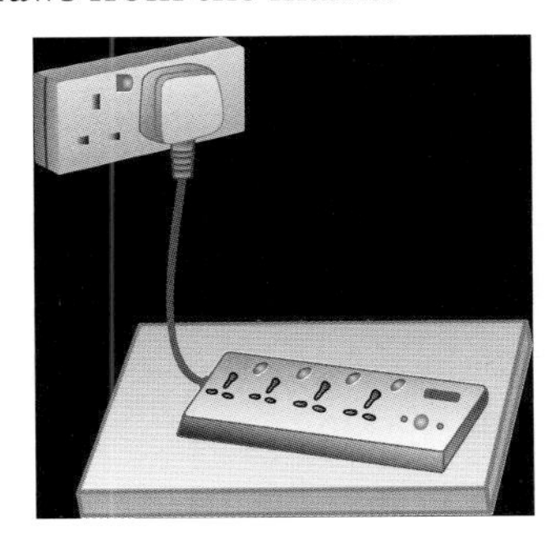

Appliance	Current in A
television	0.5
lamp	0.1
computer	0.3
radio	0.1
iron	4.0
hairdryer	2.5
heater	7.5

a) Explain why appliances are connected to the mains supply in parallel. [2]

b) Calculate the electrical power used when the television, computer, lamp and radio are all plugged in to the adaptor together. [3]

c) Explain what will happen if you plug the iron, the hairdryer and the heater into the adaptor. [2]

d) Explain why we are advised to plug heaters directly in to a mains socket, rather than using an adaptor. [1]

e) Explain why it is important to earth the metal heater. [2]

f) The hairdryer is double insulated. What does this mean? [2]

14 The circuit shown has four ammeters in it.

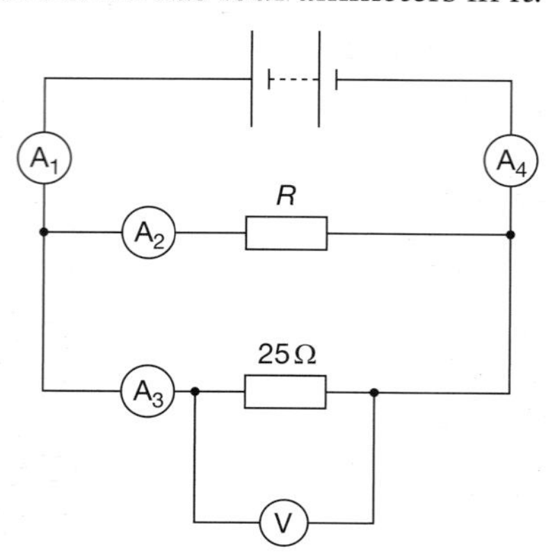

a) Ammeter A1 reads 0.6 A, and ammeter A2 reads 0.2 A. Determine what ammeters A3 and A4 read. [2]

b) Is the resistance, *R*, larger or smaller than 25 Ω? Explain your answer. [2]

c) Calculate the voltage shown on the voltmeter. Show your working. [3]

d) State the voltage of the battery. [1]

15 A student decides to investigate how the current through a filament lamp changes with the voltage applied across it.

The student uses a 12 V battery, a filament lamp, an ammeter, a voltmeter, a variable resistor, a switch and some connecting wires.

a) Draw a suitable circuit diagram for the experiment. [3]

b) The student obtained the following results.

Voltage in V	0.0	1.0	2.0	3.0	4.0	6.0	8.0	10.0
Current in A	0.0	0.45	0.80	1.05	1.25	1.50	1.75	1.90

i) Plot a graph of current against voltage. [4]

ii) Use the graph to find the current when the voltage is 7.0 V. [1]

iii) Calculate the resistance of the filament lamp for voltages of 2.0 V and 6.0 V. [3]

iv) Explain why the resistance of the lamp changes as the current increases. [2]

v) Calculate the power used by the lamp when the voltage across it is 10.0 V. [3]

16 The diagram shows three lamps connected in parallel to a 12 V battery.

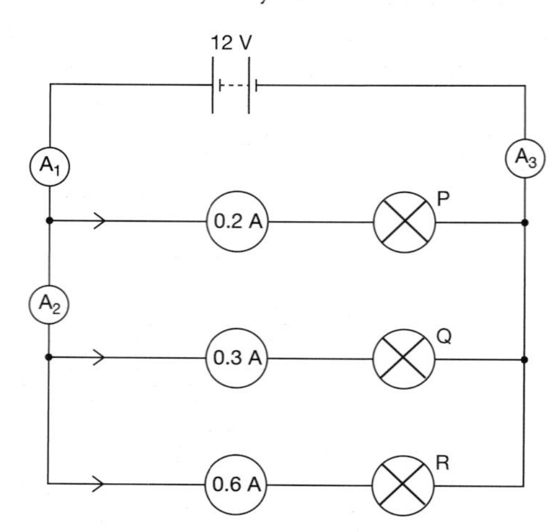

a) Calculate the currents through the ammeters A_1, A_2 and A_3. [3]

b) Calculate the resistance of lamp P. [3]

c) Calculate the power dissipated by lamp R. [3]

17 A filament lamp is connected to a 12 V battery. The current through the lamp is recorded by a data logger when the lamp is switched on. The graph shows how the current in the filament changes with time.

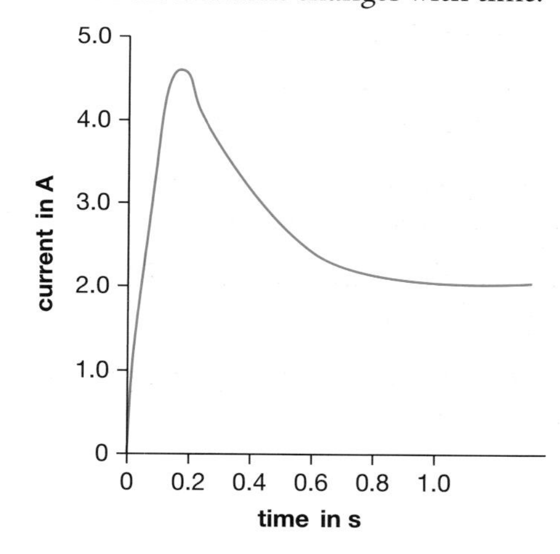

a) Describe how the current in the filament changes with time when it is switched on. [2]

b) Use the graph to determine:

i) the maximum current in the lamp

ii) the current after a time of 1 second. [2]

c) The resistance of the filament increases as its temperature rises. Use this information to explain the shape of the graph. [2]

d) The lamp is marked 12 V 24 W. Explain what these markings mean. [2]

18 A strain gauge is a device that detects the way an object changes shape when it is squashed, stretched or bent. The diagram shows how one is made. A thin piece of wire is set into a piece of flexible plastic. When the plastic is bent, the wire stretches. This causes the resistance of the wire to increase.

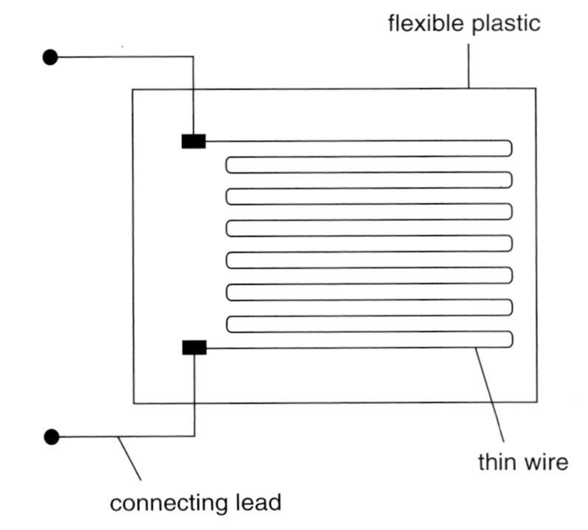

a) Draw a diagram to show how the resistance of the strain gauge can be measured using a battery, ammeter and voltmeter. [2]

b) Before the gauge is stretched it is connected to a 6 V battery. A current of 30 mA flows through it. Calculate the gauge's resistance. [3]

c) Explain why the resistance of the gauge increases when the wire is stretched. [1]

d) State what happens to the current in the circuit when the wire is stretched. [1]

19 A student is investigating the relationship between voltage and current.

a) State the equation linking voltage, current and resistance. [1]

b) The meters show the current in a resistor and the voltage across it.

current in mA

voltage in V

13.00

i) Copy the results table and add the readings on the two meters to the last column. [1]

Current in mA	0.20	0.60	1.01	1.14	1.81	2.22	
Voltage in V	1.0	3.0	5.0	7.0	9.0	11.0	

ii) Use the data in the table to draw a graph of current against voltage. [5]

iii) Circle the anomalous point on the graph. [1]

iv) How did you decide that this point was anomalous? [1]

v) Use your graph, or the table, to find the resistance of the resistor that the student used. [2]

c) The student wants to investigate the effect of changing light intensity on a circuit. She sets up equipment outside in a garden for an experiment lasting 24 hours. She uses the circuit shown in the diagram.

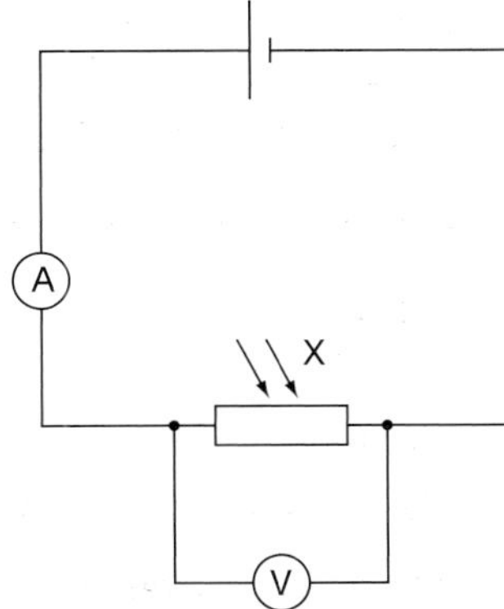

i) Give the name of the component labelled X. [1]

ii) List the variables that the student should measure. [2]

iii) Explain why the student might need help to take all the readings for this investigation. [2]

iv) The graph below shows some results of the student's investigation.

Copy the graph and label both axes with appropriate quantities and units. [2]

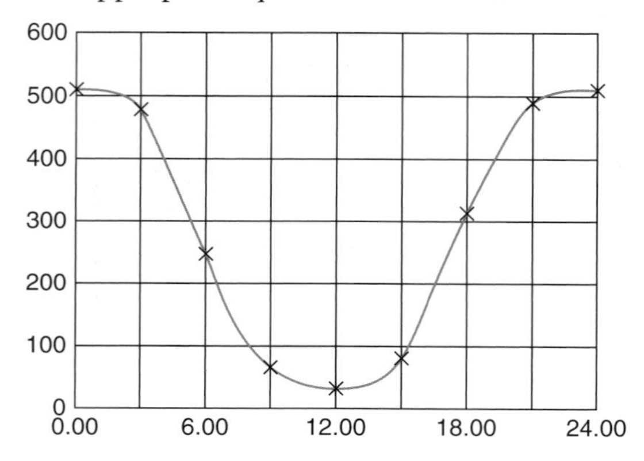

20 A student rubs a plastic rod with a cloth.

a) i) The plastic rod gains a negative charge. Explain how the rod becomes charged. [2]

ii) Describe experiments the student could do to show that there are two types of electric charge. [4]

b) The teacher uses a Van de Graaff generator to give another student a negative charge.

Use ideas about electric charge to explain why the student's hair stands on end. [3]

21 Burning coal in power stations produces soot, ash and waste gases.

The diagram shows an electrostatic precipitator, which collects the soot and ash particles.

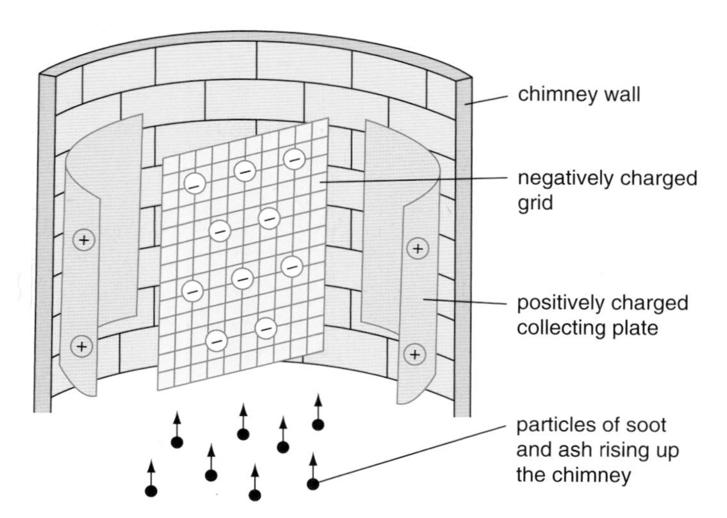

Explain how the electrostatic precipitator prevents the ash and soot escaping to the atmosphere. [4]

22 The diagram shows a car being painted by a process of electrostatic paint spraying.

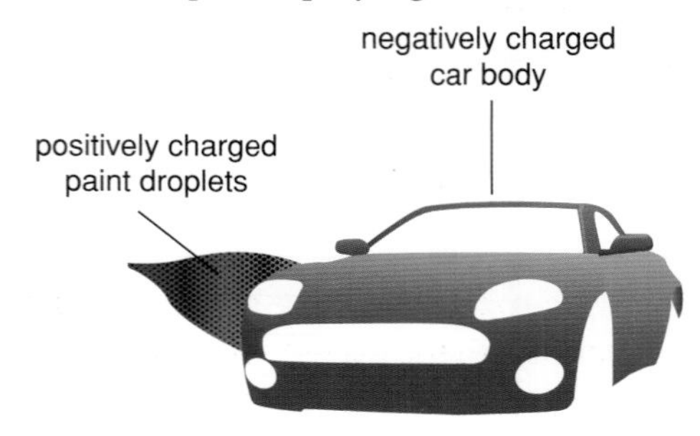

a) Explain why it is important that all the paint droplets are positively charged. [2]

b) Explain why the car must be charged negatively. [2]

23 The diagram shows a thundercloud.

a) Explain why the tree below has a positive charge. [2]

b) The base of the cloud carries a charge of 15 C. The voltage between the cloud and the ground is 800 000 kV.

i) A flash of lightning discharges the cloud completely in 0.04 s. What is the average current that flows in this time? [3]

ii) Calculate the maximum energy that could be delivered by this lightning flash. [3]

EXTEND AND CHALLENGE

1 a) The table shows the current in three different electrical appliances when connected to a 230 V a.c. supply.

appliance	current in A
kettle	11.5
lamp	0.05
toaster	4.2

i) Which appliance has the greatest resistance? Explain how the data shows this.

ii) The lamp is connected to the mains supply using thin, twin-cored cable, consisting of live and neutral connections. State **two** reasons why this cable should not be used to connect the kettle to the mains supply.

b) i) Calculate the power rating of the kettle when it is operated from the 230 V a.c. mains supply.

ii) Calculate the current flowing through the kettle when it is connected to a 115 V mains supply.

iii) The kettle is filled with water. The water takes 90 s to boil when working from the 230 V supply. Explain how the time it takes to boil changes when the kettle operates on the 115 V supply.

2 The graph shows how the current through a type of filament lamp depends on the voltage applied to it. Three of these lamps are connected into the circuit shown.

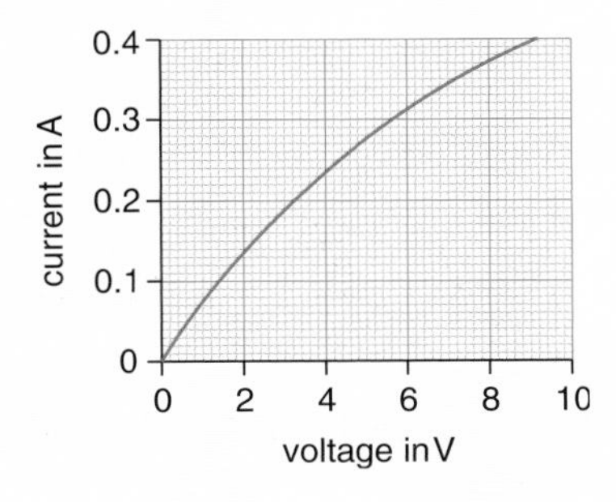

a) The circuit diagram shows 0.2 A flowing through lamp B. Use the graph to find the voltage across it.

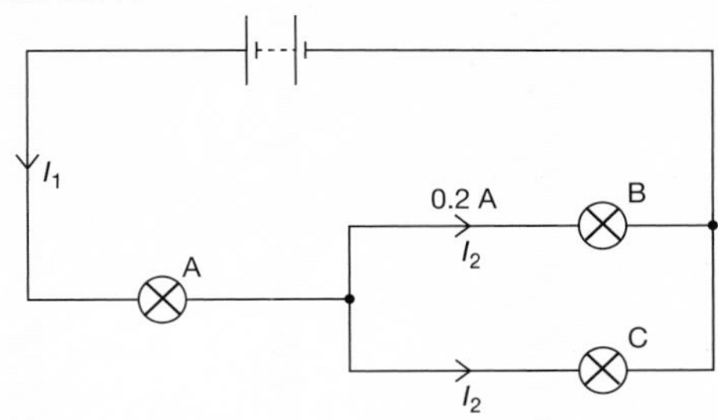

b) Calculate the resistance of lamp B.

c) Work out the currents in:

i) lamp C

ii) lamp A.

d) Use your answer to part **c) ii)** to calculate the voltage across lamp A.

e) Now calculate the voltage of the battery

3 In this question you are asked to work out the unknown values I and R in the circuit diagram below. The questions below take you through the process of solving this problem.

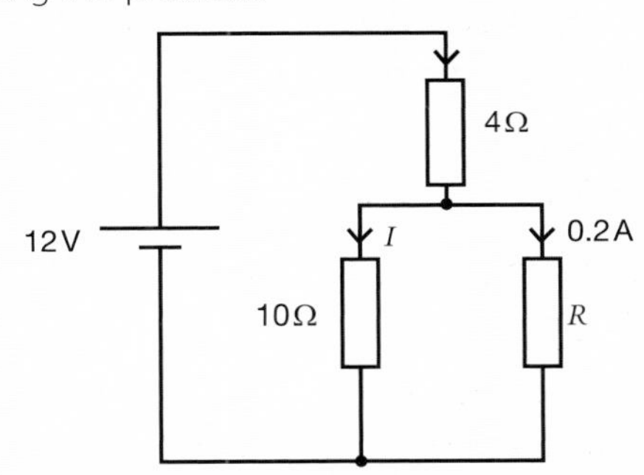

a) Write down an expression for the current flowing through the 4 Ω resistor.

b) Now explain why it is possible to write the following equation relating the voltage of the battery to the voltage across the resistors:

$$12 = 4(I + 0.2) + 10.I$$

c) Solve the equation in part **b)** to calculate the value of I.

d) Now calculate the value of R.

3 Waves

WRITE • CONSIDER

Write down five examples of the use of waves. In each case explain how we use them and how information and energy is carried by the waves.

Sea waves store a lot of kinetic energy. They also carry some information – about the weather conditions at sea. Energy to set these waves in motion may have travelled thousands of miles, but matter has not been transferred this distance. Water waves are some of the waves we can see. We make use of many types of waves in our everyday lives, to communicate, to cook and in medicine.

By the end of this section you should:

- be able to explain the difference between longitudinal and transverse waves
- know the definitions of amplitude, wavefront, frequency, wavelength and period of a wave
- know that waves transfer energy and information without transferring matter
- know and use the relationship between wave speed, wavelength and frequency
- be able to use the relationship between frequency and time period
- be able to use the relationships above in different contexts
- be able to explain the Doppler effect
- be able to explain that all waves are reflected and refracted
- know that light is part of a continuous electromagnetic spectrum
- know the order of the electromagnetic spectrum by wavelength and frequency
- be able to explain the uses and the detrimental effects of the electromagnetic spectrum
- know that light waves are transverse waves and that they can be reflected and angle of refracted
- be able to use the law of reflection
- be able to draw ray diagrams to show reflection and refraction
- be able to investigate the refraction of light
- know the relationship between refractive index and the angle of incidence and angle of refraction
- be able to investigate refractive index
- be able to describe the role of total internal reflection in optical fibres
- be able to explain the meaning of the critical angle
- know the relationship between critical angle and refractive index
- know that sound waves are longitudinal and can be reflected and refracted
- know the range of human hearing
- understand how an oscilloscope and a microphone can display sound waves
- be able to investigate the frequency of sound with an oscilloscope
- understand that the loudness of a sound relates to the amplitude of the source vibration

3.1 Introducing waves

Waves do two important things; they carry energy and information. When you watch television you are taking advantage of radio waves. These waves carry energy and information from the transmitting station to your house. Light and sound waves then carry energy and information from the television set to your eyes and ears.

Waves in solids and liquids

When you drop a stone into a pond you see water ripples spreading outwards from the place where the stone landed (with a splash). As the ripples spread, the water surface moves up and down. These ripples are examples of **waves**.

The water ripples transfer energy and information. The energy moves outwards from the centre but the water itself does not move outwards. The shape of the waves provides us with the information about where the stone landed (if we did not see it land).

Figure 1.1 Information is a major part of our lives – waves carry this information.

Figure 1.2 Water ripples spreading

Transverse and longitudinal waves

A good way for us to visualise waves is to use a stretched 'slinky' spring. When two students stretch a slinky across the floor in a laboratory, they can transmit waves that travel slowly enough for us to see.

Transverse waves

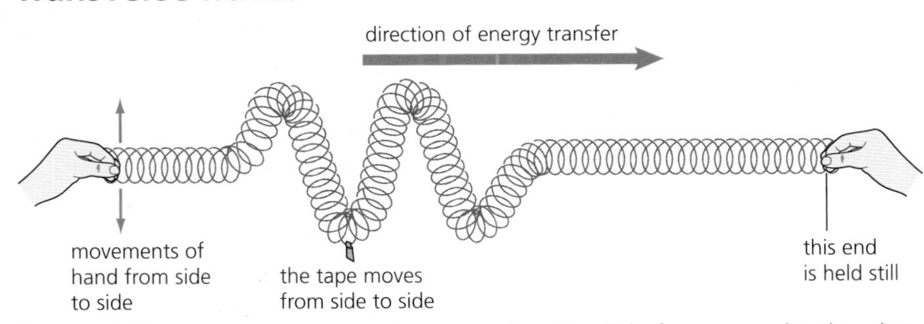

Figure 1.3 The transverse waves transfer energy along the slinky from one end to the other. The coloured tape shows that the pulses cause the slinky to vibrate from side to side, just the same as the student's hand.

Figure 1.3 shows a **transverse wave**. One student holds the end of the slinky stationary. The other end of the slinky is moved from side to side. A series of pulses moves down the slinky, sending energy from one end to the other. The student holding the slinky still feels the energy as it arrives. None of the material in the slinky has moved permanently.

Water waves and light waves are two examples of transverse wave. The water shown in Figure 1.2 moves up and down. Energy is carried outwards by the wave, but water does not pile up at the edge of the pond.

Longitudinal waves

Figure 1.4 shows a **longitudinal wave**. Energy is transmitted along the slinky by pulling and pushing the slinky backwards and forwards. This makes the slinky vibrate backwards and forwards. The vibration of the slinky is parallel to the direction of energy transfer. The coils are pushed together in some places (areas of compression). In other places the coils are pulled apart (areas of rarefaction). As energy is transferred, none of the material of the slinky moves permanently.

Sound is an example of a longitudinal wave. When a guitar string is plucked, energy is transferred through the air as the string vibrates backwards and forwards. However, the air itself does not move away from the string with the wave and there is not a vacuum left near the guitar.

TIP

Can you explain the difference between transverse and longitudinal waves?

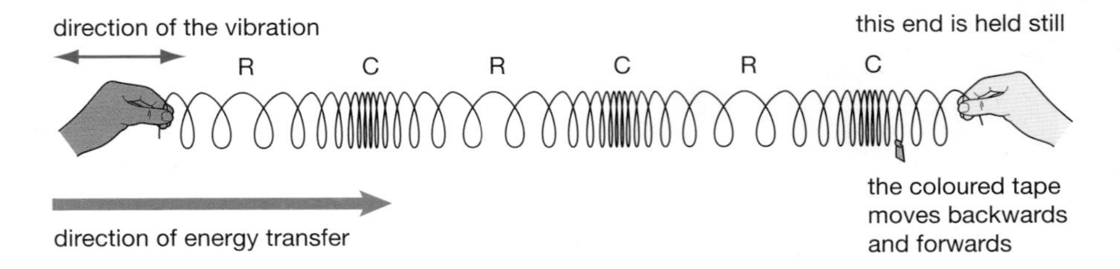

Figure 1.4 The coloured tape on the slinky shows that the pulses on the slinky move the coils backwards and forwards in the same direction as the student's hand moves.

Figure 1.5 Earthquakes produce shock waves, which may be either longitudinal or transverse waves. Buildings are damaged by the transfer of energy, not by the movement of material.

Describing waves

Waves are described by the terms, amplitude (A), wavelength (λ), frequency (f) and time period (T). λ, the symbol for wavelength, is a Greek letter pronounced lambda.

TIP

Make sure you can define amplitude, wavelength, time period and frequency.

Figure 1.6 shows a transverse water wave. This diagram helps us to understand the meaning of these terms.

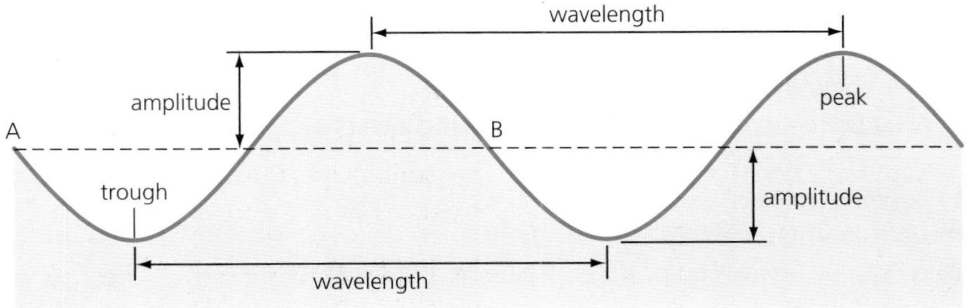

Figure 1.6

- In Figure 1.6 the amplitude is the distance from a peak to the middle or from a trough to the middle.
- In Figure 1.6 a wavelength, λ, is the distance between two peaks or two troughs (or two equivalent points such as A and B).
- Frequency is measured in hertz (Hz). A frequency of 1 Hz means a source is producing one wave per second. If a student, using a slinky, moves his hand from side to side and back twice each second, he produces two complete waves each second. The frequency of the waves is 2 Hz.
- The time period, T, of a wave is the time taken to produce one wave.
- The frequency and time period of a wave are linked by this equation:

$$T = \frac{1}{f}$$

$$\text{Time period} = \frac{1}{\text{frequency}}$$

where period is in seconds, s
frequency is in hertz, Hz.

Example. Calculate the period of a wave with a frequency of 10 Hz.

$$T = \frac{1}{10}$$
$$= 0.1\,\text{s}$$

You can remember this as follows: if a source produces 10 waves each second, the source takes 0.1 s to produce one wave.

TIP

Make sure you can recall an equation linking wave speed to frequency and wavelength.

Wave speed and the wave equation

The **wave speed** is the speed at which energy is transferred (or the wave moves) through the medium.

All waves obey this equation:

$$v = f\lambda$$

wave speed = frequency × wavelength
where wave speed is in metres per second, m/s
frequency is in hertz, Hz
wavelength is in metres, m.

TIP

A wavefront is a surface that is affected in the same way by a wave at any given time. All the points in a wavefront move in phase.

Example. Sound travels at a speed of 330 m/s in air. Calculate the wavelength of a sound wave with a frequency of 660 Hz.

$$v = f\lambda$$
$$330 = 660 \times \lambda$$
$$\lambda = \frac{330}{660}$$
$$= 0.5\,\text{m}$$

STUDY QUESTIONS

1. Use diagrams to illustrate the nature of
   a) transverse waves
   b) longitudinal waves.
2. What do the terms 'area of compression' and 'area of rarefaction' mean?
3. Describe how you would use a slinky to show that waves transfer both energy and information.
4. In Figure 1.7 waves go past a cork travelling from left to right.
   a) In which way does the water move as the waves travel?
   b) Describe the motion of the cork.

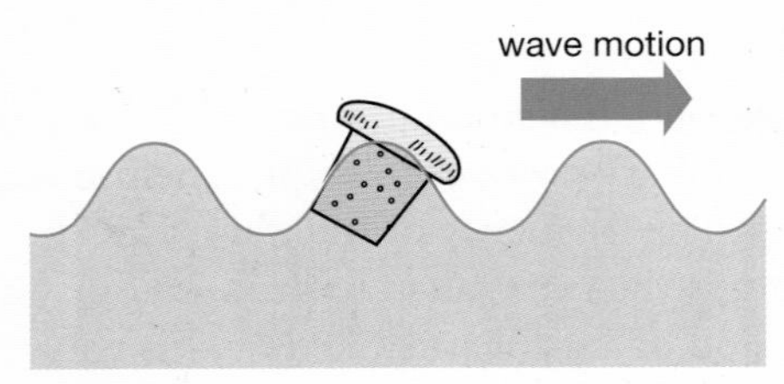

Figure 1.7

5. Figure 1.8 shows wave pulses travelling at the same speed along two ropes. How are the wave pulses travelling along rope A different from the wave pulses travelling along rope B?

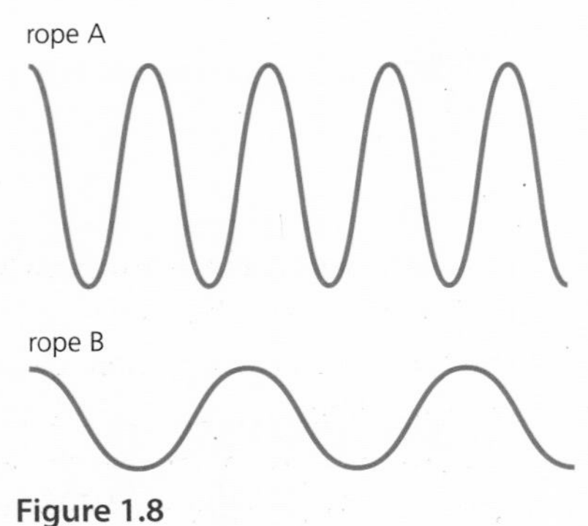

Figure 1.8

6. Figure 1.9 shows transverse waves on a rope of length 8 m.

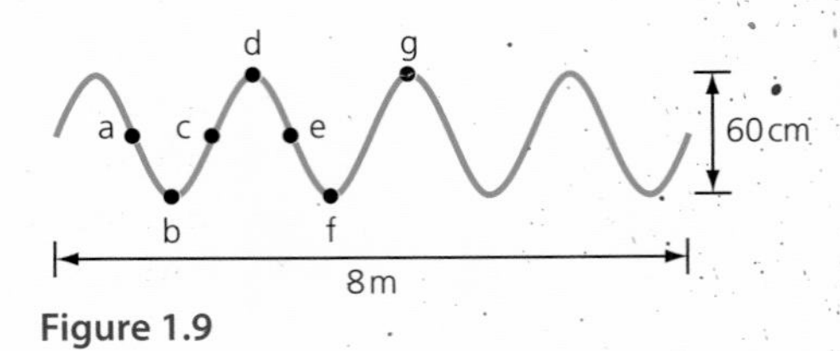

Figure 1.9

   a) What is the name given to each of these horizontal distances?
      i) ae
      ii) bf
      iii) dg
   b) What is the name given to each of these vertical distances?
      i) ed
      ii) ef
   c) Calculate:
      i) the wavelength of the wave
      ii) the amplitude of the wave
      iii) the horizontal distance ag.
7. The time periods of two waves are:
   a) 0.25 s
   b) 0.01 s.

   Calculate the frequency of each wave.
8. A stone is thrown into a pond and waves spread outwards. The waves travel with a speed of 0.4 m/s and their wavelength is 8 cm. Calculate the frequency of the waves.
9. Make a sketch to show the wavelength of a longitudinal wave.

3.2 Ripple Tanks

We can learn more about the properties of waves by studying ripples that move over the surface of water. Figure 2.1 shows a ripple tank. A motor is used to produce waves continuously, by vibrating a dipper up and down. A beam can be used to produce straight waves, and a small sphere is used to produce circular waves. By shining a light over the top of the glass tank, we can see the pattern of crests and troughs on the screen below the tank.

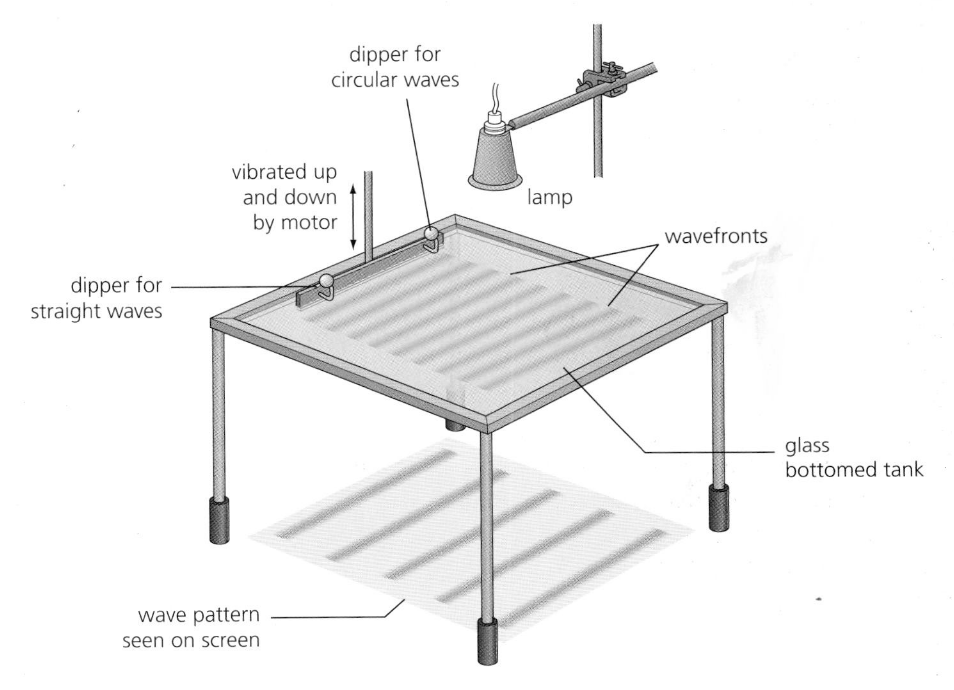

Figure 2.1

We use the word **wavefront** to describe a line where the water is affected by a wave in the same way at the same time. For example, the circular crests of the waves seen in Figure 1.2 are parts of the same wavefront: where there is a crest, all the water has been lifted to its maximum displacement. In Figure 2.2, we can see straight wavefronts produced by the straight beam. As the wavefront passes, all the water along the line of the wavefront, rises and falls together.

Reflection

Figure 2.2 shows what happens when straight water wavefronts, in a ripple tank, approach a solid barrier (a metal plate). The waves are reflected off the barrier as shown, so that they leave the barrier (the red wavefronts) at the same angle as they approached (the blue wavefronts).

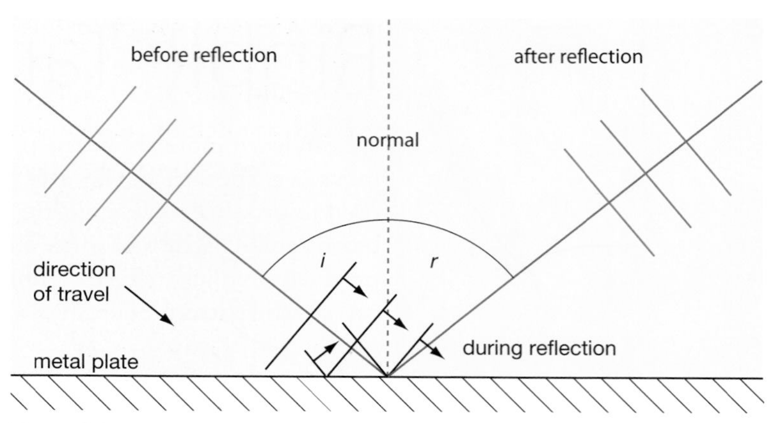

Figure 2.2

When describing reflection, we define an angle of incidence, i, and angle of reflection, r; these are shown in Figure 2.2. The normal is a line at right angles to the barrier. The angle of incidence i, is the angle between the direction of travel (the green line) for the incident waves and the normal. The angle of reflection, r, is the angle between the direction of travel (the green line) for the reflected waves and the normal.

The angle of incidence always equals the angle of reflection.

$$i = r$$

All types of waves show the property of reflection. For example, we see light waves reflected in mirrors, and when we hear an echo, sound waves have been reflected. The reflection of light is studied further in Chapter 3.4.

Refraction

A ripple tank can also be used to show the refraction of water waves. When water waves travel from deep water to shallow water they slow down, and their wavelength also reduces. Figure 2.3 shows the side view of what happens to water ripples as they travel from deep water to shallow water. (The frequency of the waves stays the same.)

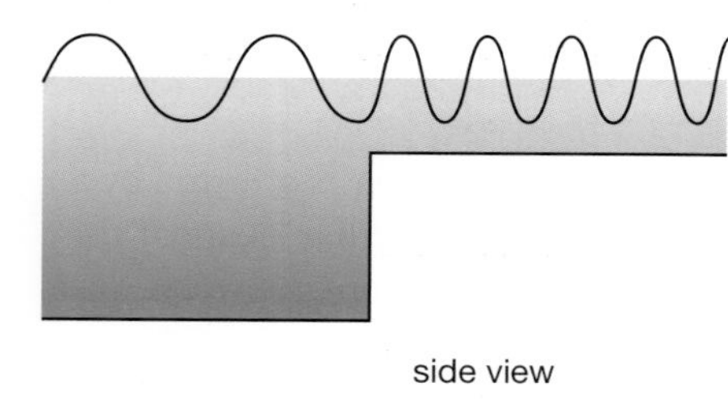

Figure 2.3

Figure 2.4 shows the ripple tank from the top. Here some waves travel into some shallow water. The waves in the deep water travel faster than the waves in shallow water.

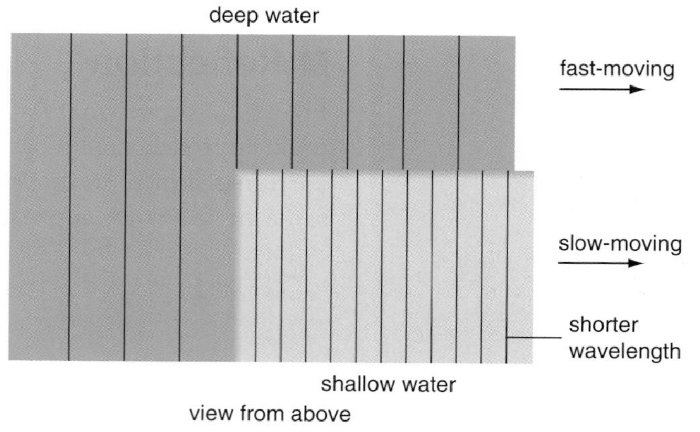

Figure 2.4

Figure 2.5 shows what happens when the water waves meet a region of shallow water at an angle. Now the waves slow down and change direction. This is called **refraction**. All waves show the property of refraction. The refraction of light is studied further in chapter 3.5.

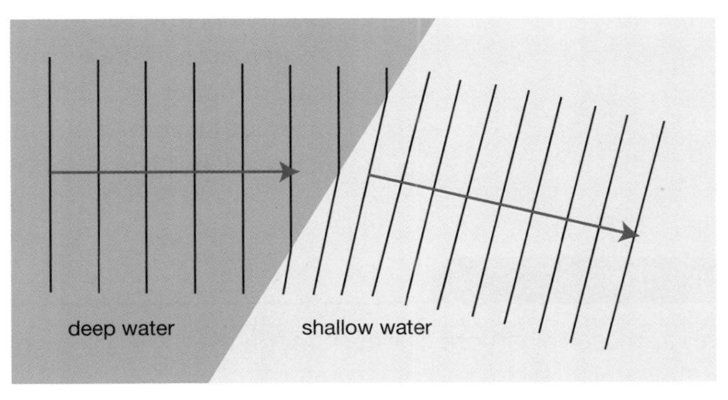

Figure 2.5

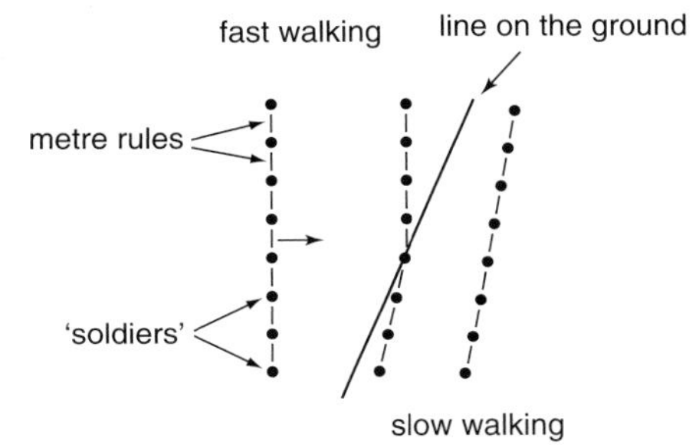

Figure 2.6

Figure 2.6 shows how refraction can be demonstrated by students acting as a platoon of soldiers. The idea is that a platoon (of eight soldiers in this example) walks briskly towards a line, with their arms linked firmly by meter rules. When they cross the line, each in turn slows down. The result is that the platoon slows and their row changes direction. This is a model or an **analogy** of refraction.

Doppler effect

You will be familiar with the Doppler effect. This is the name given to the apparent change in the frequency of a moving source of sound (or other type of wave). When you hear the siren from a fire engine as you stand in a street, you hear one pitch (frequency) of sound as the fire engine approaches you, and a lower pitch of sound after the fire engine has passed you and is travelling away from you. Figure 2.7 helps you to understand why the sound changes pitch.

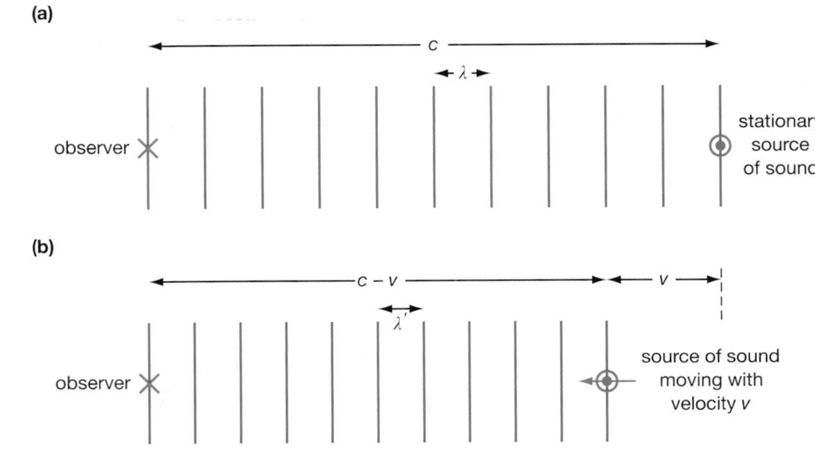

Figure 2.7

MATHS TIP

You are not expected to be able to derive this formula, but you meet this formula again in Section 8. From diagrams 2.7a) and 2.7b), you can see that:

$$\frac{\lambda'}{\lambda} = \frac{c - v}{c}$$

but $\lambda' = \lambda - \Delta\lambda$, where $\Delta\lambda$ is the change in wavelength.

So

$$\frac{\lambda - \Delta\lambda}{\lambda} = \frac{c - v}{c}$$

and

$$\frac{\Delta\lambda}{\lambda} = -\frac{\Delta f}{f} = \frac{v}{c}$$

In Figure 2.7a), a source of sound is emitting waves which are heard by the observer, who is a distance c metres away from the source; c is the speed of sound in m/s. A one-second burst of sound stretches from the source to the observer. In Figure 2.7b), the source is moving towards the observer with a velocity v. Now, the one-second burst of sound is squashed into a length $(c - v)$ metres. This means that the wavelength is reduced, and the observer hears a higher frequency. If the source moves away from the observer, the waves are stretched out into a length of $c + v$ metres; the wavelength is increased and the observer hears a lower frequency.

STUDY QUESTIONS

1 Sea waves approach a harbour wall as shown in the diagram. Copy the diagram and show the path of the waves after they have hit the harbour wall.

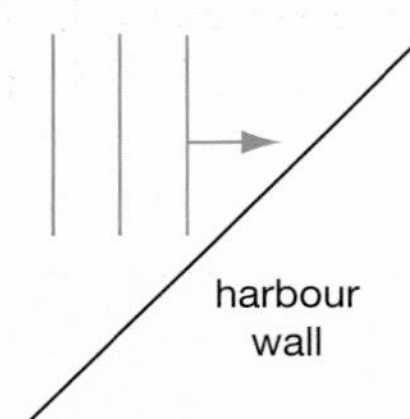

2 **a) i)** Sea waves approach a shallow beach as shown in the diagram. Copy the diagram and show the direction of the waves as they cross into the shallow area of water.

ii) Explain why the waves change direction as they move into shallow water.

iii) Wavefronts approaching a beach are usually parallel to the beach. Use your answers to parts **i)** and **ii)** to explain why.

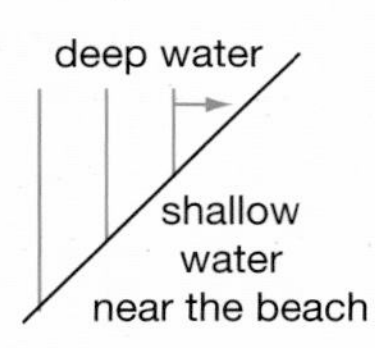

b) The diagram below shows water waves moving from shallow water to deep water. Copy the diagram to show the wavefronts in the deep water. [Hint: the waves travel faster in deep water and have a longer wavelength; and the wavefronts must join the waves in the shallow water.]

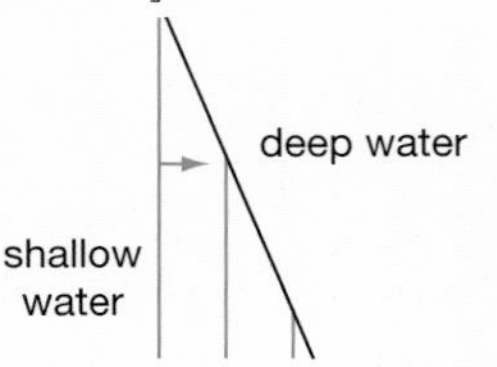

3 Explain why the apparent pitch of a fire engine's siren changes as the vehicle passes you. Draw diagrams to show how the wavefronts of sound are affected by the siren's motion towards you, and then away from you.

3.3 Electromagnetic waves

Figure 3.1 These sunbathers are exposed to several kinds of electromagnetic waves.

These people on the beach are exposed to several types of electromagnetic waves. Can you name four types of electromagnetic wave, some of which could be damaging their bodies?

Can you explain what happens to the sunbathers as time passes from morning to afternoon?

The electromagnetic spectrum

You will already have heard of radio waves and light waves. These are two examples of **electromagnetic waves**. There are many sorts of electromagnetic wave, which produce very different kinds of effect. Figure 3.2 shows you the full electromagnetic spectrum. The range of wavelengths in the spectrum stretches from 10^{-12} m for gamma rays to about 2 km for radio waves.

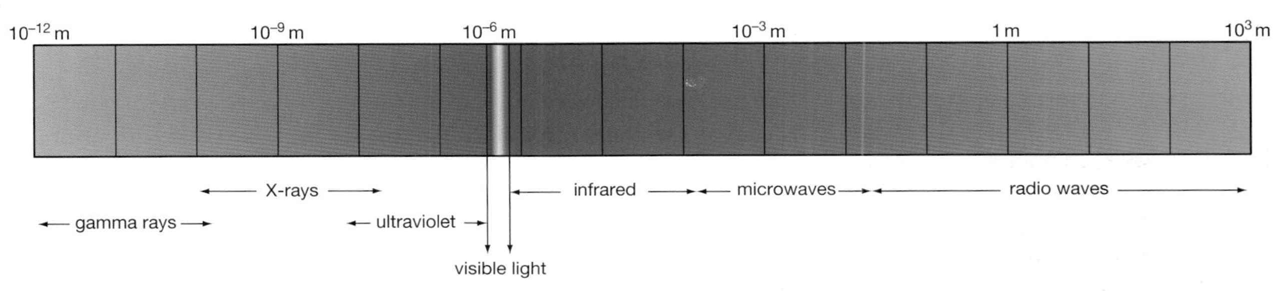

Figure 3.2 The range of wavelengths in the electromagnetic spectrum.

TIP

Electromagnetic waves are transverse waves, carrying energy and information in their electric and magnetic force fields.

Electric and magnetic fields

Many waves that you are familiar with travel through some material. Sound waves travel through air; water ripples travel along the surface of water. Electromagnetic waves can travel through a vacuum; this is how energy reaches us from the Sun. The energy is carried by changing electric and magnetic fields. These changing fields are at right angles to the direction in which the wave is travelling. So electromagnetic waves are transverse waves (Figure 3.3).

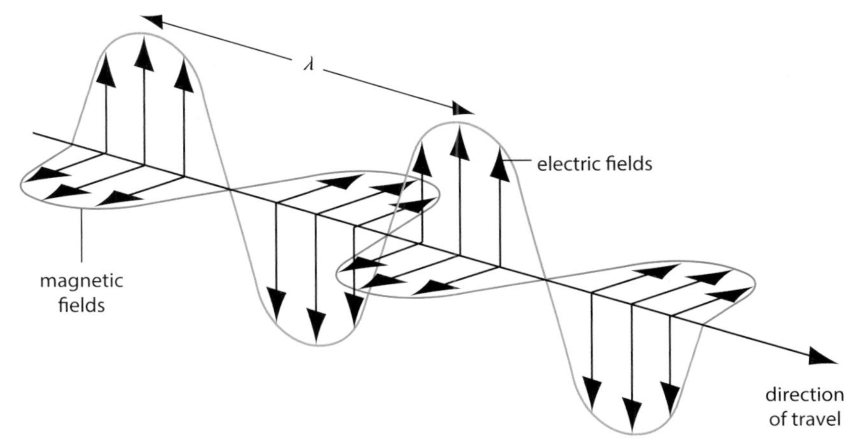

Figure 3.3 In an electromagnetic wave, energy is carried by oscillating electric and magnetic fields. These fields are at right angles to the direction in which the wave travels.

TIP

In a vacuum, all electromagnetic waves travel at the speed of light.

Wave speed

Electromagnetic waves show the usual wave properties. They can be described by frequency and wavelength. In a vacuum all electromagnetic waves travel at the same speed of 3×10^8 m/s. However, electromagnetic waves travel at slower speeds when they travel in a material. For example, light travels at a slower speed in glass.

medium

Radio waves

Radio waves with wavelengths of a few hundred metres are used to transmit local and national radio broadcasts. Radio waves with wavelengths of a few centimetres are used to transmit television signals and international phone calls. These short-wavelength radio waves are also known as microwaves. Microwaves are also used for mobile internet connection and satellite navigation. When you make a phone call to America, your radio signals are sent into space by large aerial dishes (see Figure 3.4). These signals are received by a satellite in orbit around the Earth. The signals are then relayed to another dish in America.

Microwaves

Figure 3.4 A large satellite communications dish.

Microwaves have wavelengths between radio waves and infrared waves. As well as satellite transmissions, microwaves are used for cooking. Microwaves of a particular wavelength are chosen that are absorbed by water molecules. Microwave cooking provides the advantage that the waves can penetrate further than infrared waves, thereby cooking the inside as well as the outside. This is much faster than the conventional method of cooking, where heat is conducted to the inside. Microwave ovens must be made of metal in order to trap the waves inside them. Microwaves of certain wavelengths are a hazard and could burn us internally by heating up our body tissues. The metal case of the oven protects us from the harmful effects of microwaves.

Infrared waves

Figure 3.5 An image taken with an infrared camera.

Infrared waves have wavelengths between about 10^{-4} m and 10^{-6} m. Anything that is warmer than its surroundings will transfer energy by giving out infrared radiation. You transfer some thermal energy by radiation. You can feel the infrared radiation given out, or *emitted*, by an electric fire. Prolonged exposure of the skin to infrared in sunlight can cause skin burns. We can limit sunburn by keeping out of the sun and wearing protective clothing.

Infrared photography can be used to measure the temperature of objects. The hotter something is, the more infrared radiation it gives out. We can also use infrared photography to 'see' objects at night, by detecting radiation from warm objects.

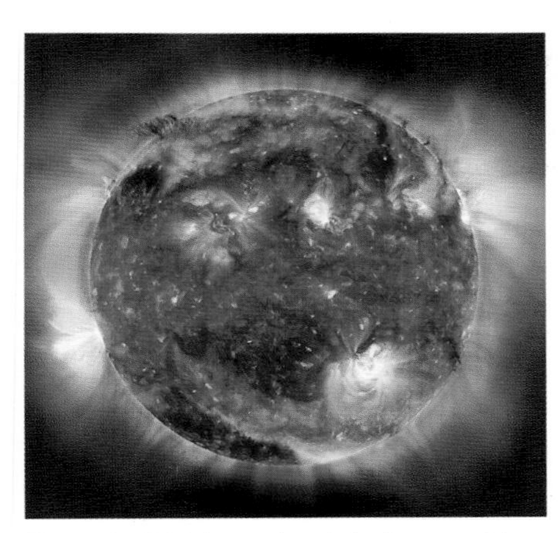

Figure 3.6 This is an ultraviolet image of the Sun. Very hot objects like the Sun produce these waves. Astronomers often 'see' what space looks like in wavelengths other than visible light. Imagine if our eyes detected ultraviolet rather than visible light – how different would the world look?

Light waves

Light waves form the part of the electromagnetic spectrum to which our eyes are sensitive ('visible' light). Red light has a wavelength of 7×10^{-7} m and violet light a wavelength of 4×10^{-7} m. Common uses of light waves include transmitting information along optical fibres (see Chapter 3.4) and taking photographs with cameras. Very bright light can damage the eye. It is more comfortable to wear sunglasses on a bright day and you must avoid looking directly at the Sun.

Ultraviolet waves

Ultraviolet waves have wavelengths shorter than visible light – approximately 10^{-8} m or 10^{-9} m. While infrared waves are emitted from hot objects (1000 °C or so), ultraviolet waves are emitted from even hotter objects 4000 °C and above). The Sun and other stars are sources of ultraviolet 'light'. Exposure to ultraviolet waves can damage your eyes, including snow-blindness, which is sunburn of the retina. Prolonged exposure to ultraviolet light from the Sun can damage your skin and cause skin cancer. We can protect ourselves from the harmful effects of ultraviolet by limiting our time in the sun, wearing protective clothing and using suncreams to block ultraviolet rays.

Some materials fluoresce when exposed to ultraviolet light. This can have applications in deterring crime (see Figure 3.7).

TIP

Ultraviolet lamps can cause some materials to fluoresce.

Figure 3.7 This man has been sprayed with 'smartwater'. Each batch of the water has a unique code which shows up under UV light. The code determines where the material has come from. Such materials are now widely used to mark property.

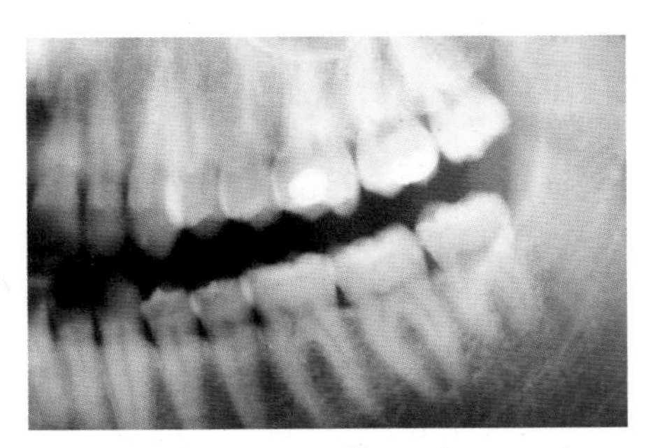

Figure 3.8 Dentists use X-ray photography to check on the health of teeth. X-rays allow dentists to see tooth disease or decay, which are not visible in an ordinary visual check-up.

X-rays

X-rays have wavelengths of about 10^{-10} m. These rays can cause damage to body tissues, so your exposure to them should be limited. X-rays are widely used in medicine. X-rays of short wavelengths will pass through body tissues but will be absorbed by bone. Such rays can be used to take a photograph to see if a patient has broken a bone or has tooth decay.

Slightly longer wavelength X-rays are used in body scanning. These X-rays are absorbed to different degrees by body tissues, so doctors can build up a picture of the inside of a patient's body. This allows doctors to investigate whether a patient has a disorder in the body.

TIP

Make sure you know and understand the uses and hazards of the seven types of electromagnetic waves listed here.

Gamma rays

Gamma rays are very short-wavelength electromagnetic waves that are emitted from the nuclei of atoms. Gamma rays are very penetrating and can be very harmful. They can cause mutations in cells leading to cancer. These rays have many industrial, agricultural and medical uses (see Chapter 3.7).

People who work with X-rays and gamma rays make sure to limit their exposure by keeping well away from the source. They work behind protective screens and radiation badges monitor their radiation dose (see Chapter 3.7).

The discovery of infrared and ultraviolet

This is an example of discovery by investigation. In 1800 William Herschel decided to investigate the temperature produced by each colour of the visible spectrum. Figure 3.9 shows how a prism can be used to split white light up into all the colours of the rainbow. Herschel put the bulb of a blackened thermometer in turn into each colour, noting that as he did so, the temperature increased towards the red end of the spectrum. He then put the thermometer just beyond the red end of the spectrum and noted that, although he could see no light, the temperature was even higher. He had discovered invisible heat rays, which we now call infrared rays. 'Infra' means below – the rays are outside the red end of the spectrum.

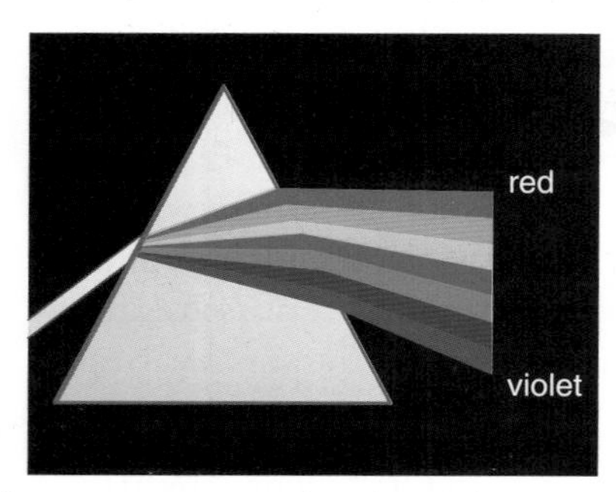

Figure 3.9 A prism splits white light up into the colours of the rainbow.

Inspired by Herschel, a year later in 1801, Johann Ritter investigated the other end of the spectrum. He investigated the effect of light on silver chloride, which was known to be blackened by sunlight. Ritter discovered that the chemical blackened more quickly as he moved towards the violet end of the spectrum. To his surprise, he discovered that the blackening effect was most pronounced just beyond the violet end. So he had discovered ultraviolet rays – 'ultra' means beyond.

STUDY QUESTIONS

1 a) Name the parts of the electromagnetic spectrum that have frequencies higher than ultraviolet waves.
  b) Name the parts of the electromagnetic spectrum that have wavelengths longer than light waves.

2 a) Give two reasons why microwave ovens are made from steel.
  b) Which electromagnetic waves cause:
    i) skin burns
    ii) skin cancer?
  c) Give three examples of how electromagnetic waves are used in your life every day.

3 In the table below you can see some data showing typical values of wavelengths and frequencies for different types of radio wave. The speed of radio waves is 300 000 000 m/s.
  a) Copy the table. Then use the equation $v = f \times \lambda$ to fill in the missing values.
  b) Which wave would you use for
    i) a local radio station,
    ii) television broadcasts to the USA?

4 For each of the following types of electromagnetic wave, give an example of a use of the wave and a hazard posed by each: microwave, infrared, ultraviolet, gamma ray.

Type of radio wave	Wavelength in m	Frequency in MHz
Long	1500	
Medium	300	
Short	10	
VHF		100
UHF		3000

(VHF = very high frequencies, UHF = ultra high frequencies)

3.4 Reflection of light

Light is a transverse wave, which can be reflected in the same way that water waves are reflected.

Figure 4.1 How does this trick work?

TIP

All types of waves can be reflected. Your image in a mirror and the echo of a sound are both examples of reflection.

Reflection of light rays

You may use a mirror every day for shaving or putting on make-up. Mirrors work because they reflect light. In Figure 4.2 you can see an arrangement for investigating how light is reflected from a mirror. A **ray box** is used to produce a thin beam of light. Inside the ray box is a light bulb; light is allowed to escape from the box through a thin slit.

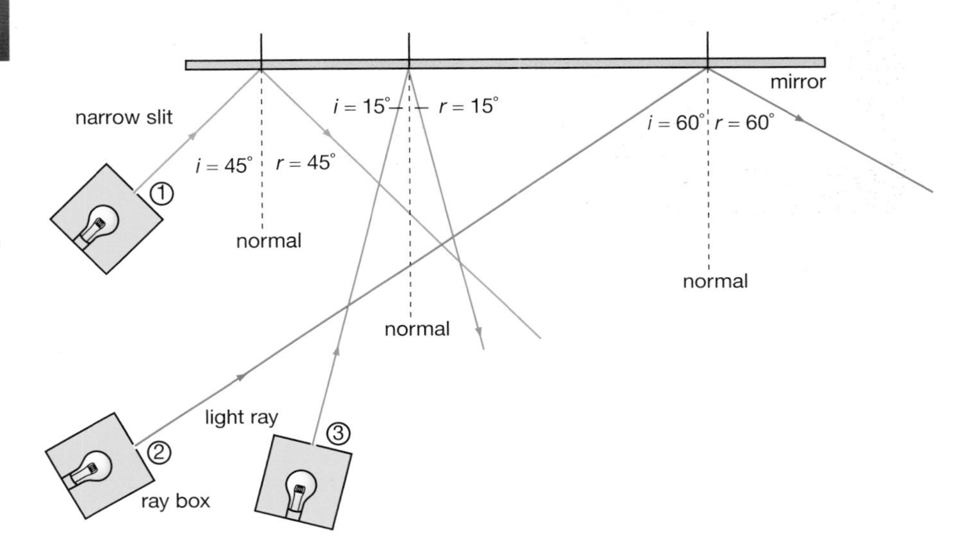

Figure 4.2 Reflection of light rays from a mirror.

PRACTICAL

Make sure you understand how to investigate angles of incidence and reflection using ray boxes. **Remember that ray box bulbs can become very hot.**

Before the light ray strikes the mirror it is called the **incident ray**. The **angle of incidence**, i, is defined as the angle between the incident ray and the normal. The **normal** is a line at right angles to the surface of the mirror. After the ray has been reflected it is called the **reflected ray**. The angle between the normal and this ray is called the **angle of reflection**, r. Light waves reflect in the same way as water waves.

There are two important points about reflection of light rays, which can be summarised as follows:

The angle of incidence always equals the angle of reflection; $i = r$.
The incident ray, the reflected ray and the normal always lie in the same plane.

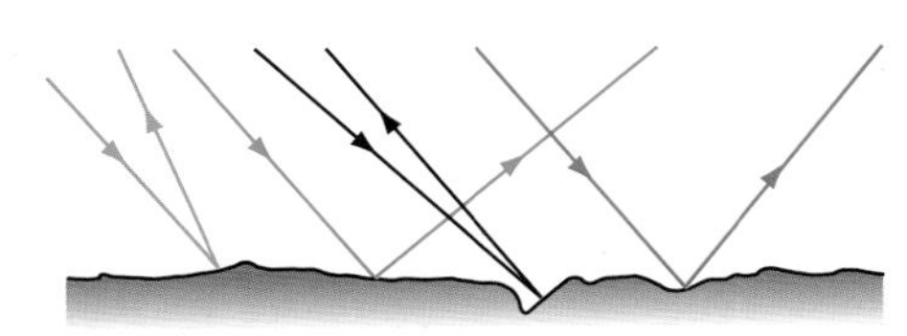

Figure 4.3 Light is reflected from a rough surface in all directions.

All surfaces can reflect light. Shiny smooth surfaces produce clear images. Figure 4.3 shows that light is reflected in all directions from a rough surface so that no clear image can be produced.

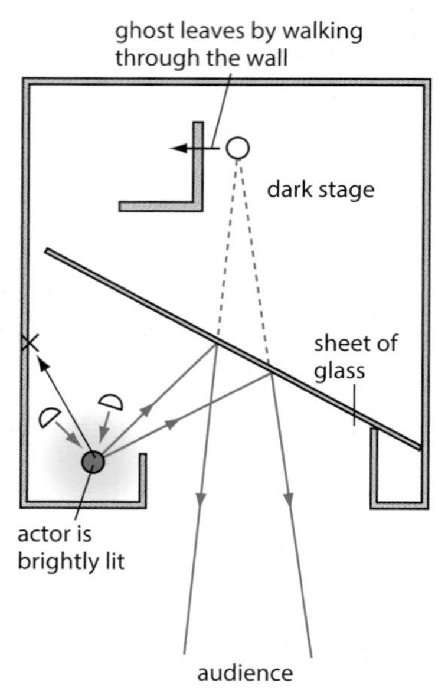

Figure 4.4 Pepper's Ghost.

Figure 4.4 shows how it is possible to produce the illusion of a ghost in a play. The technique is known as 'Pepper's Ghost'. A large sheet of glass is placed diagonally across the stage. The audience can see a wall through the glass on the darkened stage. An actor is hidden from the audience in the wings. He is brightly illuminated so that his image is reflected by the glass for the audience to see. When the actor walks off up the stage, his image (the ghost) appears to leave by walking through the wall.

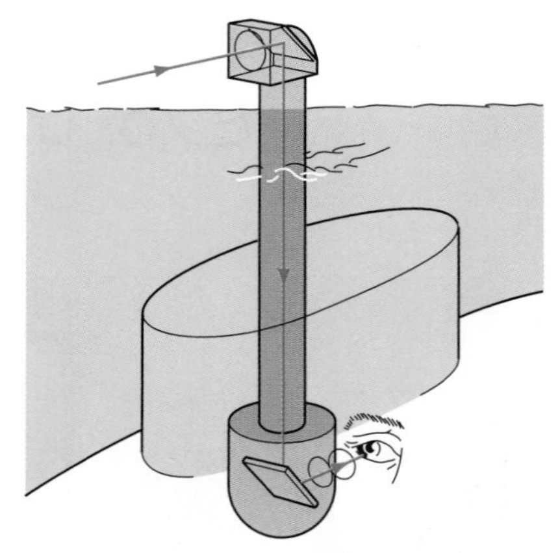

Figure 4.5 Submarines use periscopes so that crew can see above the surface. You can make a simple periscope using two mirrors.

STUDY QUESTIONS

1 At what angle must the mirrors be fitted into the periscope in Figure 4.5?

2 John runs towards a mirror at 5 m/s. At what speed does his image approach him?

3 Copy the diagrams and complete the paths of the rays after reflections in the mirrors. Use a protractor to draw your diagrams accurately.

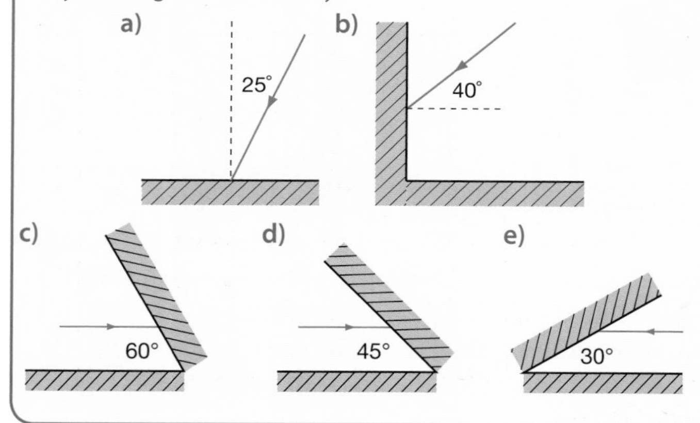

4 **a)** Draw a diagram to show a ray of light being reflected off a mirror. Show the angle of incidence and the angle of reflection.

b) Describe an experiment to show that the angle of incidence always equals the angle of reflection.

5 This question refers to Figure 4.4.

a) The actor moves to position X. Draw a diagram to show the reflection of 2 light rays from this new position. Use a protractor to ensure the angle of incidence and the angle of reflection are the same.

b) Use your diagram to suggest where the actor might appear to be now.

3.5 Refraction of light

Figure 5.1 This photograph shows light being refracted (changed in direction) by water in a glass.

Refraction of light

When a light ray travels from air into a transparent material such as glass or water, the ray changes direction – it is refracted. Refraction happens because light travels faster in air than in other substances (Figure 5.2).

The amount by which a light ray bends when it goes from air into another material depends on two things:

- what the material is
- the angle of incidence.

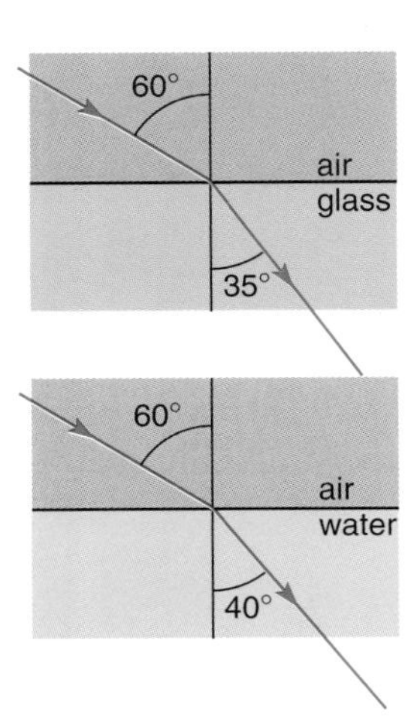

Figure 5.2 Light travels more slowly in glass than it does in water. So a light ray bends more when it goes into glass.

Refraction by a rectangular block

Figure 5.3 shows you an experiment to study how light rays bend when they go into a block of glass. The angle *i* between the normal and the incident ray, AB, is the angle of incidence. The angle *r* between the normal and refracted ray, BC, is the angle of refraction.

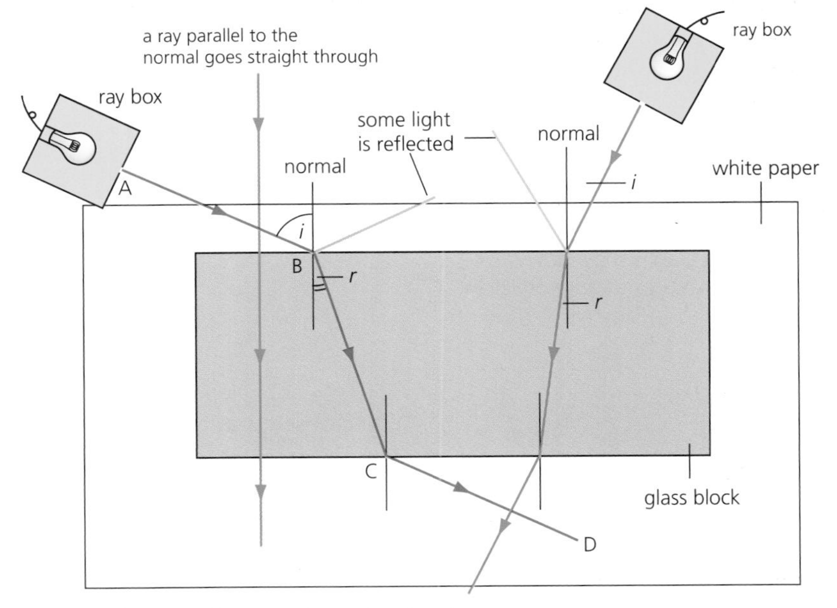

Figure 5.3 Refraction of light rays by a glass block.

Method

1. Set up the ray box, adjusting the slit to produce a narrow ray.
2. Position the block.
3. Mark the path of the rays AB and CD by putting a series of pencil marks along their lines.
4. Trace around the block, then remove it.
5. Use a ruler to draw the rays AB, BC, CD. Show that the rays AB and CD are parallel.
6. Use a protractor to mark the two normals.

PRACTICAL

Make sure you understand how to trace the incident rays, how to measure the refracted rays and how to measure the angles of incidence and refraction.

7 Finally, measure the angle of incidence i, and the angle of refraction r. Record the angles in a suitable table.

8 Repeat the experiment to determine six pairs of i and r.

TIP

Refraction occurs as light changes speed in passing from one medium to another.

Taking it further

Devise an experiment to investigate the refraction of light by a triangular prism and a semi-circular glass block.

The experiment should show these points:

- The light ray is bent towards the normal when it goes into the glass. The angle of incidence is greater than the angle of refraction.
- When the light ray leaves the block of glass it is bent away from the normal.
- If the block has parallel sides, light comes out at the same angle as it goes in.

Refractive index

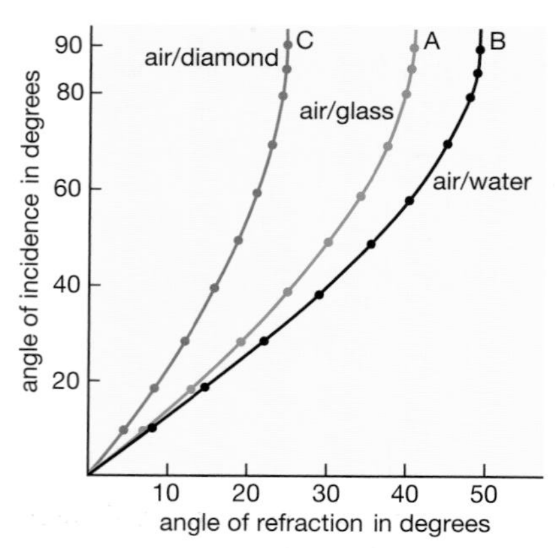

Figure 5.4 These graphs show how r depends on i when light travels from air into A glass, B water and C diamond.

You can use the apparatus in Figure 5.3 to investigate a relationship between the incident and the refracted rays. Figure 5.4 shows a graph of the investigations for three materials: glass, water and diamond.

There is also a mathematical connection between the incident and refracted rays. This is illustrated in Table 1.

Table 1 Comparison of i and r for glass

Angle of incidence (i)	Angle of refraction (r)	sin i	sin r	sin i/sin r
0	0	0	0	
23	15	0.39	0.26	1.50
34	22	0.56	0.37	1.51
48	30	0.74	0.50	1.48
59	35	0.86	0.57	1.51
80	41	0.98	0.66	1.48

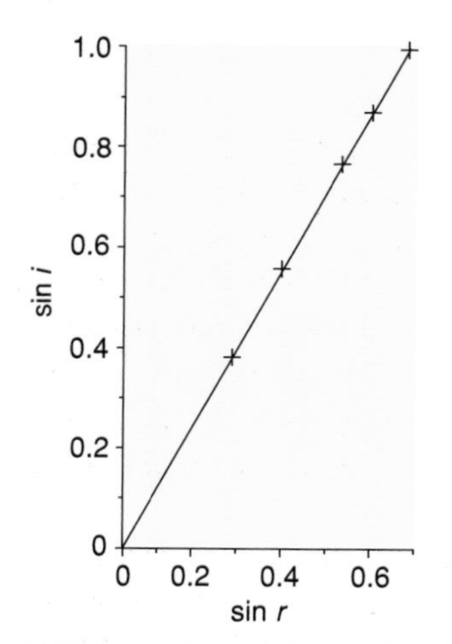

Figure 5.5 The gradient of this graph is the refractive index of glass.

You can see that the ratio of sin i/sin r is approximately constant – in this case its average value is about 1.5. The ratio, 1.5 is the **refractive index** (n) of glass. The refractive index is the **ratio** of the speed of light in air to the speed of light in glass.

$$n = \frac{\sin i}{\sin r}$$

Because n is a ratio of speeds, it has no unit.

These results may also be plotted on a graph, Figure 5.5. The gradient tells us the refractive index of the glass when sin i is plotted against sin r.

REQUIRED PRACTICAL

Make sure you are familiar with this practical.

Figure 5.6 James demonstrates refraction in the pool. Why does the pole look bent?

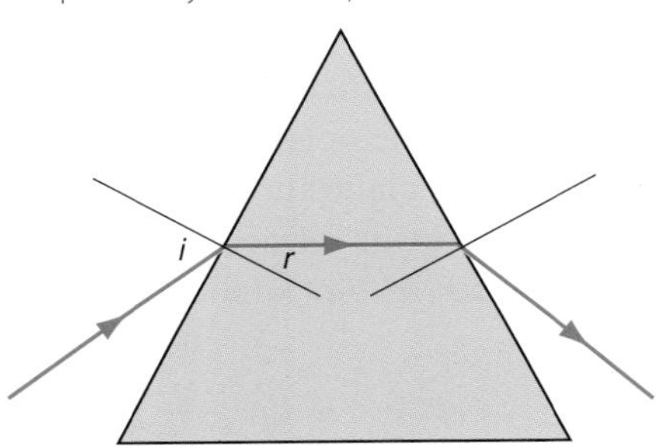

Figure 5.7 Light is refracted by a triangular prism.

Investigating the refractive index of glass

Method

Follow the procedure outlined in the practical work above.

1 Add two further columns to your table and use a calculator to add values of sin i and sin r, as shown in Table 1.

2 Draw a graph of sin i on the y-axis against sin r on the x-axis.

3 Measure the gradient. Comment on how your answer compares with the refractive index for glass (usually about 1.5).

4 Discuss the nature of the errors in your experiment.

Example. A light ray is incident on still water at an angle of 45° to the normal. Calculate the angle of refraction. The water has a refractive index of 1.33.

$$1.33 = \frac{\sin 45}{\sin r}$$

$$\sin r = \frac{\sin 45}{1.33}$$

$$\sin r = \frac{0.71}{1.33}$$

$$\sin r = 0.53$$

$$r = 32°$$

Refraction by prisms

You learnt some simple rules about refraction from the experiment shown in Figure 5.3. You can apply these rules to predict what will happen to light rays going into any shape of block. In Figure 5.7 you can see a light ray passing through a triangular glass **prism**. Notice that as the ray goes into the prism it is bent towards the normal; as it leaves the prism it is bent away from the normal.

STUDY QUESTIONS

1 a) Explain what is meant by the word ***refraction***.
b) Explain why light refracts when it travels from air into water.

2 When light travels from air into glass, light bends towards the normal. When light travels from glass into air, it bends away from the normal.
a) Use this rule to show the paths of light rays through each of these glass shapes.

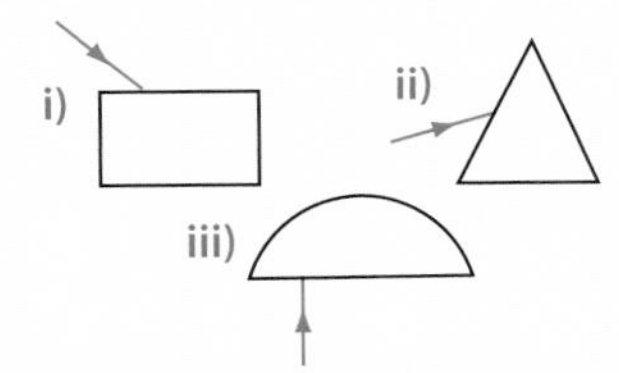

b) Use your knowledge of refraction, to sketch a diagram to show the path rays through the glass of water in Figure 5.1. Use your diagram to explain why the blue and yellow colours of the background are reverse by the refraction of the light rays.

3 A student uses an experimental arrangement similar to the one shown in Figure 5.3 to measure the refractive index of a block of glass. He measures the angle of incidence as 64° and the angle of refraction as 34°.
a) Use the student's results to calculate the refractive index of the glass.
b) Discuss how the student could improve his experiment to provide a more accurate answer.

4 Water has a refractive index of 1.33. A ray is incident at an angle of 30° on to a flat water surface.
Calculate the angle of refraction for the ray entering the water.

3.6 Total internal reflection

Figure 6.1 Diamonds are very popular jewels, because they sparkle so brightly. Diamonds are skilfully cut, so that light is reflected back inside the crystal. This has the effect of concentrating the light so that it comes out brighter from some faces than others. This gives us the sparkle.

In the last chapter you learnt that when a light ray crosses from glass into air, it bends away from the normal. However, this only happens if the angle of incidence is small. If the angle of incidence is too large, all of the light is reflected back into the glass. This is called **total internal reflection**.

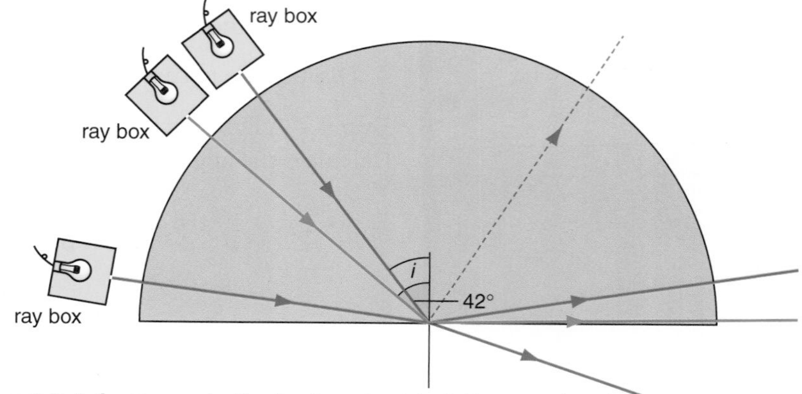

Figure 6.2 Refraction and reflection in a glass block. The critical angle is 42°, the angle between the blue ray and the normal.

Figure 6.2 shows how you can see this effect for yourself in the lab. Three rays of light are directed towards the centre of a semicircular glass block. Each ray crosses the circular edge of the block along the normal, so it does not change direction. However, when a ray meets the plane surface there is a direction change. For small angles of incidence, the ray is refracted (solid red emerging ray). Some light is also reflected back into the block (shown as a dashed red on the diagram).

PRACTICAL

Make sure you understand how to determine the critical angle for glass.

Critical angle

At an incident angle of 42° the ray is refracted along the surface of the block (blue ray). This is called the **critical angle**. If the angle of incidence is greater than this critical angle then all of the light is reflected back into the glass (green ray). The critical angle varies from material to material. While it is 42° for glass, for water it is about 49°.

Total internal reflection in prisms

An ordinary mirror has one main disadvantage: the silver reflecting surface is at the back of the mirror. So light has to pass through glass before it is reflected by the mirror surface. This can cause several weaker reflections to be seen in the mirror, because some light is internally reflected off the glass/air surface (Figure 6.3(a)).

We can avoid the extra reflections by using prisms. In Figure 6.3(b) the light ray AB meets the back of the glass prism at an angle of incidence of 45°. This angle is greater than the critical angle for glass, so the light is totally reflected. There is only one reflection because there is only one surface. Total internal reflection by prisms is also put to use inside binoculars and cameras.

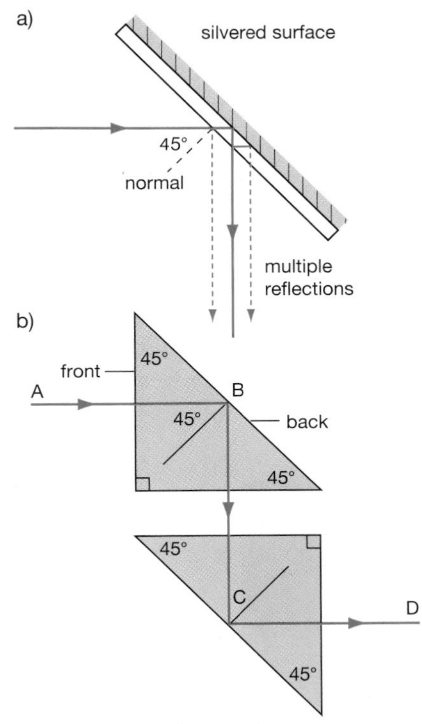

Figure 6.3 **(a)** Reflection from a mirror. **(b)** Reflection from two prisms to make a periscope.

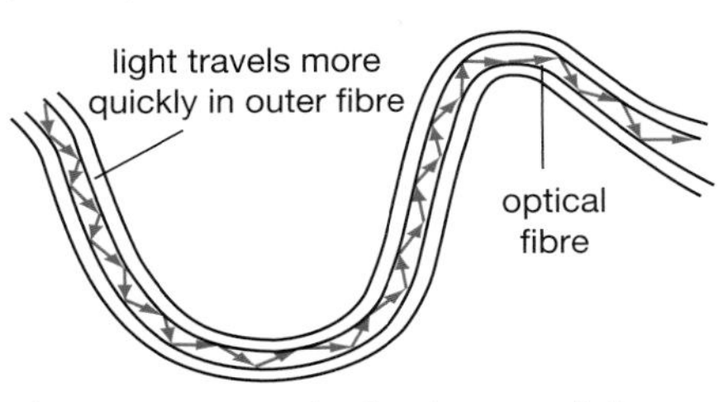

Figure 6.4 Internal reflection traps light inside the optical fibre.

Critical angle and refractive index

We can calculate the critical angle, c, for a material if we know its refractive index, n, using the equation:

$$\sin c = \frac{1}{n}$$

Example. Calculate the critical angle for a sample of Perspex, with a refractive index of 1.4.

$$\sin c = \frac{1}{1.4}$$

$$= 0.71$$

$$c = 46^\circ$$

Optical fibres

Optical fibres are used for carrying beams of light. The fibres usually consist of two parts. The inner part (core) carries the light beam. The outer part provides protection for the inner fibre. It is important that light travels more quickly in the outer part. Then the light inside the core is trapped due to total internal reflection (Figure 6.4).

Surgeons use a device called an **endoscope** to examine the inside of patients' bodies. This is made of two bundles of optical fibres. One bundle carries light down inside the patient, and the other tube allows the surgeon to see what is there.

Optical fibres are also used in telephone networks. A small glass fibre, only about 0.01 mm in diameter, is capable of carrying hundreds of messages, coded as light pulses, at the same time. These fibres have largely replaced the old copper cables in telephone systems.

STUDY QUESTIONS

1 Turn back to the last chapter. Explain how the graphs in Figure 5.4 can help you work out critical angles. State the critical angles of glass, water and diamond.

2 Below, you can see a ray of light entering a five-sided prism (pentaprism) in a camera. The ray undergoes three internal reflections before emerging. Copy the diagram and mark in the ray's path.

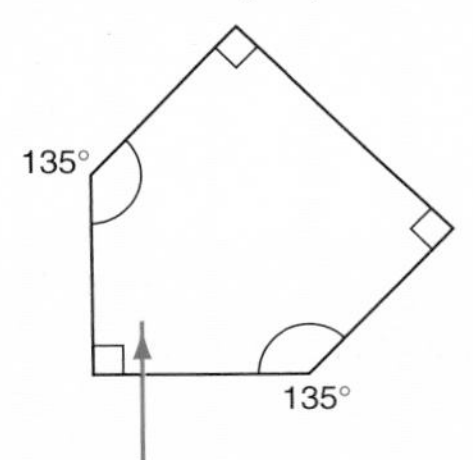

3 State two uses for internal reflection. Illustrate your answer with diagrams.

4 The diagram below shows sunlight passing through a prismatic window that is used to light an underground room.

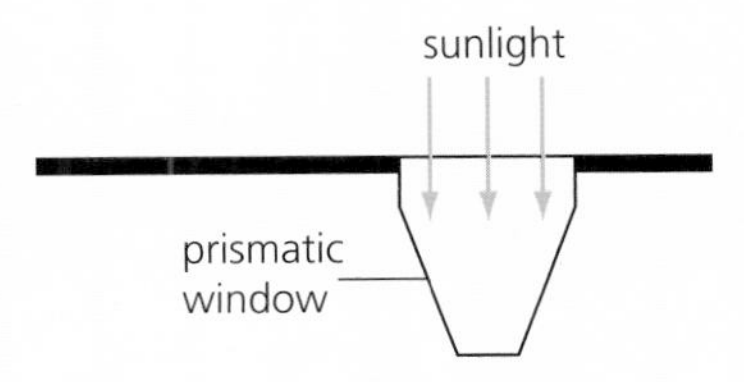

a) Copy the diagram and show the path of the rays through the window.

b) Explain why the window spreads out the light.

5 Diamond has a high refractive index of 2.4. Use the equation: $\sin c = 1/n$ to calculate the critical angle for diamond.

6 Describe an experiment to measure the critical angle of a transparent material.

3.7 Sound waves

The ear

Sound waves travel as compressions and rarefactions through air in the same way that compressions and decompressions move along a slinky. Your ear detects the changes in pressure caused by these waves. When a compression reaches the ear it pushes the eardrum inwards. When a decompression arrives, the eardrum moves out again. The movements of the eardrum are transmitted through the ear by bones. Then nerves transmit electrical pulses to the brain.

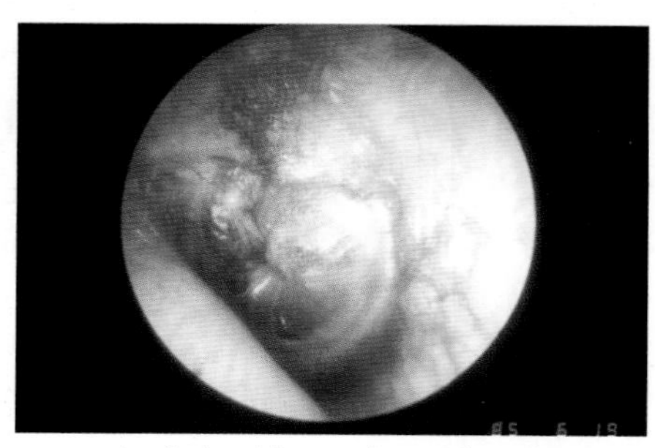

Figure 7.1 This is a photograph of a healthy eardrum. The vibrations carried in sound waves set the membrane in motion.

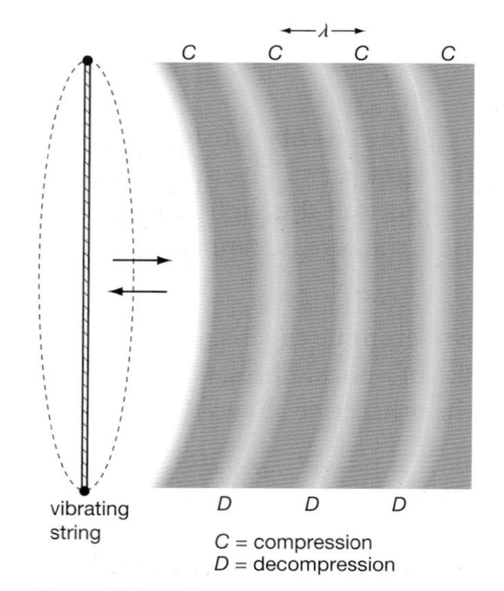

Figure 7.2 A vibrating guitar string.

Making and hearing sounds

Vibrations

When you pluck a guitar string the instrument makes a noise. If you put your finger on the string you can feel the string vibrating. Sounds are made when something vibrates. The vibrations of a guitar string pass on energy to the air. This makes the air vibrate. Our ears can hear frequencies in the range 20 Hz to 20 000 Hz.

Longitudinal waves

Sound is a longitudinal wave. Molecules in air move backwards and forwards along the direction in which the sound travels. In Figure 7.2, when the guitar string moves to the right it compresses the air on the right hand side of it. When the string moves to the left the air on the right expands. This produces a series of **compressions** and **decompressions**. In a compression the air pressure is greater than normal atmospheric pressure. In a decompression (rarefaction) the air pressure is less than normal atmospheric pressure.

Speed of sound

The speed of sound depends on which material it travels through. Sound waves are transmitted by molecules knocking into each other. In air, sound travels at about 340 m/s. In solids and liquids, where molecules are packed more tightly together, sound travels faster (Table 1). In a vacuum there are no molecules at all. Sound cannot travel through a vacuum.

Table 1

Material	Speeds of sound in m/s
air	340
water	1500
steel	5000

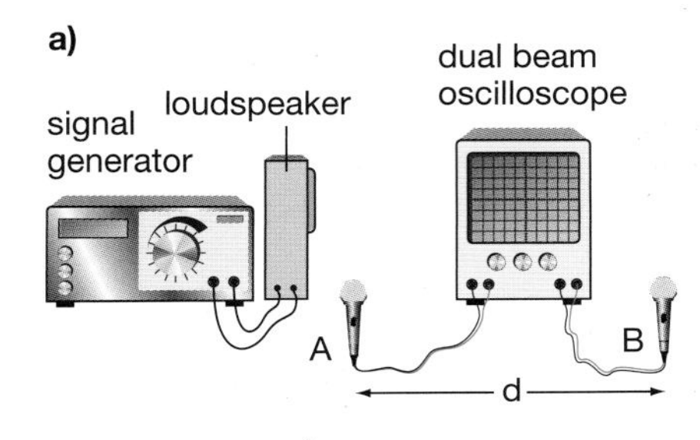

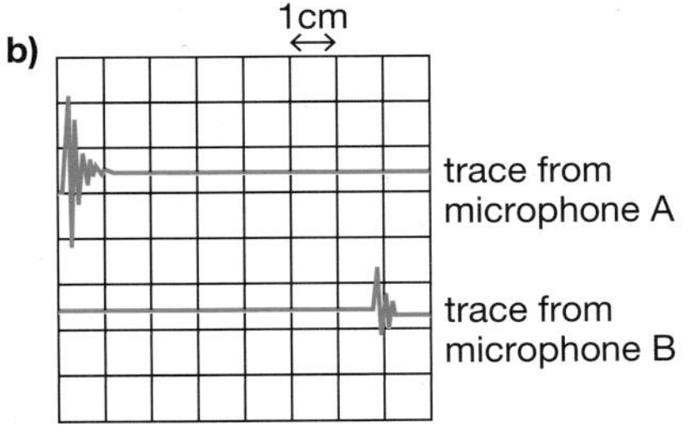

Figure 7.3

TIP

The speed of sound in air is 330 m/s.

Measuring the speed of sound in a laboratory

It is possible to measure wave speed using the equation:

$$\text{speed} = \frac{\text{distance travelled}}{\text{time taken}}$$

The problem with measuring the speed of sound in air is that sound travels quickly. So we must find a way to measure short times accurately in the laboratory. Figure 7.3 shows how we can do this.

- A loudspeaker is connected to a signal generator which produces short pulses of sound.
- Two microphones are placed near the loudspeaker but separated by a short distance, d. Each microphone is connected to an input of a dual beam oscilloscope.
- The oscilloscope can measure the time difference, t, between the sound reaching microphone A and microphone B.

Example. A student sets the microphones a distance of 220 cm apart. The time base on the oscilloscope is set to 1 ms per centimetre: this means that a distance of 1 cm on the horizontal scale corresponds to a time of 1 ms (0.001 s). Using Figure 7.3(b), you can measure that the sound reaches microphone B 6.5 ms after the sound reaches microphone A.

$$\text{speed} = \frac{d}{t}$$

$$= \frac{2.2}{0.0065} = 340\,\text{m/s}$$

Remember: you must convert distance into metres and time into seconds.

PRACTICAL

Make sure you can explain how to measure the speed of sound.

Measuring the speed of sound outside

- When outside, the echoes from a tall building can be used to measure the speed of sound in air.
- Stand 40 m in front of a tall building and clap your hands or bang two blocks of wood together – watch your fingers.
- Each time you hear an echo, clap or bang the blocks together again.
- Have another student use a stopwatch to time how long it took to hear 10 echoes.

The results obtained by two students are given in the table.

Table 2

Trial number	Time in seconds
1	2.4
2	2.8
3	2.3

1 Calculate the mean value of the time for 10 echoes.

2 How long does it take the sound to travel 80 m?

3 Calculate the speed of sound given by these results.

So the speed of sound is given by:

$$v = \frac{d}{t}$$

$$v = \frac{80\,\text{m}}{0.25\,\text{s}} = 320\,\text{m/s}$$

TIP

When calculating the speed of sound by reflection, make sure you remember that the sound travels there and back – so in this example the distance travelled is $2 \times 40\,\text{m} = 80\,\text{m}$.

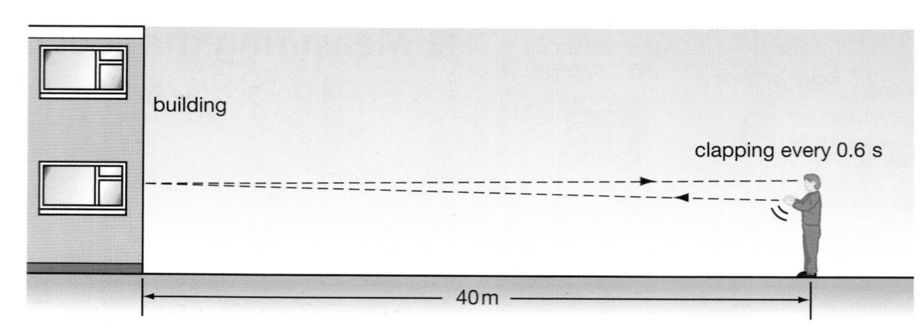

Figure 7.4 Measuring the speed of sound through air.

TIP

You will not be examined on ultrasound. All you need to know is that ultrasound is a very high pitched sound wave.

Ultrasound

Ultrasound waves are high-frequency sound waves. Like all sound waves, these waves are longitudinal, and their vibrations produce compressions and decompressions (expansions) in the material they travel through. Ultrasound has such a high frequency (above 20 000 Hz) that our ears cannot detect it. The high-frequency waves have a short wavelength.

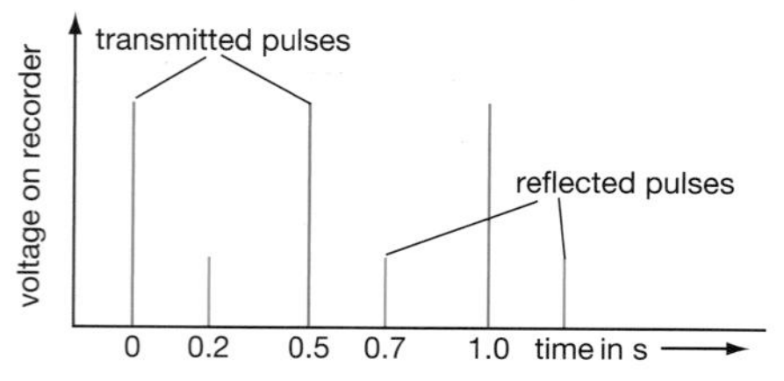

Figure 7.5 Ultrasound pulses for a depth of 150 m.

Ultrasonic depth finding

Ships use beams of ultrasound (or sonar) for a variety of purposes. Fishing boats look for fish; destroyers hunt for submarines; explorers chart the depth of the oceans. Figure 7.6 demonstrates the idea. A beam of ultrasonic waves is sent out from the bottom of a ship. The waves are reflected from the sea bed back to the ship. The longer the delay between the transmitted and reflected pulses, the deeper the sea is (Figure 7.5).

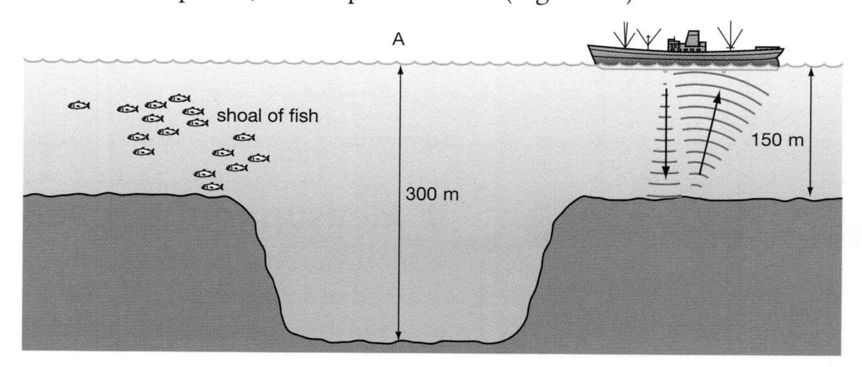

Figure 7.6 Ultrasonic depth finding.

STUDY QUESTIONS

1. The frequency of a particular sound from your mouth is 250 Hz. Calculate the wavelength of this sound (use Table 1 to find the speed).
2. A student is using the apparatus in Figure 7.3. She moves the microphones closer together so that the horizontal distance between the traces on the oscilloscope is 4.2 cm. Calculate how far apart the microphones are. (The speed of sound in air is 340 m/s.)
3. The ship in Figure 7.6 sends out short pulses of ultrasound every 0.5 s; the frequency of the waves is 50 kHz (50 000 Hz).
   a) The duration of each pulse is 0.01 s. Calculate how many complete oscillations of the ultrasound waves are emitted during that time.
   b) Use the information in Figures 7.5 and 7.6 to show that the ultrasound travels at a speed of 1500 m/s through water.
   c) Sketch a graph, similar to Figure 7.5, to show both the transmitted and reflected pulses when the ship reaches point A.
   d) What difficulties would the sonar operator face when trying to measure depths of around 500 m?
   e) Calculate the wavelength of the ultrasonic waves in water.
4. Describe an experiment that you could do to measure the speed of sound. Discuss the accuracy of your method.

3.8 Loudness, quality and pitch

Figure 8.1 Each of these musical instruments can play the same musical notes. However, you can tell the difference between a note played on a trumpet and a note played on a violin. What is it about the notes that enables you to tell the difference?

Amplitude and loudness

The loudness of a noise depends on the amplitude of the vibrations that produce the noise. When a guitar string vibrates it causes compressions and rarefactions in the air. These pressure changes cause vibrations in the air, which in turn set our eardrum vibrating. The amplitude of the vibrations in our eardrum determines the loudness of the sound. Figure 8.2 compares the amplitude of pressure changes for two voices: the larger amplitude sounds louder.

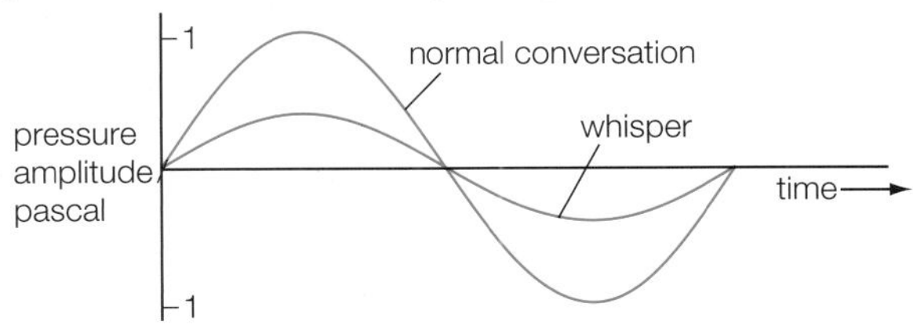

Figure 8.2 Sound waves caused by the human voice.

Pitch

We use the term **pitch** to describe how a noise or a musical note sounds to us. Men have low-pitched voices, women have voices of higher pitch. The pitch of a note is related directly to its frequency. The higher-pitched notes are the notes with higher frequency.

A data logger and microphone can be used to analyse the waveform of the sound from a tuning fork. Once the data has been collected it can be downloaded and displayed on a computer screen. The higher-pitched sound, shown in Figure 8.3(b), has more cycles per second than the lower-pitched sound, shown in Figure 8.3(a).

(a) Waveform of tuning fork of low pitch.

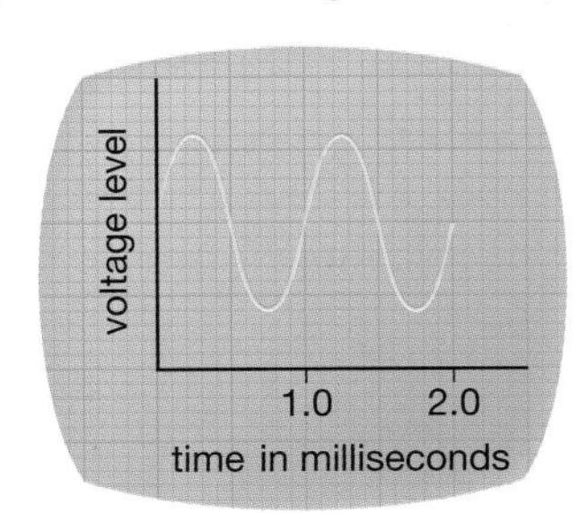

(b) Waveform of tuning fork of higher pitch.

Figure 8.3

Quality

On a piano the note that is called middle C has a frequency of 256 Hz. This note could be played on a piano or a violin, or you could sing the note. Somebody listening to the three different sounds would recognise straight away whether you had sung the note or played it on the piano or violin. The **quality** of the three notes is different. The quality of a note depends on the shape of its waveform. Although notes may have the same frequency and amplitude, if their waveforms are of a different shape you will detect a different sound (Figure 8.4).

(a) flute

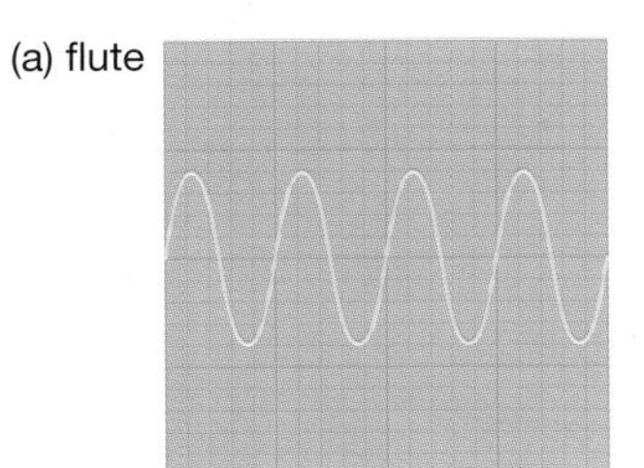

(b) trumpet

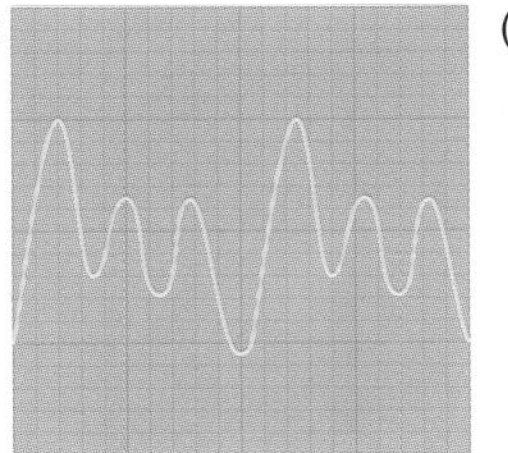

(c) piano

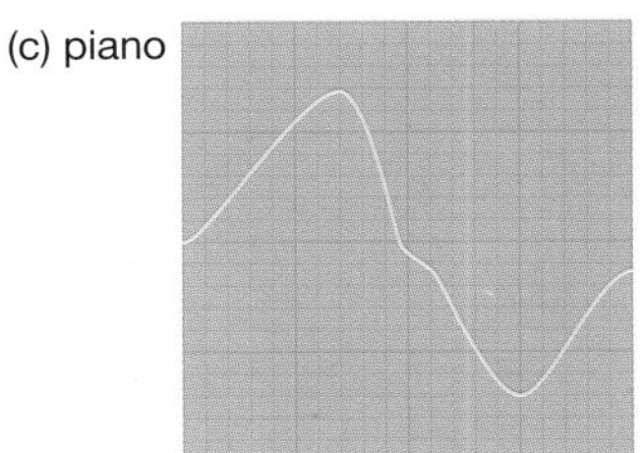

(d) violin

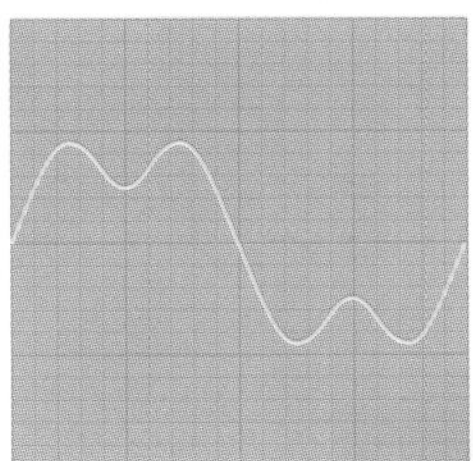

Figure 8.4 Data-logger readings from various instruments.

PRACTICAL

You can use data loggers to display the shape of sound waves, but you also need to know how to use an oscilloscope to measure the frequency of a sound wave. Make sure you understand how to use the *y*-gain and timebase controls.

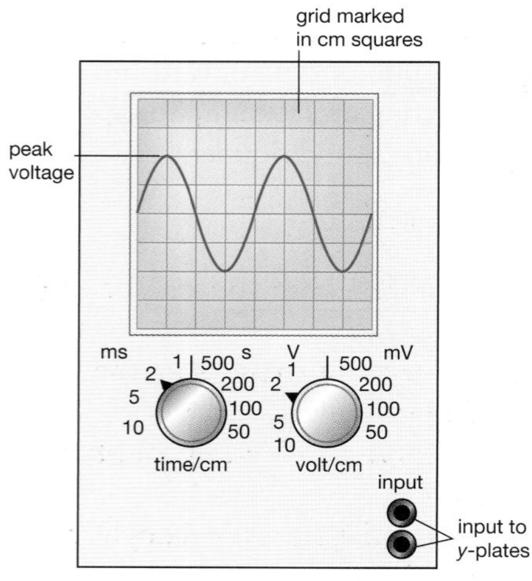

Figure 8.5 A simple waveform on an oscilloscope screen.

Using an oscilloscope to measure frequency

A microphone is a device that turns sound vibrations into an electrical signal. The vibrations of a sound move a diaphragm in a microphone backwards and forwards, which generates an a.c. voltage. If the microphone is attached to an **oscilloscope**, we can see how the a.c. voltage changes with time. This is shown in Figure 8.5.

Here are some notes on how the oscilloscope works.

The size of the voltage is measured with the *y*-axis. The *y*-gain control (on the right-hand side in Figure 8.5) adjusts the sensitivity of the oscilloscope. Here the *y*-gain is set at 2 V/cm. This means that each centimetre displacement in a vertical direction measures 2 V. So here the peak voltage is 4 V, as the peak is 2 cm above or below the zero voltage level.

The *x*-axis allows us to measure the time period (and frequency) of the voltage. The timebase (left-hand control in Figure 8.5) is set at 2 ms/cm. This means that each centimetre on the *x*-axis measures a time of 2 ms. One complete cycle of this voltage takes 4 squares. This is 8 ms or 8/1000 s. Since the frequency of the voltage is equal to 1/time period, the voltage frequency is 1000/8 = 125 Hz.

You should practise using an oscilloscope to measure the frequency for yourself.

STUDY QUESTIONS

1 Look at Figure 8.4. The data logger collected information about the four musical instruments for the same length of time in each case.
   a) Which note has the highest frequency?
   b) Which instrument is played most loudly?
   c) Which instruments are played most softly?
   d) Which two instruments are playing notes of the same pitch?

2 a) Use Figure 8.3 to calculate the frequencies of the sounds emitted from the two tuning forks.
   b) Using the same scales as Figure 8.3, sketch the forms of notes from tuning forks of frequencies
      i) 2000 Hz
      ii) 750 Hz.

3 Look at the signal in Figure 8.5. Draw another diagram to show how the trace looks when you make these two separate changes:
   a) increase the *y*-gain to 1 V/cm
   b) increase the timebase setting to 1 ms/cm.

4 The diagram below shows a waveform on an oscilloscope screen; the timebase is set to 1 ms/cm and the *y*-gain control to 2 V/cm. Calculate the peak voltage and frequency of the a.c. supply

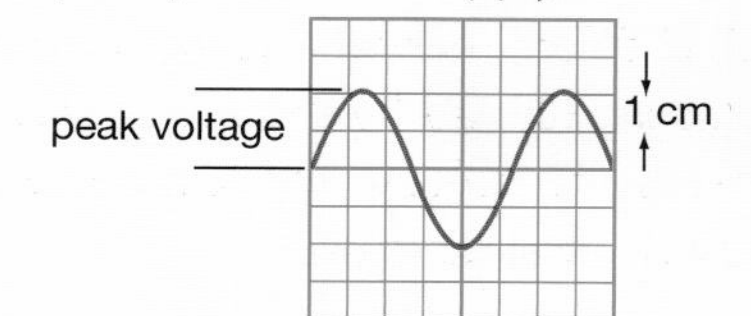

Summary

I am confident that:

✓ I know and understand the properties of waves.

- I know that waves carry information and energy
- I can describe transverse and longitudinal waves.
- I can describe the meaning of the terms: amplitude, wavelength, time period, frequency and wavefront.
- I can recall and use the equations:
  wave speed = frequency x wavelength
  $$v = f \times \lambda$$
  $$\text{frequency} = \frac{1}{\text{time period}}$$
  $$f = \frac{1}{T}$$
- I know that there is a change in the observed frequency and wavelength of a wave when its source is moving relative to an observer. The frequency appears to increase when the source is moving towards the observer, and to decrease when the source is moving away. This is called the Doppler effect.
- I know that all waves reflect and refract.

✓ I can recall the electromagnetic spectrum.

- I understand that all electromagnetic waves travel at the speed of light in a vacuum (3×10^8 m/s)
- I know that electromagnetic waves are transverse waves.
- I know the spectrum in the order of its increasing frequency: radiowaves, microwaves, infrared, visible light, ultraviolet, X-rays, gamma rays.
- I can recall uses of all the types of electromagnetic wave.
- I can recall the detrimental effects of microwaves, infrared, ultra violet and gamma rays, and the precautions we take to reduce the risks of exposure to these radiations.

✓ I can recall the properties of light.

- Light can be reflected and refracted.
- Light is a transverse wave.
- I can draw ray diagrams to show refraction and reflection.
- I know how to draw a normal to a surface.
- I know that the angle of incidence equals the angle of reflection.
- I can describe experiments to investigate the refraction of light.
- I know that refractive index can be calculated using the equation:
  $$n = \frac{\sin i}{\sin r}$$
  where i is the angle of incidence and r is the angle of refraction.
- I can describe an experiment to measure the refractive index of a glass block.
- I can describe total internal reflection, and know about the critical angle.
- I can use the equation relating critical angle, c to refractive index, n:
  $$\sin c = \frac{1}{n}$$

✓ I can recall and describe the properties of sound.

- Sound can be reflected and refracted.
- Sound is a longitudinal wave.
- The range of human hearing is 20 Hz to 20 000 Hz.
- I can describe experiments to measure the speed of sound in air.
- I understand how to use an oscilloscope to measure the frequency of sound.
- I know that the pitch of a sound is related to its frequency.
- I know that the loudness of a sound is related to the amplitude of the vibrations producing the sound.

Sample answers and expert's comments

1 The diagram shows water waves in a ripple tank. The waves are travelling in deep water. The waves cross a boundary into some shallow water, where they travel more slowly than in the deep water.

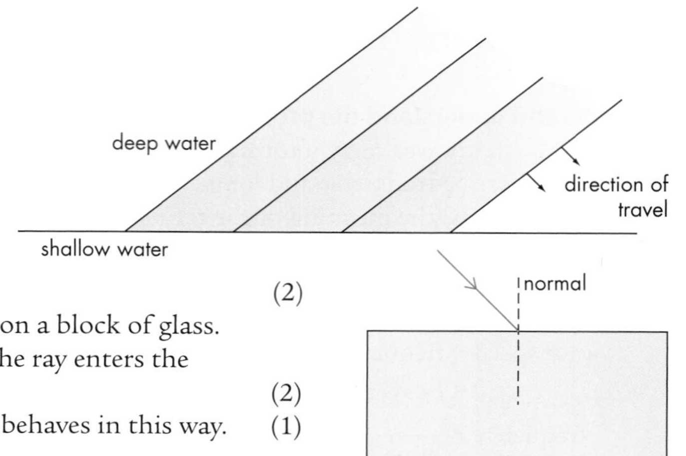

a) Copy and complete the diagram to show what happens to the waves after they travel into the shallow water. (2)

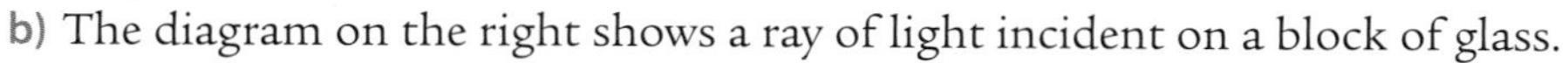

b) The diagram on the right shows a ray of light incident on a block of glass.

i) Copy and complete the diagram to illustrate how the ray enters the glass and leaves on the other side of the block. (2)

ii) Explain, in terms of the speed of light, why the ray behaves in this way. (1)

(Total for question = 5 marks)

Student response Total 4/5	Expert comments and tips for success
a) deep water; shallow water ✔ ○	There is no change of direction, but the wavelength is drawn smaller in the shallow water, so 1 mark is awarded (see Figure 2.5).
b) i) normal ✔ ✔	The ray is drawn correctly at both boundaries, bending towards the normal on entry and away from the normal on leaving the block.
ii) Light travels faster in air than in glass. ✔	This is correct.

2 The diagram shows the waveform of a sound wave displayed on an oscilloscope.

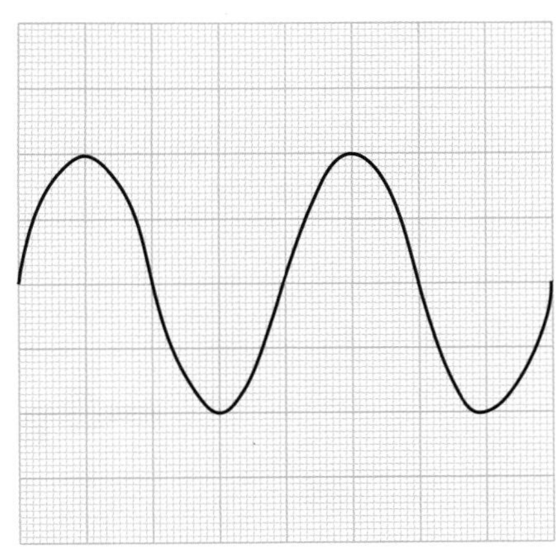

The settings on the oscilloscope are: y-axis 0.2 V/cm

x-axis 5 ms/cm

Each large square on the screen measures 1 cm by 1 cm.

a) Determine the voltage that corresponds to the amplitude of the wave on the screen. (2)

b) i) Calculate the time period of the note. (3)

ii) Calculate the frequency of the note. State the unit of frequency. (2)

c) Make a copy of the diagram of the screen. Add the waveform of a second sound wave which has twice the frequency and the same loudness as the first note. (2)

(Total for question = 9 marks)

Student response Total 4/9	**Expert comments and tips for success**
a) *It is 2 squares high* ✔ *so 2 V* ○	The answer correctly identifies that the amplitude is measured from the centre to the top. This gets a mark. But the student has forgotten to use the scale of 0.2 V/cm. So the correct answer is 0.4 V (0.2 x 2)
b) i) *4 cm* ✔	The answer shows the student understands that the time period is linked to one complete cycle. This earns a mark. The scale (time base here) has not been used, so 2 marks are lost. The time period is 4× 5 = 20 ms
ii) *The frequency is the number waves per second.*	This is a correct definition, but it was not requested, so does not earn a mark. The question asked for a calculation and the unit – neither of which are given. $f = \frac{1}{T} = \frac{1}{0.02} = 50\text{ Hz}$
c) 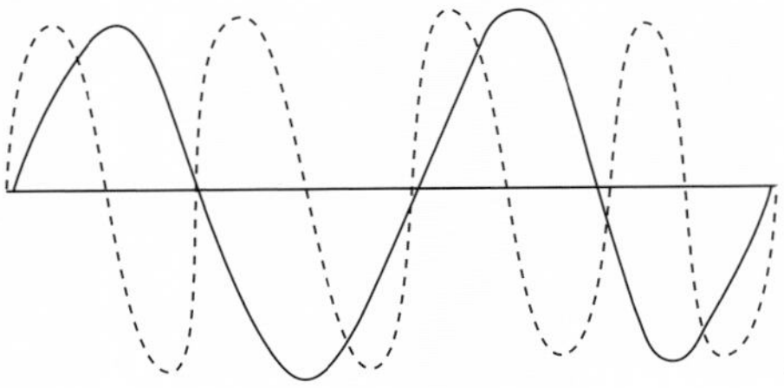 ✔✔	The diagram correctly shows twice the number of waves per second. The peak voltage is the same, so the amplitude of the two waves is the same, and therefore the loudness.

3 The diagram shows how the depth of a liquid in an oil storage tank can be checked using sound waves. Some of the sound emitted by the transmitter is reflected back from the surface of the liquid. The time interval between the transmitted and reflected pulses is measured using the time base on an oscilloscope.

oil storage tank

transmitter/receiver that sends and receives high frequency sound waves

a) State the maximum frequency of sound which can be heard by humans. State the unit of frequency. (2)

b) The oscilloscope trace is shown below.

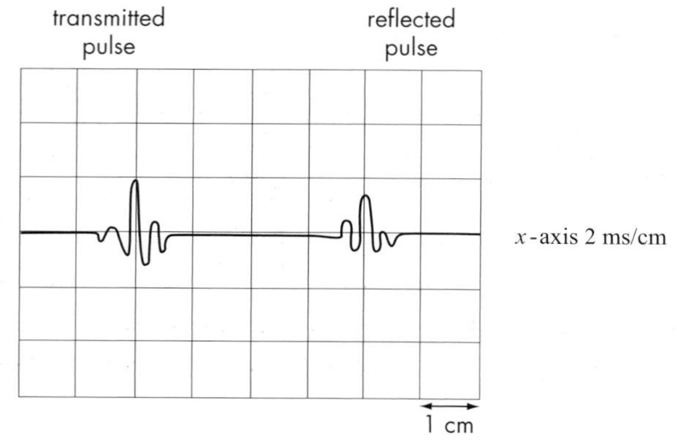

Use the oscilloscope trace to find the time taken for the sound to travel from the transducer to the surface and back. (2)

c) The speed of the sound through the liquid is 1200 m/s. Calculate the depth of the liquid. (3)

Student response Total 5/7	**Expert comments and tips for success**
a) 20 000 ✓	This is correct, but you must include the unit, Hz (hertz), so a mark is lost.
b) 5 cm = 10 ms ✓ ✓	10 ms is the correct answer, but this is poorly expressed. The oscilloscope time base is 2 ms/cm, so the time is: 2 ms/cm x 5 cm = 10 ms
c) $d = v \times t$ = 1200 x 0.01 ✓ = 12 m ✓	12 m is the correct answer for the distance travelled. But sound travels to the surface and back. So the depth = $\frac{12}{2}$ = 6 m

Exam-style questions

The diagram shows a wave on the sea.

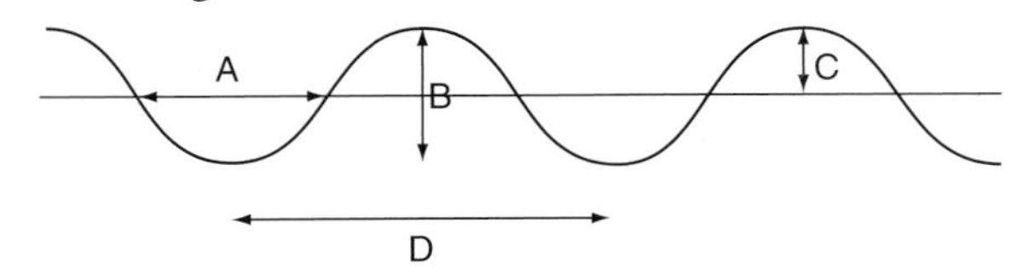

1 Which letter shows the wavelength of the wave? [1]

2 Which letter shows the amplitude of the wave? [1]

3 Which of the following decreases as the pitch of a sound decreases?

A amplitude

B speed

C frequency

D wavelength [1]

4 A sound took 0.6 seconds to travel 900 m in water. Based on that timing, which of the following is the speed of sound in water, in m/s?

A 540 m/s **B** 1500 m/s **C** 1800 m/s

D 3000 m/s [1]

Below are listed four parts of the electromagnetic spectrum.

A infrared

B ultraviolet

C microwave

D radio wave

5 Which of the above is used for transmitting phone calls via satellite? [1]

6 Which of the above is the most likely to cause skin cancer? [1]

A seismic wave has a speed of 6.5 km/s; it has a frequency of 0.04 Hz [1]

7 Which of the following is the wavelength of the seismic wave?

A 260 m **B** 26 km **C** 160 km

D 6500 km [1]

8 Which of the following is the time period of the seismic wave?

A 25 s **B** 6.5 s **C** 4.0 s **D** 0.04 per second [1]

9 Which of the following correctly describes the nature of light and sound waves? [1]

	light	sound
A	longitudinal	longitudinal
B	longitudinal	transverse
C	transverse	longitudinal
D	transverse	transverse

10 A car drives past you blowing its horn. Which of the following correctly describes the pitch of the horn that you hear?

A The apparent pitch gets lower.

B The apparent pitch stays the same.

C The apparent pitch gets higher.

D The apparent pitch goes from low to high and back to low. [1]

11 A teacher and two students are measuring the speed of sound.

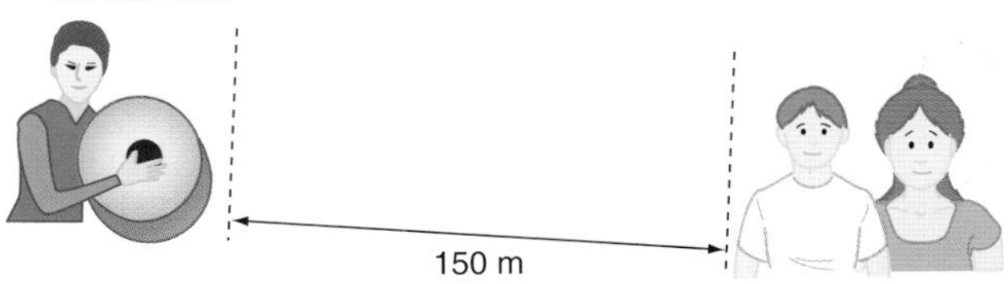

The teacher makes a sound by hitting two cymbals together. Each student starts a stopwatch when they see the teacher hit the cymbals. They each stop their stopwatch when they hear the sound.

a) Describe how a sound wave moves through the air. [3]

b) The students repeat the experiment and record their readings in a table.

Student	Time in s
Andrew	0.44, 0.46, 0.44, 0.48, 0.43
Kefe	0.5, 0.6, 0.4, 0.4, 0.6

i) State the precision of Andrew's readings. [1]

ii) State the equation linking speed, distance travelled and time taken. [1]

c) The teacher was standing 150 m from the students. Use the experimental data recorded by each student to calculate:

i) the mean measured time for each student [2]

ii) the speed of sound calculated by each student. [2]

Write each answer to an appropriate number of significant figures.

d) The students look in a data book and find that the speed of sound in air is given as 341 m/s.

The students discuss their results:

Andrew: 'My experiment was more accurate because my answer was closest to 341 m/s.'

Kefe: 'No, you did not allow for reaction time. My result is the best that you can get with this method.'

Andrew: 'No, reaction time didn't matter because I had to react twice and it cancelled out.'

Evaluate these conclusions. [5]

12 A student is investigating refraction of light.

a) Explain the word 'refraction'. [2]

b) The diagram shows a ray of light travelling from air to glass.

Copy the diagram and add labels to show the angle of incidence, *i*, and the angle of refraction, *r*. [2]

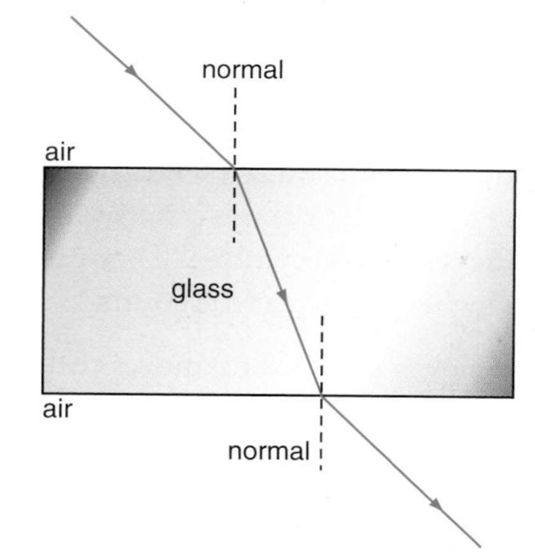

c) The student wants to find the refractive index of the glass.

i) State the equation linking refractive index, angle of incidence and angle of refraction. [1]

ii) The student has available the following pieces of apparatus: a rectangular glass block, a protractor and a ray box.
Describe how the student should carry out the experiment. You should include:

- what the student should measure
- how the measurements should be made
- how the student should use a graph to find the refractive index. [6]

13 Light from an object forms an image in a plane mirror.

a) State which statement in the box below is correct. [1]

Light from the object passes through the image in a plane mirror. Light waves are longitudinal. The angle of incidence equals the angle of reflection. The incident ray is always at right angles to the reflected ray.

b) i) Copy the diagram and use words from the box to label the numbers on the diagram. [2]

mirror	normal	ray	reflection

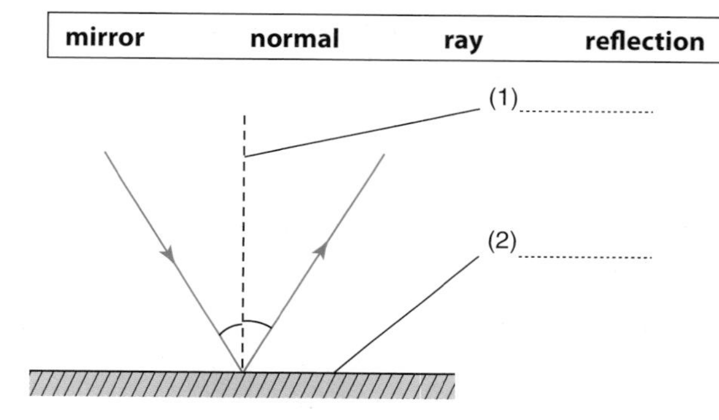

ii) Write *r* on your diagram to show the angle of reflection. [1]

14 a) The diagram shows a ray of light directed at a semicircular glass block.

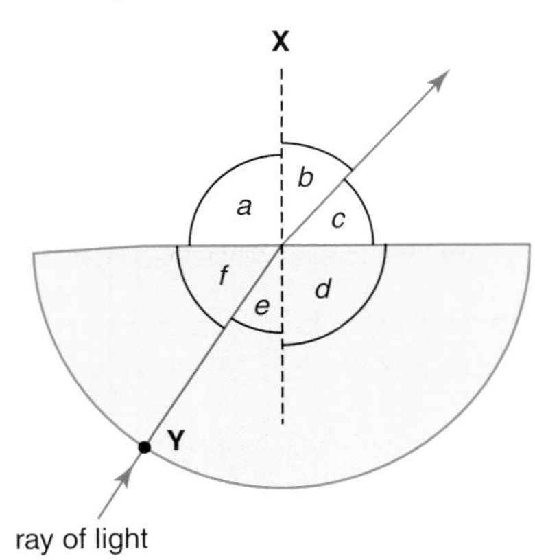

i) Name line X. [1]

ii) State which letter, a, b, c, d, e or f, is the angle of incidence. [1]

iii) Name angle b. [1]

iv) State an equation that relates angle of incidence, angle of refraction and refractive index of glass. [1]

v) At point Y light passes from air to glass but refraction does not take place.
How can you tell this from the diagram? [1]

vi) Why does refraction not take place at point Y? [1]

b) Glass with a critical angle of 42° was used to make the blocks shown below.

i) Copy and complete the diagram to show how the ray of light passes through the rectangular block and out into the air. [3]

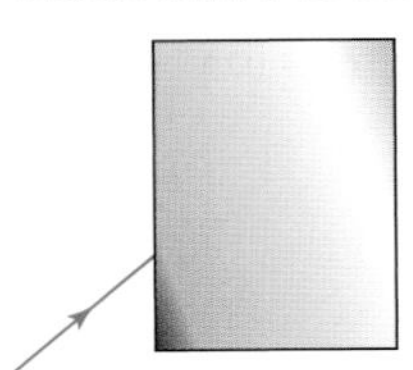

ii) Copy and complete the diagram to show how the ray of light passes through the triangular glass block and out into the air. [2]

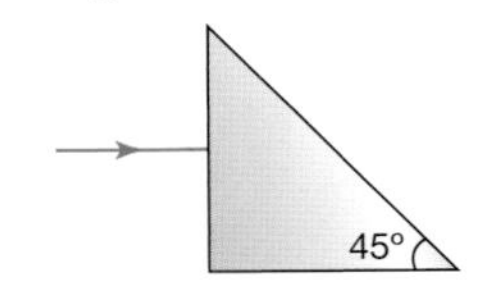

15 This question is about radiations in the electromagnetic spectrum.

radio waves	micro-waves	infrared	A	ultraviolet	B	gamma rays

a) Name the two parts of the spectrum that are missing (A and B). [1]

b) Which electromagnetic radiation is used for heating and night vision equipment? [1]

c) Which electromagnetic radiation is used for cooking and satellite transmission? [1]

d) Which end of the electromagnetic spectrum has the highest frequency? [1]

e) Exposure to excessive electromagnetic radiation can be harmful to the human body. For two named types of radiation, describe:

i) a harmful effect [2]

ii) how the risks of exposure can be reduced. [2]

16 The diagram shows a ray of light incident at point **A** inside a glass optical fibre. The ray of light is totally internally reflected at **A** and eventually passes out into the air from point **B**.

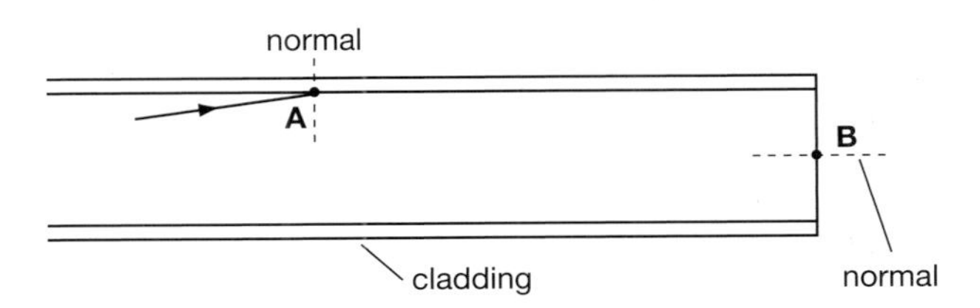

a) Copy then draw on the diagram the path taken by the ray of light:

i) between A and B [1]

ii) as it passes into the air from B. [2]

b) Explain:

i) why the light is totally internally reflected inside an optical fibre [1]

ii) why the light is able to emerge at **B**. [1]

17 A student investigates total internal reflection of light inside semicircular blocks.

a) The student draws this diagram of the apparatus, which shows the path followed by a ray of light in a plastic block.

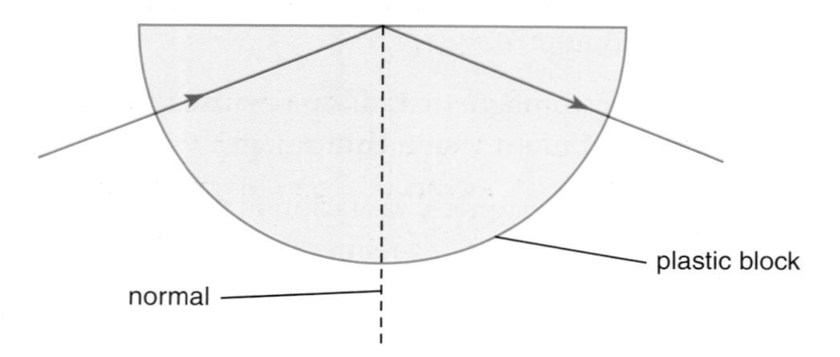

i) Copy the diagram. Mark and name the two angles that the student should measure. [1]

ii) Describe how the student should measure and record the angles. [3]

iii) State how these two angles are related. [1]

b) The student repeats the experiment with a glass block rather than a plastic one. The light emerges from a different place, as shown in the next diagram.

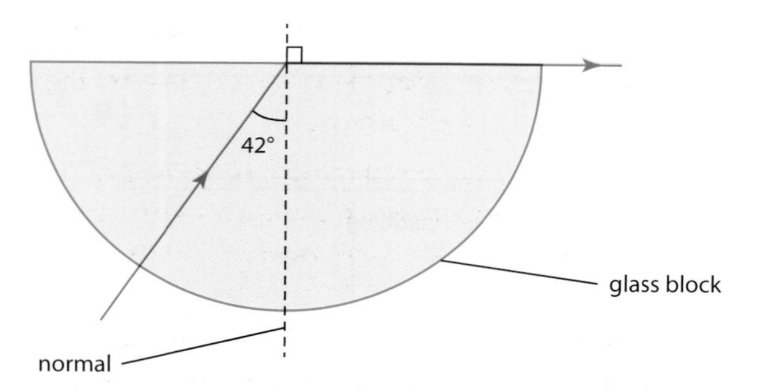

i) State the name given to the angle marked 42° in the diagram. [1]

ii) Calculate the refractive index of the glass. [3]

18 Pete and Annie use the microphone and an oscilloscope to display the different waveforms of their voices. Trace A shows the form of Annie's voice, and trace B shows the form of Pete's voice.

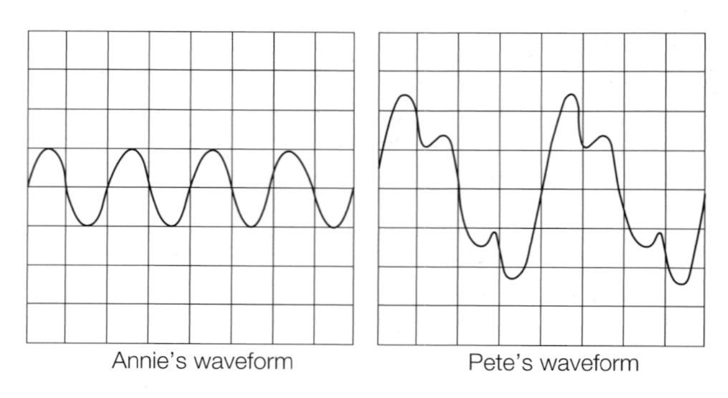

The oscilloscope is adjusted to these settings for both singers:

y-axis 0.5 V/cm

x-axis 2 ms/cm

Each square on the grid measures 1 cm by 1 cm.

a) i) Which singer has the louder voice? Explain your answer. [2]

ii) Which singer's voice has the higher pitch? Explain your answer. [2]

b) Calculate the voltage that corresponds to the amplitude of the sound from Pete. [2]

c) i) Calculate the time period of the sound from Annie. [2]

ii) Calculate the frequency of this sound. State the unit. [2]

19 Explain what is meant by the Doppler Effect. Give an example of the Doppler Effect in your answer. [4]

EXTEND AND CHALLENGE

1 Read the article below about whales, then answer the questions.

Whale song

Not only are blue whales the largest animals in the world, but they are also the noisiest. They give out low-frequency sounds that allow them to communicate over distances of thousands of kilometres.

The whales are helped in their long-distance communications because the sounds they give out are trapped in the upper surface layers of the ocean. Sounds that hit the surface of the sea at a shallow angle are reflected back. Sounds that travel downwards are turned back upwards. At greater depths sound travels faster, because the water is more compressed. This causes sound waves to be refracted as shown in the diagram. This is similar to the refraction of light waves on a hot day, which allows us to see a mirage. Only in the top layers of the ocean does sound travel a long way. The sounds are trapped, rather like light waves in an optical fibre.

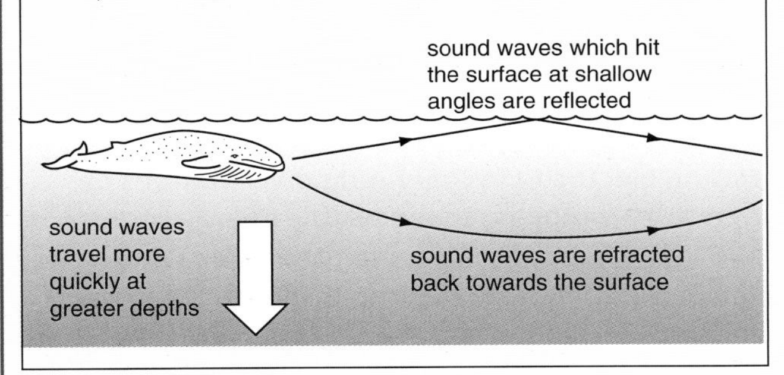

a) Explain why the reflection and refraction of waves shown in the diagram help the whales to communicate over large distances.

b) Explain why creatures living at the bottom of the sea cannot communicate over long distances.

c) Wallis, an amorous bull whale, is prepared to swim for a day to find a mate. After singing for half an hour, he gets a response from Wendy 7 minutes after he stops. How far away is she, and will Wallis bother to make the journey? (Assume Wendy replies as soon as Wallis finishes his serenade.)

- Wallis can swim at 15 km/h.
- Sound travels at about 1500 m/s in water.

d) Why are there long pauses in whale conversations?

e) Explain how the refraction of sound waves shown in the diagram tells us that sound travels faster at greater depths. Draw a diagram to show what would happen to whale sounds if sound travelled more *slowly* at greater depths.

f) On a hot day, it is common to see a mirage on a road as you drive along. The mirage is an image of the sky. Draw a diagram to explain how mirages occur. (Hint: light travels slightly faster in the hot air just above the road surface.)

2 An earthquake produces seismic waves which travel around the surface of the Earth at a speed of about 6 km/s. The graph shows how the ground moves near to the centre of the earthquake as the waves pass.

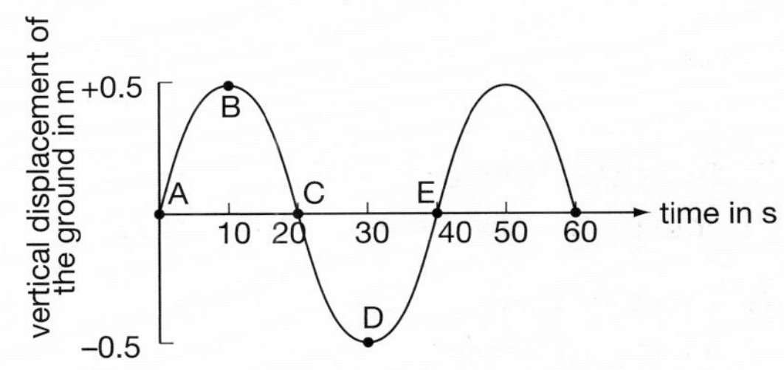

a) Calculate the time period of the waves.

b) Calculate the frequency of the waves.

c) Calculate the wavelength of the seismic waves.

d) Explain why the ground is moving most rapidly at times C and E.

e) When is the ground accelerating at its greatest rate?

f) Use the graph to estimate the vertical speed of the ground at the time marked C.

g) Make a sketched copy of the graph. Add to it a second graph, to show the ground displacement caused by a second seismic wave of the same amplitude but twice the frequency.

h) Discuss whether high-frequency or low-frequency seismic waves will cause more damage to buildings.

i) The diagram shows seismic waves passing a house. The waves produce ground displacements that have a vertical component YY_1 and a horizontal component XX_1. Which component is more likely to make the house fall down? Explain your answer.

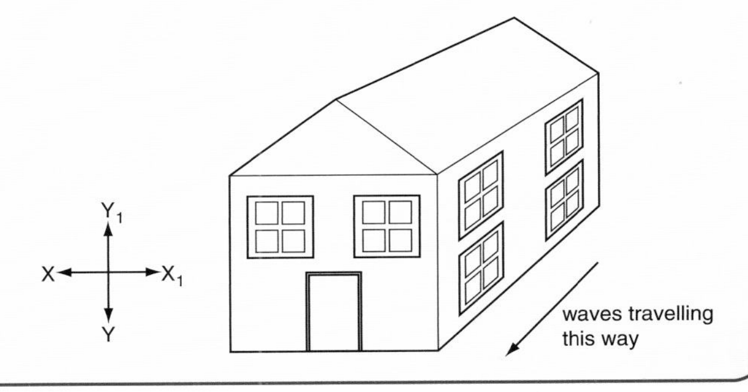

4 Energy resources and energy transfer

CALCULATE • PRESENT

1 Find out the electrical energy requirement for your country each year.
2 What fraction of electricity is generated in your country using fossil fuels?
3 Find out the total electrical energy requirement each year for the world.

The UK will stop generating electricity in coal-fired power stations by 2025. Increasingly the country is moving to renewable energy sources. In 2009 two per cent of the UK's electricity was generated using wind power, which has grown to eleven per cent of production in 2017.

By the end of this section you should:

- **be able to describe energy transfers from one energy store to another**
- **be able to use the principle of energy conservation**
- **know and use the relationship between efficiency, useful energy output and total energy input**
- **be able to use Sankey diagrams**
- **be able to describe how thermal energy transfer may take place by conduction, convection and radiation**
- **be able to explain the role of convection in every day phenomena**
- **be able to explain how emission and absorption of radiation are related to surface and temperature**
- **be able to investigate thermal energy transfer by conduction, convection and radiation**
- **be able to explain ways of reducing unwanted energy transfer**
- **know and be able to use the definitions of work and power**
- **know and be able to use the relationships for gravitational potential energy and kinetic energy**
- **understand how conservation of energy produces a link between gravitational potential energy, kinetic energy and work**
- **be able to describe the energy transfers involved in generating electricity**
- **be able to describe the advantages and disadvantages of methods of large-scale electricity production from various renewable and non-renewable resources.**

4.1 Energy

Figure 1.1

When you fill a car up with fuel, do you think about the principle of the conservation of energy?

- State the type of energy store in the fuel.
- Where did that energy store come from in the first place?
- When the fuel is used, its energy is transferred to other stores of energy. Can you name three other stores of energy?

Energy stores and systems

We can begin to understand energy by studying changes in the way energy is stored when a **system** changes. A ‘system’ is an object or a group of objects that interact. Here are some situations that you should be familiar with.

- **Throwing an object upwards**
  When you throw a ball upwards, just after the ball leaves your hand it has a store of kinetic energy. When the ball reaches its highest point, it has a store of gravitational potential energy. Just before you catch it again, it has a store of kinetic energy.
- **Boiling water in a kettle**
  When you turn on your electric kettle, the water in the kettle gets hotter. There is now more internal (or thermal) energy stored in the hot water than there was in the cold water.
- **Burning coal**
  When we burn coal, there is a chemical reaction. Coal has a store of chemical energy which is transferred to thermal energy as it burns. A coal fire can warm up a room.
- **A car using its brakes to slow down**
  A moving car has a store of kinetic energy. When the car slows to a halt, it has lost this store of kinetic energy. The brakes exert a frictional force on the wheels, and the brakes get hot. The store of kinetic energy in the car has been transferred to a store of thermal energy in the brakes. This energy is then transferred to the surroundings.
- **Dropping an object which does not bounce**
  Just before the object hits the ground, it has a store of kinetic energy. Once the object has stopped moving, all of its kinetic energy has been transferred to a store of thermal energy in the object and the surroundings, so they warm up a little. (You might hear a noise, but the energy carried by the sound is also transferred to the internal energy of the surroundings.)

Figure 1.2

- **Accelerating a ball with a constant force**
  We have a store of chemical potential energy in our muscles. When we throw a ball, our store of chemical potential energy decreases, and the ball’s store of kinetic energy increases. The hand applies a force to the ball and does work to accelerate it.

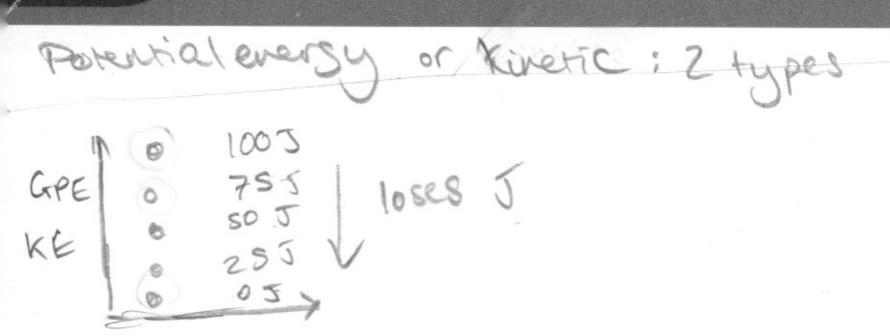

- **Holding two magnets with north poles facing**
  When you hold two magnets with like poles facing, you can feel a force which repels the magnets from each other. When the magnets are close together, there is a store of magnetic potential energy. When you release the magnets, they move apart. The magnets' store of magnetic potential energy has reduced, and their store of kinetic energy has increased.

Energy stores

In the simple everyday events and processes that were described above, we identified objects that had gained or lost energy. For example, objects slow down or get hotter. We saw that the way energy is stored changes.

We use the following labels to describe the stores of energy you will meet:

- kinetic
- chemical
- thermal (or internal)
- gravitational
- magnetic
- electrostatic
- elastic
- nuclear.

Counting the energy

Energy is a quantity that is measured in joules, J. Large quantities of energy are measured in kilojoules, kJ, and megajoules, MJ.

$$1\,\text{kJ} = 1000\,\text{J}\ (10^3\,\text{J});\ 1\,\text{MJ} = 1\,000\,000\,\text{J}\ (10^6\,\text{J})$$

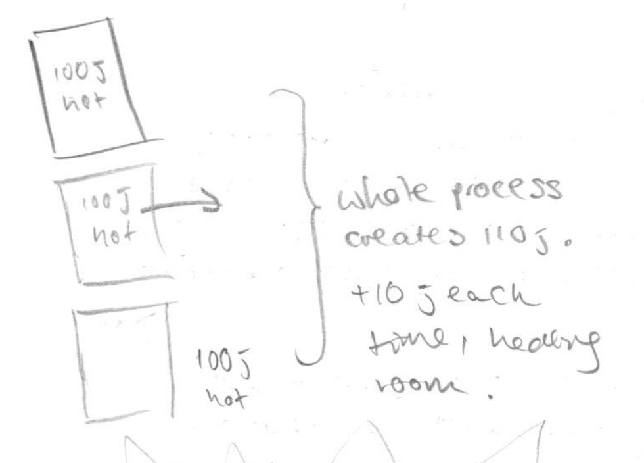

The reason that energy is so important to us is that there is always the same energy at the end of a process as there was at the beginning. If we add up the total energy in all the stores at the end, the amount is the same as it was at the beginning.

Figures 1.3, 1.4 and 1.5 show some examples of counting the energy.

TIP
You will often find the word 'potential' used to describe an energy store. A rock on the top of a hill has a store of gravitational (potential) energy. It may not look as if it has an energy store, but the rock has the potential to roll downhill – when it has a store of kinetic energy (which is more obvious to the observer).

	gravitational energy store	kinetic energy store
(b) ball stationary above the ground	100 J	0
(a) ball moving upwards	0	100 J

Figure 1.3 A ball is thrown upwards from the ground. In **(a)**, the ball has 100 J of kinetic energy and zero gravitational potential energy. In **(b)**, the ball has 100 J of gravitational potential energy and zero kinetic energy.

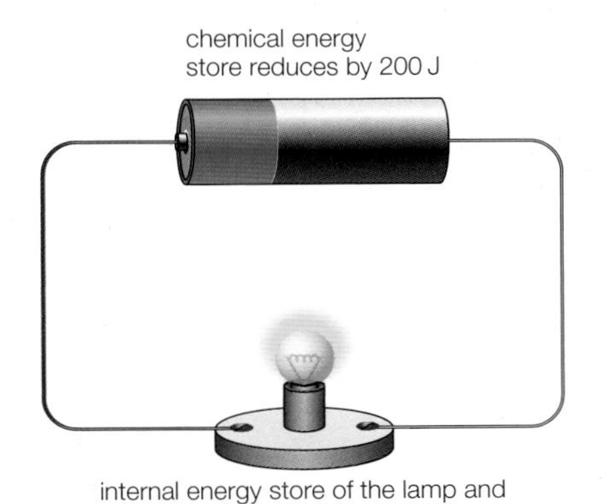

Figure 1.5 A cell passes a charge through a lamp. This charge flows for a few minutes. After this time, the store of chemical energy in the cell has decreased by 200 J. The store of thermal energy in the lamp and the surroundings has increased by 200 J.

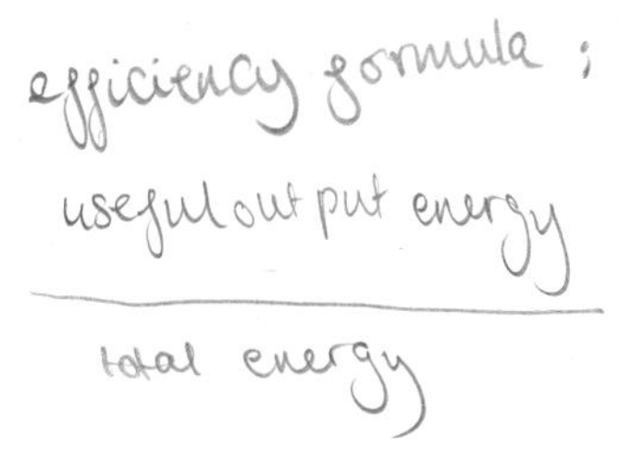

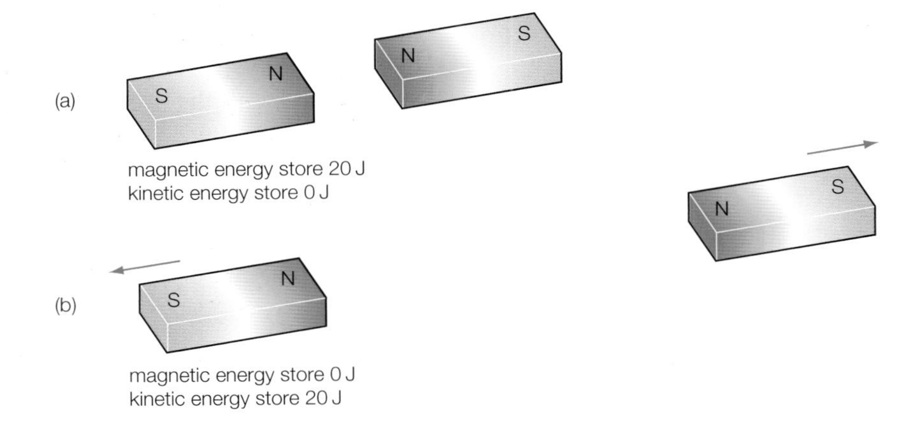

Figure 1.4 (a) Two magnets have been pushed close together. There is a store of 20 J of magnetic (potential) energy. **(b)** The magnetic store of energy has reduced to zero, and each magnet has 10 J of kinetic energy, making a total store of 20 J.

The principle of conservation of energy

The principle of conservation of energy states that the amount of energy always remains the same. There are various stores of energy. In any process, energy can be transferred from one store to another, but energy cannot be destroyed or created.

Transferring energy from one store to another

Light, sound and electricity are useful, but they are not stores of energy. They are ways of transferring energy from one store to a different energy store. You cannot go into a shop to buy a box of 'electrical energy', but you can buy a cell or battery. In a circuit, the chemical energy stored in a cell or battery causes electric charge to flow.

In a torch, the chemical energy stored in the battery causes an electric current (a flow of charge). The electric current causes the temperature of the bulb to increase so much that the bulb lights up. The light cannot be stored but it is useful. When the light strikes an object and is absorbed, the thermal energy of the object increases.

If we drop a bunch of keys onto a table, the collision will make the air vibrate and we hear a sound. The sound wave transfers energy; it is not an energy store. The energy will transfer to the air and surrounding objects causing an increase in their store of thermal energy.

Useful and wasted energy transfers

When we use an energy store, we usually want to transfer the energy into another useful store. For example, we may use a crane to lift a heavy load for us. What we want to do is to use the chemical energy stored in the petrol to transfer energy into the gravitational store of the load we are lifting. Unfortunately, when we use an engine there are unwanted energy transfers. You know from experience that a car engine makes a noise and gets hot. So there is an unwanted transfer to the thermal store of the surroundings.

TIP

The word dissipate means to spread out, or waste. Thermal energy spreads out and is hard to transfer back in to a useful store.

Efficiency

When a machine wastes energy we say that it is inefficient. We can express **efficiency** as a percentage or a decimal. In the case of a petrol engine, only about 25% of the available energy is usefully used to lift a load or drive a vehicle forwards. The other 75% of the stored chemical energy is transferred wastefully to the thermal store of the surroundings. We calculate efficiency using the equation:

$$\text{efficiency} = \frac{\text{useful energy output}}{\text{total energy input}}$$

A steam engine is rather less efficient than a petrol engine. Calculate the efficiency of a steam engine that gets 18 kJ of useful energy transfer from 150 kJ of chemical energy stored in the coal in its furnace:

$$\text{efficiency} = \frac{18\ \text{kJ}}{150\ \text{kJ}} = 0.12 \text{ or } 12\%$$

Figure 1.6 shows how we can show the same information in a **Sankey diagram**. Such a diagram shows how the energy put into a machine is used usefully or dissipated as thermal energy.

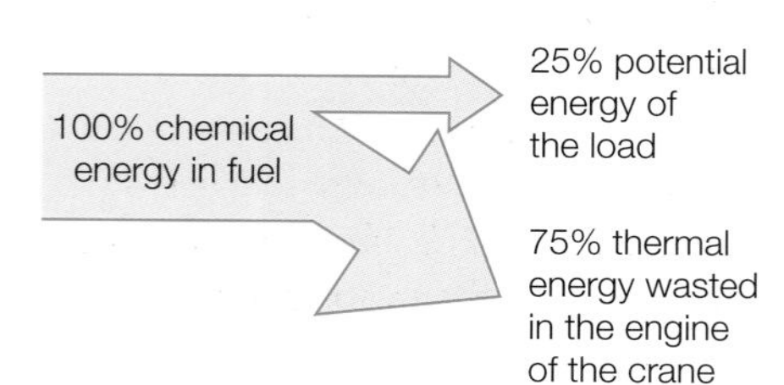

Figure 1.6 A Sankey diagram to show the fate of the chemical energy store used to lift a load.

STUDY QUESTIONS

1 In the question below, state the energy transfers that occur. State the starting and end points only.
   a) An electric motor lifts a load.
   b) A firework rocket flies into the air and explodes.
   c) A car brakes and comes to a halt without skidding.
   d) A car brakes and skids to a halt.
   e) A golfer hits a ball down the fairway.

2 Explain what is meant by the principle of conservation of energy.

3 In the diagram below, a ball falls to the ground. In positions A to E some values of its kinetic energy (KE) and gravitational potential energy (GPE) are shown. Copy the diagram and fill in the missing values (you may ignore air resistance.)

	KE	GPE
A •	0	50 J
B •	10 J	
C •		25 J
D •	35 J	
E •		0

4 A student compares two engines, A and B. Both machines have the same energy input. She discovers that A wastes less energy as heat than B does. Which engine is more efficient?

5 **a)** A car's engine is supplied with a kilogram of diesel, which stores 45 MJ of chemical energy. The engine is 36% efficient. How much useful energy is available to drive the car forwards?
   b) Draw a Sankey diagram to show how the energy in the diesel is transferred to other stores.

6 **a)** An electric crane, with an efficiency of 30%, transfers 240 J of useful gravitational potential energy in lifting a load. How much energy was transferred to the motor to do this job?
   b) Draw a Sankey diagram for this process to show how the energy put into the motor was transferred to other stores of energy.

7 Explain what is wrong with this statement: 'A battery stores electrical energy for the lights, horn and starter motor'.

4.2 Conduction and convection

Figure 2.1 When you walk on a carpet in bare feet, your feet feel warm. But if you walk into the kitchen, which has a tiled floor your feet feel cold. The tiles are better conductors than the carpet, so thermal energy is transferred quickly from your feet and they feel cold.

Insulators and conductors

Materials that conduct thermal energy quickly are known as conductors, and materials that only allow a slow transfer of energy are know as insulators. All metals are very good conductors, building materials such as concrete, brick and glass are reasonably good conductors. Air, wood and plastics are classed as insulators. Metals are good conductors because they have free electrons, which can carry energy quickly. Insulators transfer energy slowly, as one atom can only transfer energy to its immediate neighbours.

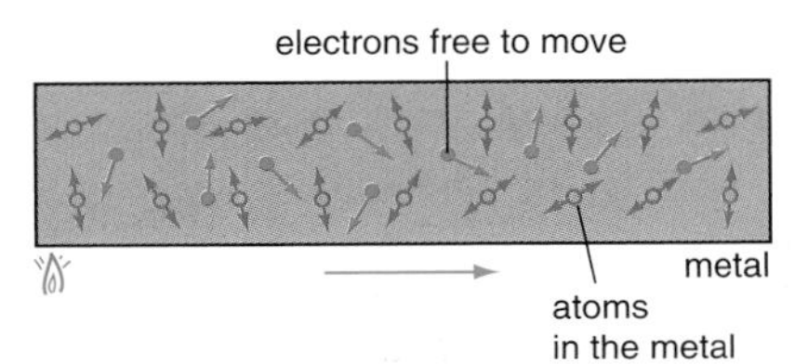

(a) Thermal energy is transferred quickly by fast-moving electrons.

insulator

hot

cold

(b) Thermal energy is transferred slowly by atoms bumping into each other.

Figure 2.2 Thermal energy transfer by conduction.

Keeping warm

Fat, air and wool are all insulators. These three materials are important for keeping warm-blooded animals (including us) at the right temperature. Marine animals such as whales and seals, have a thick layer of blubber to keep them warm. Birds trap air in feathers, and we trap layers of air in our clothing.

Practical: The conductivity of metals

Figure 2.3 shows how you can compare the conductivities of four metals. In this case, aluminium, copper, brass and iron have been used.

Figure 2.3

Method

1 The four metals are placed centrally on a tripod.
2 Water is placed into hollows at the end of each conductor.
3 A Bunsen burner is placed in the centre so that all the metal sheets are heated equally. The water on the best conductor will boil first. NB the sheets will become very hot – handle with tongs to avoid burning.

Figure 2.4 This robin is keeping warm by trapping a layer of air in its feathers.

Questions

1 Explain how this is made a fair test.
2 The time taken for the water to boil on the four conductors is given below. Match each of the conductors A, B, C, D to the metals mentioned above.

Conductor	Time for water to boil in seconds
A	10
B	16
C	36
D	54

warm less dense air rising

colder more dense air sinking

Figure 2.5 Convection currents near a person in a room.

Convection currents

When air is heated it expands and its density becomes less. When air is surrounded by cold denser air, it rises. This is the same principle as a submerged cork rising to the surface in water.

Figure 2.5 shows how air circulates around you if you stand lightly clothed in a cold room. Your body transfers thermal energy to the air surrounding you. The air rises as it is less dense, and colder denser air takes its place. This is a convection current. Although air is a poor conductor, it is very efficient at transferring thermal energy by convection. We avoid thermal energy transfer by convection, by trapping layers of air in our clothing.

Sea breezes

Convection currents also play an important part in our weather systems. Winds are convection currents on a large scale. You have probably noticed when you have been sunning yourself on the beach that there is usually a sea breeze (Figure 2.6). In the daytime the land warms up quickly. This causes a rising air current and so a breeze blowing in from the sea. Seabirds are quick to take advantage of these rising currents of air, and can soar upwards without flapping their wings.

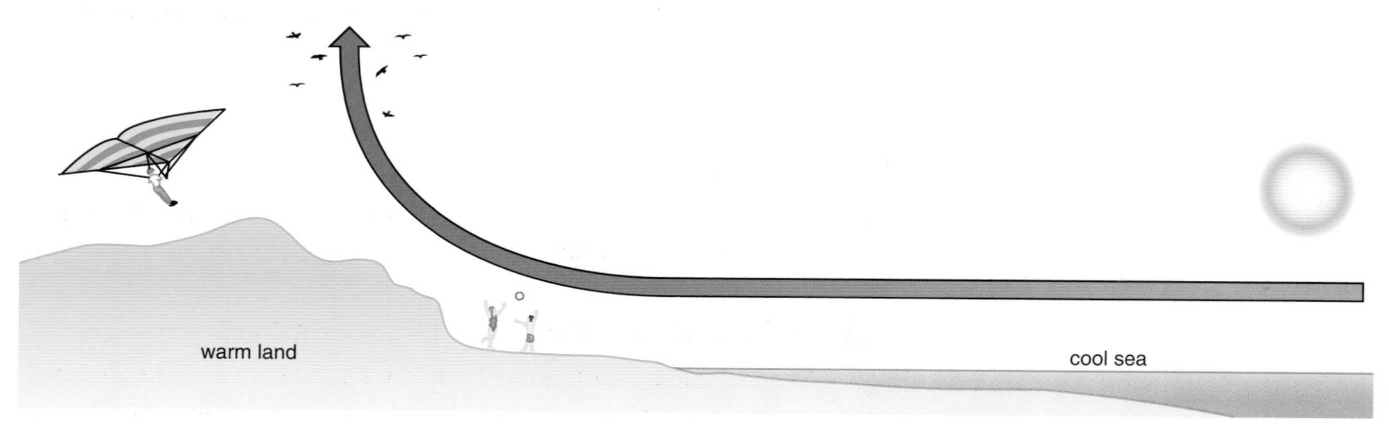

Figure 2.6 The sea takes lots of thermal energy to warm up, so it remains cooler than the land. The land warms quickly, and warm air rises above the land.

REQUIRED PRACTICAL

Make sure you can explain how you can see convection currents in water.

Practical: Observing convection currents

Method

1 Fill a beaker (or a flask) with water, and place the beaker onto a tripod and gauze.

2 Carefully drop a crystal of potassium manganate(VII) into the centre of the beaker.

3 Heat the centre of the beaker slowly under the crystal. Observe the path of the dissolved potassium manganate(VII) which shows the direction of the currents.

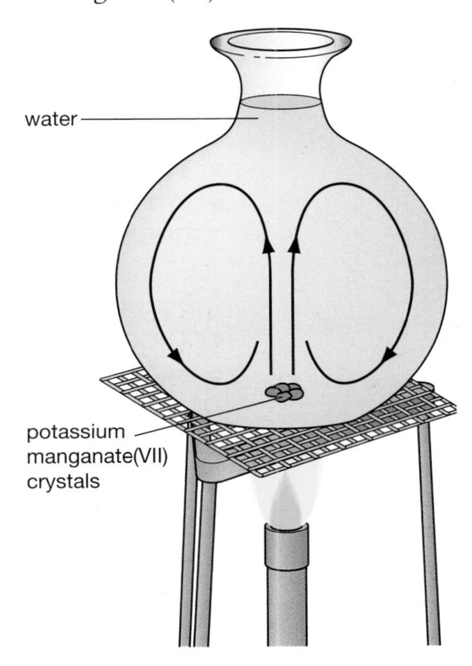

Figure 2.7 A laboratory experiment to show convection currents in water.

Questions

1 Draw the path of the convection currents.

2 Explain the paths of the convection currents in terms of the density of water.

Safety advice

1 Wear safety glasses.

2 Be careful – potassium manganate(VII) is harmful if swallowed. Do not allow it to stain your clothing.

■ Avoiding unwanted energy transfer

When we heat our homes, the energy stored inside the house is dissipated through the roof, walls, windows or doors of the house to warm up the air outside. We want to make sure the energy escapes as slowly as possible.

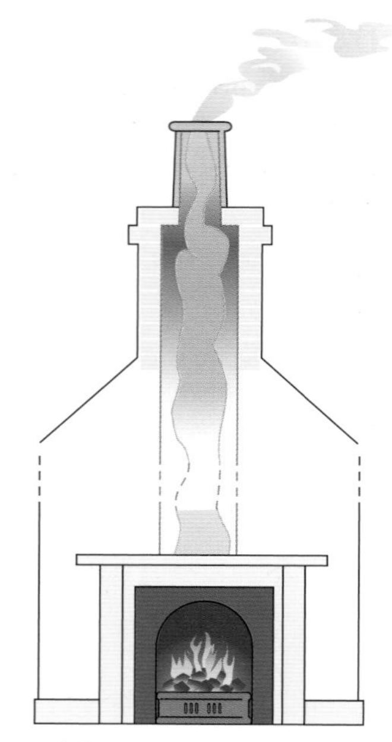
Figure 2.8

Here are some ways in which we reduce unwanted energy dissipation at home.

Chimneys

Figure 2.8 shows a coal fire burning in a sitting room. Some of the energy from the burning coal is transferred to the air outside the house. This is wasted energy. By having the chimney inside the house, thermal energy can be transferred into the bedrooms upstairs. This is useful energy.

Walls

The rate at which energy is transferred through the walls of a house depends on four factors.

- The temperature difference between inside and outside. (Our heating bills are larger in winter than in summer.)
- The area of the walls. (Large houses cost more to heat than small houses.)
- The thermal conductivity of the walls. Brick and glass are quite good thermal conductors. Energy flows out of a warm house through the walls and windows. The higher the thermal conductivity of a material, the higher the rate of energy transfer by conduction across the material.
- The thickness of the walls (or windows) is important. The thicker the walls, the slower the rate of thermal energy transfer.

TIP

Make sure you understand how houses are insulated to keep them warm.

Figure 2.9 This house was built in 1820. The walls are 70 cm thick. This keeps the house warm by reducing the thermal energy transferred through the walls.

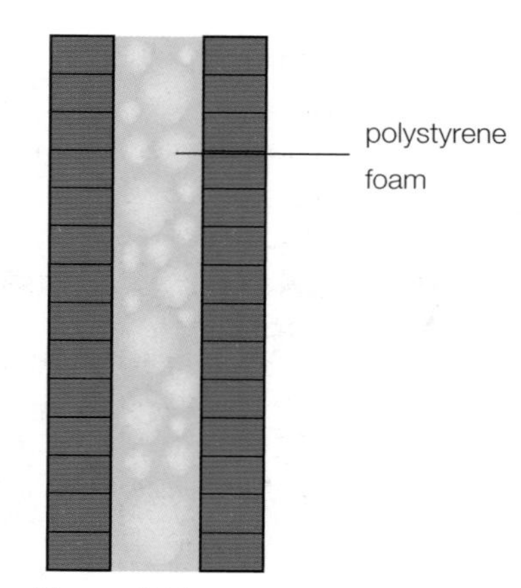

Figure 2.10

Modern houses are built with two layers of brick as shown in Figure 2.10. Then the house is insulated with cavity wall insulation, between the two layers of brick. The foam which insulates the walls is full of trapped air. The air is a good insulator; it has a much lower thermal conductivity than brick or glass.

Loft insulation and carpets

The most efficient way to reduce energy waste from our house is to insulate the loft. A thick layer of loft insulation reduces thermal energy transfer by conduction through the roof. We also use insulating carpets to reduce thermal energy transfer by conduction through the floor.

Double glazing

A thin pane of glass in a window transfers energy out of the house. We use double glazing to reduce thermal energy transfer through the windows. A layer of gas trapped between two panes of glass provides good insulation (Figure 2.12).

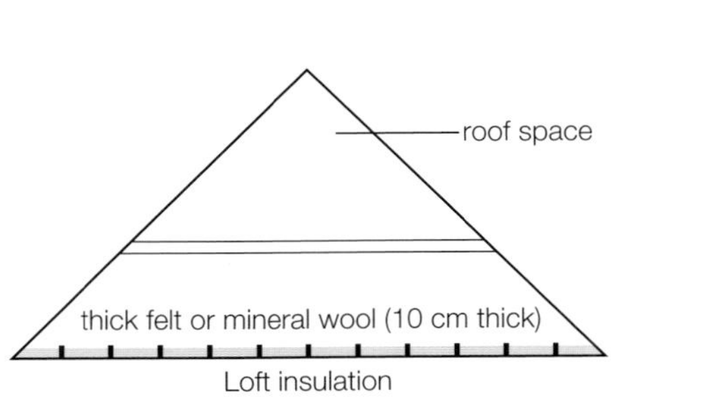

Figure 2.11

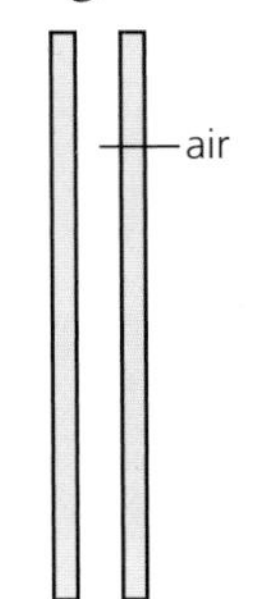

Figure 2.12 Double glazing.

STUDY QUESTIONS

1 The kitchen and lounge of a house are at the same temperature. When a girl walks bare foot on the carpet in the lounge, her feet feel warm. When she walks on the tiled floor in the kitchen, her feet feel cold. Explain why this happens.

2 The diagram below shows a piece of wood with a ring of brass round part of it. A piece of paper has been stuck partly across the wood and partly across the brass. A flame is now applied gently underneath the paper as the wood is turned slowly. The paper over the wood turns black and burns.
The paper over the brass discolours slightly and does not burn. Explain why.

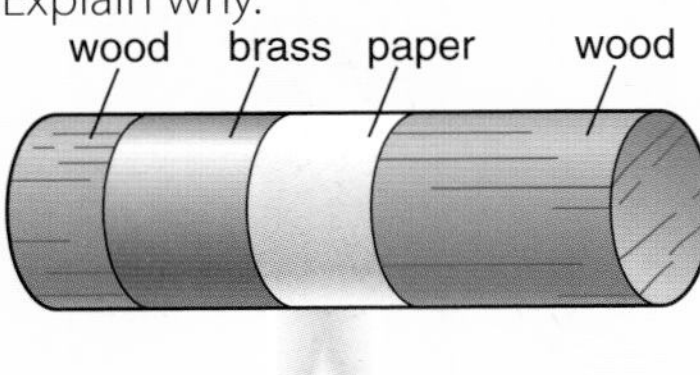

3 **a)** Explain why a woollen sweater keeps you warm.
b) Explain why it is dangerous to put your bare hand on to metal when the temperature is −15 °C, but it is safe to make a snowball.

4 Thermal energy cannot be transferred by convection in a solid. Explain why.

5 Figure 2.6 shows the direction of a sea breeze in daytime. At night, the land cools down more quickly than the sea. Draw a diagram to show the direction of the breeze at night.

6 Over a hundred years ago in tin mines, fresh air used to be provided for the miners through two ventilation shafts shown below.
a) To improve the flow of air, a fire was lit at the bottom of one of the shafts. Explain how this worked.
b) One day, Mr Trevethan, the owner of the mine, had an idea. To make the ventilation even better he lit another fire at the bottom of the second shaft.
Comment on this idea.

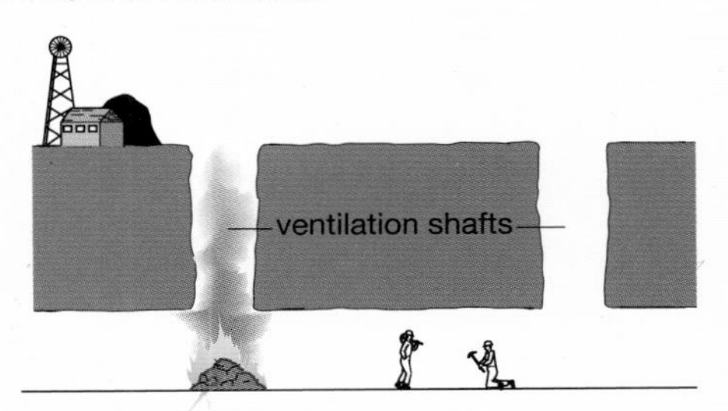

7 The diagram shows 2 test tubes A and B. In A, ice has been forced down to the bottom of the tube by some metal gauze. In B, some ice floats at the top.

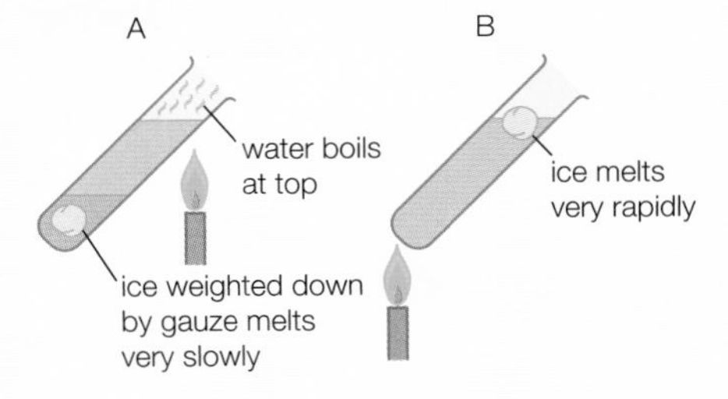

When A is heated at the top, water boils at the top, and the ice melts slowly.
When B is heated at the bottom, the ice melts quickly at the top.
Explain these observations.

8 Name 4 ways to reduce thermal energy transfers from your home. In each case, explain how the energy transfer is reduced.

4.3 Radiation

Figure 3.1 The Sun emits various types of electromagnetic wave.

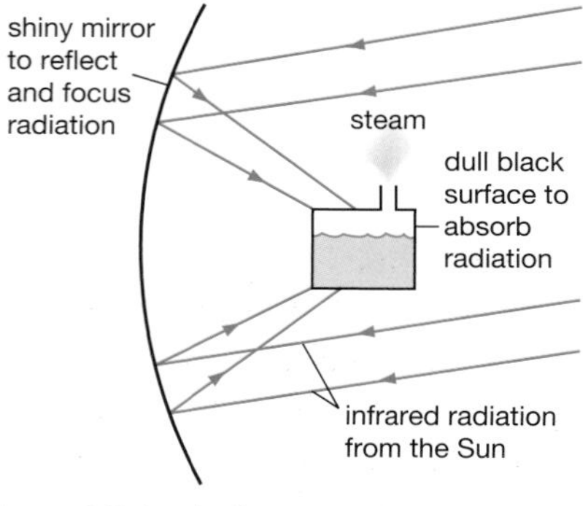

Figure 3.2 A solar furnace.

TIP

Don't forget that thermal energy can be transferred by three methods: conduction, convection and radiation.

Radiation is the third way by which thermal energy can be transferred. Electromagnetic energy from the Sun is transferred to us by radiation. The Sun emits electromagnetic waves, which travel through space at high speed. Light is such a wave, but the Sun also emits a lot of infrared waves (or infrared radiation). Infrared waves have longer wavelengths than light waves. It is the infrared radiation that makes you feel warm when you lie down in the sunshine.

Good and bad absorbers

Infrared radiation behaves in the same way as light. It can be reflected and focused using a mirror. Figure 3.2 shows the idea behind a solar furnace. The shiny surface of the mirror is a poor absorber of radiation, but a good reflector. Radiation is absorbed well by dull black surfaces. So the boiler at the focus of the solar furnace is a dull black colour.

Good and bad emitters

Figure 3.3 shows an experiment to investigate which type of teapot will keep your tea warm for a longer time. One pot has a dull black surface, the other is made out of shiny stainless steel. Radiation that is emitted from the hot teapots can be detected using a thermopile and a sensitive ammeter. When you do the experiment you will find that, when the teapots are filled with hot water at the same temperature, the black teapot emits radiation at a greater rate than the shiny surface. So a shiny teapot will keep your tea warmer than a black teapot.

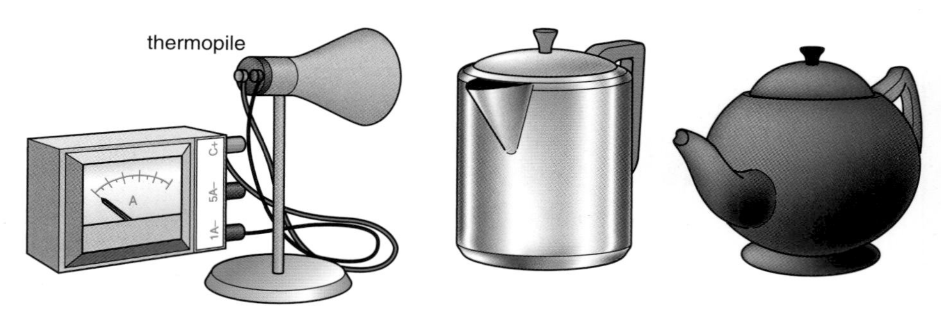

Figure 3.3 A sensitive instrument called a thermopile can detect radiation.

- Black surfaces are good absorbers and good emitters of radiation.
- Shiny and white surfaces are bad absorbers and bad emitters of radiation.

Energy transfer by radiation

Figure 3.4 shows how you can investigate how different surfaces absorb radiation.

REQUIRED PRACTICAL

Make sure you are familiar with this practical. Take care when using the heater to avoid getting burnt.

Method

1 You have two sheets of aluminium, A and B. A has been painted black on one side, and B is shiny on both sides. Use some melted wax from a candle to stick a marble on to each sheet (using the shiny side of A).

TIP

Make sure you know which surfaces are good emitters and absorbers of radiation.

2 Place the sheets equal distances from a heater and turn the heater on.

3 Observe which marble falls off the sheet first.

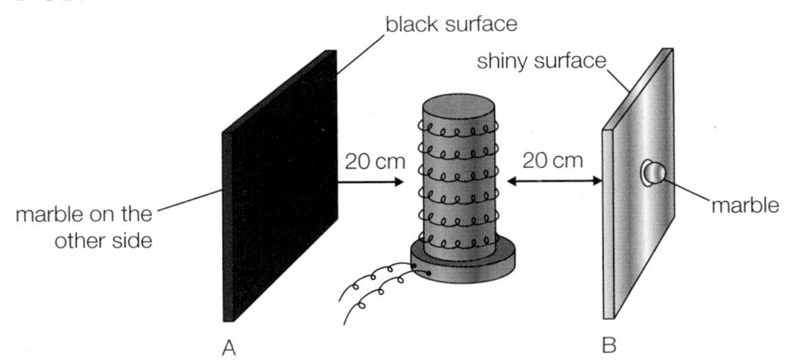

Figure 3.4

Questions

1 Discuss whether this is a fair test.

2 Which marble fell off first? Explain why.

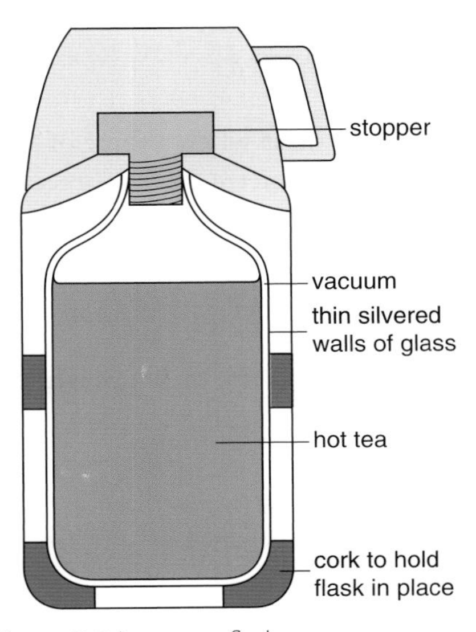

Figure 3.5 A vacuum flask.

Vacuum flask

A vacuum flask keeps things warm by reducing thermal energy losses in all possible ways. The flask is made with a double wall of glass and there is a vacuum between the two walls. Conduction and convection cannot take place through a vacuum. The glass walls are thin, so that little thermal energy is conducted through the glass to the top. Thermal energy can be radiated through a vacuum, but the glass walls are silvered like a mirror so that they are poor emitters of radiation. The stopper at the top prevents thermal energy loss by evaporation or convection currents (Figure 3.5).

STUDY QUESTIONS

1 Explain why each of the following is true, in terms of thermal energy transfer.
   a) Clean snow does not melt quickly in bright sunshine, though dirty snow does.
   b) Some casserole dishes that are used in ovens are black, but the outside of an electric kettle is shiny.
   c) After finishing a marathon, athletes may be wrapped in aluminium-coated plastic sheets.
   d) In hot countries many houses are painted white.

2 Explain three ways in which a vacuum flask helps to keep drinks warm.

3 Describe an experiment to investigate which type of surface is a good emitter of radiation. Explain how you make it a fair test.

4.4 What is work?

Figure 4.1 You agree to help a friend move a heavy load. What two factors determine how hard the job will be?

TIP

In everyday language we use the word 'work' to mean the job we do. When physicists say 'work' they mean 'force × distance moved'.

A job of work

Tony works in a supermarket. His job is to fill up shelves when they are empty. When Tony lifts up tins to put them on the shelves he is doing some work. The amount of work Tony does depends on how far he lifts the tins and how heavy they are.

We define **work** like this:

$$\text{work done} = \text{force} \times \text{distance} = F \times d$$

Work is measured in **joules** (J). 1 joule of work is done when a force of 1 newton moves something through a distance of 1 metre, in the direction of the applied force.

$$1\,\text{J} = 1\,\text{N} \times 1\,\text{m}$$

Example. Calculate the work Tony does when he lifts a tin with a weight of 20 N through a height of 0.5 m.

$$\begin{aligned} W &= F \times d \\ &= 20\,\text{N} \times 0.5\,\text{m} \\ &= 10\,\text{J} \end{aligned}$$

Tony does the same amount of work when he lifts a tin with a weight of 10 N through 1 m.

The unit of work, J, is the same as that used for energy, J. When work is done, energy is transferred. In the case of the tins, the work done by Tony is transferred to increase the tin's store of gravitational potential energy.

$$\text{work done} = \text{energy transferred}$$

TIP

A store of energy can be used to do work to transfer energy to another store.

You do work when you throw a ball. The work done transfers energy from the body's chemical store to the ball's kinetic store. When you stretch a rubber band, the work done in stretching it transfers energy from the body's chemical store to the band's elastic potential store.

Work and energy transfers

There are four ways in which energy can be transferred

- by mechanical work
- by electrical work
- by heating
- by radiation.

The word radiation includes light and electromagnetic waves. Radiation also includes 'mechanical radiation' such as sound or shock waves.

Figure 4.2 summarises how energy is transferred when a crane's engine lifts a load.

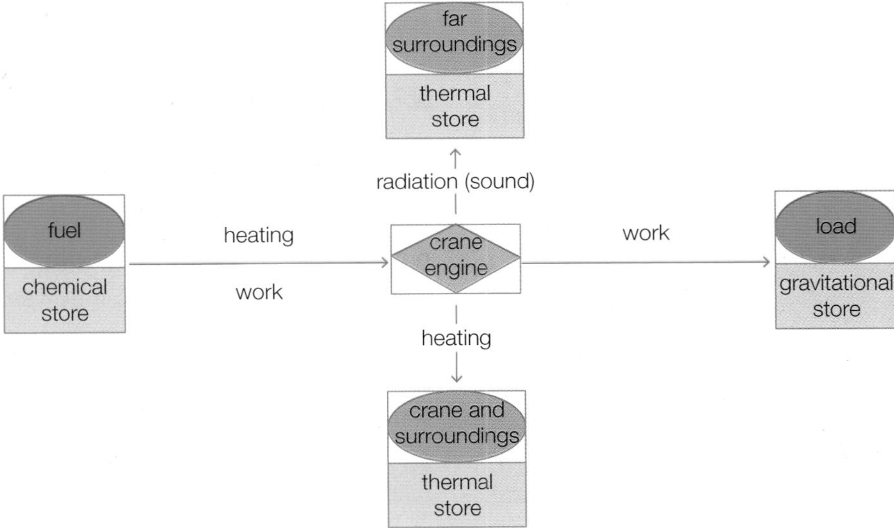

Figure 4.2

- The fuel burns in the engine's cylinders making hot high pressure gas. Pistons then do work to drive the engine.
- The motor turns and does work to lift the load – energy is transferred to the gravitational potential energy store of the load.
- The engine gets hot and energy is transferred to the thermal store of the crane by heating (by conduction and convection).
- The engine makes a noise and energy is transferred to the thermal store of the surroundings by sound. Sound waves cause random vibrations in atoms and molecules, thus increasing the store of thermal energy.

Does a force always do work?

Does a force always do work? The answer is no. In Figure 4.3 Martin is helping Salim and Teresa to give the car a push start. Teresa and Salim are pushing from behind; Martin is pushing from the side. Teresa and Salim are doing some work because they are pushing in the right direction to get the car moving and to increase its store of kinetic energy. Martin is doing nothing useful to get the car moving. Martin does no work because he is pushing at right angles to the direction of movement.

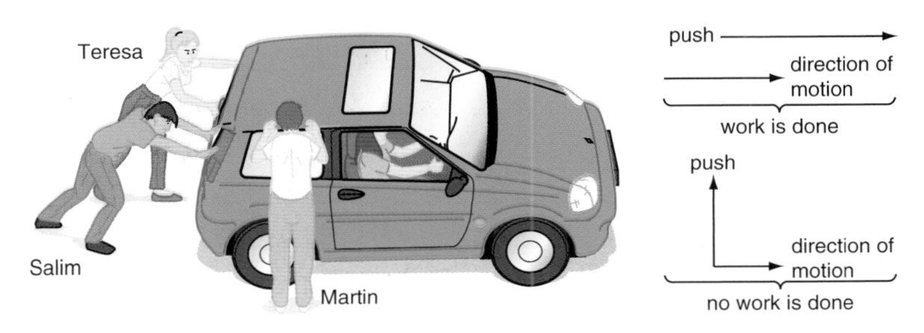

Figure 4.3

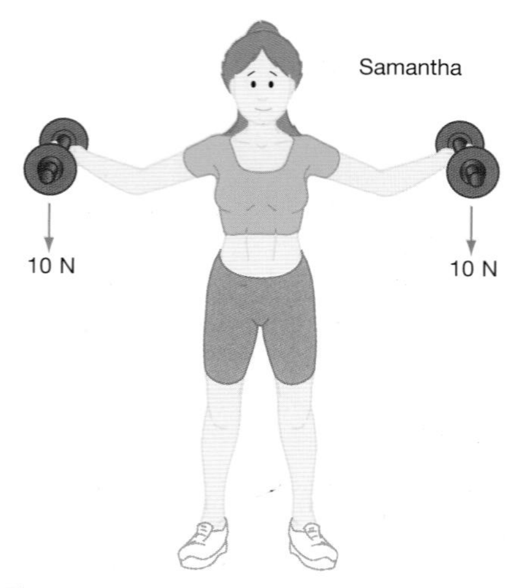

Figure 4.4

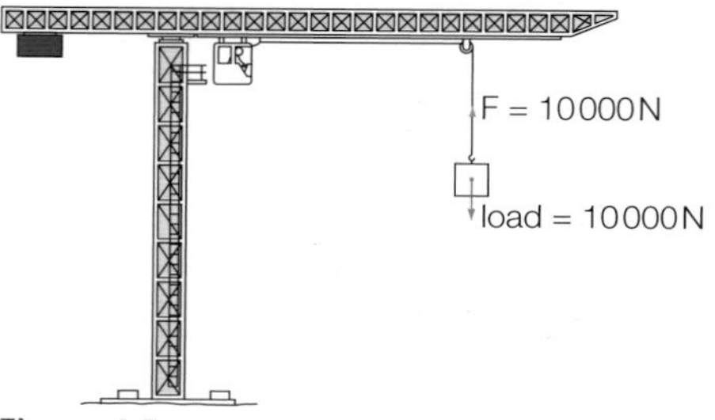

Figure 4.5

In Figure 4.4 Samantha is doing some weight training. She is holding two weights but she is not lifting them. She becomes tired because her muscles use energy, but she is not doing any work because the weights are not moving. To do work you have to move something, for instance lifting a load or pushing a car.

Paying for the fuel

When you want a job done you usually have to pay for it. This is because you have to buy fuel. Figure 4.5 shows a crane at work on a building site. The crane runs on diesel fuel.

Table 1 shows the amount of diesel used by the crane in Figure 4.5 for some jobs. You can see that to lift a heavier load, or to move it further, more diesel is used. When the crane does work, energy is transferred into the gravitational potential energy store of the load.

Table 1

Load lifted in N	Distance moved in m	Work done in J	Diesel used in litre
10 000	1	10 000	0.01
10 000	2	20 000	0.02
20 000	2	40 000	0.04
40 000	3	120 000	0.12

STUDY QUESTIONS

1 a) Define work.
b) Give the unit of work and the unit of energy.

2 In which of the following cases is work being done?
a) A magnetic force holds a magnet on a steel door.
b) You pedal a cycle along a road.
c) A pulley is used to lift a load.
d) You hold a 2 kg mass, but without moving it.

3 Calculate the work done in each case below.
a) You lift a 20 N weight through a height of 2 m.
b) You drag a 40 kg mass 8 m along a floor using a force of 80 N.

4 Table 2 below shows some more jobs done by the crane in Figure 5.4. Copy the table and fill in the missing values.

Table 2

Load lifted in N	Distance moved in m	Work done in J	Fuel used in litre
5000	2		0.01
10 000	4		
	10	40 000	
6000			0.06
	5		0.1
25 000		90 000	

5 Jo is on the Moon in her spacesuit. Her mass including her suit is 120 kg. The gravitational field strength on the Moon is 1.6 N/kg.
a) Calculate Jo's weight in her spacesuit.
b) Jo now climbs 30 m up a ladder into her spacecraft. How much work does she do?

6 Mr Hendrix runs a passenger ferry service in the Caribbean. He has three ships, which are all the same. Sometimes he has problems with the bottoms of the ships, when barnacles stick to them. This increases the drag on the ships, and they use more fuel than usual. Explain why a larger drag makes the ships use more fuel.

4.5 Calculating energy

In Figure 5.1, how much work is done in lifting 110 kg through 2 m? Can you estimate the total kinetic energy stored of the vehicles shown in Figure 5.2? On the vampire roller coaster in Figure 5.3, you can fall a height of 15 m. If you start from a very low speed what is the speed you will reach by the bottom of the fall?

Figure 5.1

Figure 5.2

Figure 5.3 These riders get a thrill when potential energy is transferred to kinetic energy.

Gravitational potential energy

When a weightlifter lifts his weights he does some work. This work increases the gravitational potential energy (GPE store) of the weights. The total mass of the bar is m. How much work does he do when he lifts the bar a height h? (m is in kilograms and h is in metres.) The pull of gravity on the bar (its weight) is $m \times g$. We call the pull of gravity on each kilogram g; this is 10 N/kg.

$$\begin{aligned}\text{work done} &= F \times d \\ &= mgh\end{aligned}$$

The work done is equal to the increase in potential energy.

$$\text{gravitational potential energy} = mgh$$

Kinetic energy

The kinetic energy of a moving object is given by the formula:

$$\text{kinetic energy} = \tfrac{1}{2}mv^2$$

(m is the mass of the object in kg, and v is its velocity in m/s.)

Using equations to calculate the energy changes

Now that you have learnt some formulae for various types of energy, you can use them together with the principle of conservation of energy to do some calculations. Some examples are given below.

1 A ball of mass 100 g is thrown vertically upwards with a speed of 12 m/s. Calculate the maximum height it will reach.

As the ball rises, its kinetic energy store will be transferred to a gravitational potential energy store. So we can use equations for those quantities to solve the problem. (The gravitational field strength g is 10 N/kg.)

$$\begin{aligned}mgh &= \tfrac{1}{2}mv^2 \\ 0.1 \times 10 \times h &= \tfrac{1}{2} \times 0.1 \times 12^2 \\ 10h &= \tfrac{1}{2} \times 144 \\ h &= 7.2\text{ m}\end{aligned}$$

2 A car, with a mass of 1200 kg and travelling at 30 m/s, is slowed by its brakes to a speed of 20 m/s. The brakes are applied for a distance of 75 m. Calculate the force the brakes apply in slowing the car down.

TIP

Look at the calculation for the change in the kinetic energy store. Work out the first value for kinetic energy, then subtract the second value for kinetic energy.

We solve this by thinking of energy transfers. Work is done by the brakes in transferring the car's kinetic energy store to a thermal energy store in the brake blocks.

$$\text{work done by the brakes} = \text{kinetic energy transferred}$$
$$Fd = \tfrac{1}{2}mv_1^2 - \tfrac{1}{2}mv_2^2$$
$$F \times 75 = \tfrac{1}{2}(1200 \times 30^2) - \tfrac{1}{2}(1200 \times 20^2)$$
$$F \times 75 = 600 \times 900 - 600 \times 400$$
$$F \times 75 = 540\,000 - 240\,000$$
$$F = \frac{300\,000}{75}$$
$$= 4000 \text{ N}$$

Figure 5.4

3 This problem refers to energy transfers in a bungee jump. Figure 5.4 shows a jumper, Chloe, launching herself off the Victoria Falls Bridge, which crosses over the Zambezi River at a height of 128 m. Chloe has a mass of 60 kg. Clearly, the bungee rope is designed so that she does not fall into the river. Figure 5.5 shows the elastic potential energy stored in the rope as it stretches; the unstretched length of this rope is 90 m.

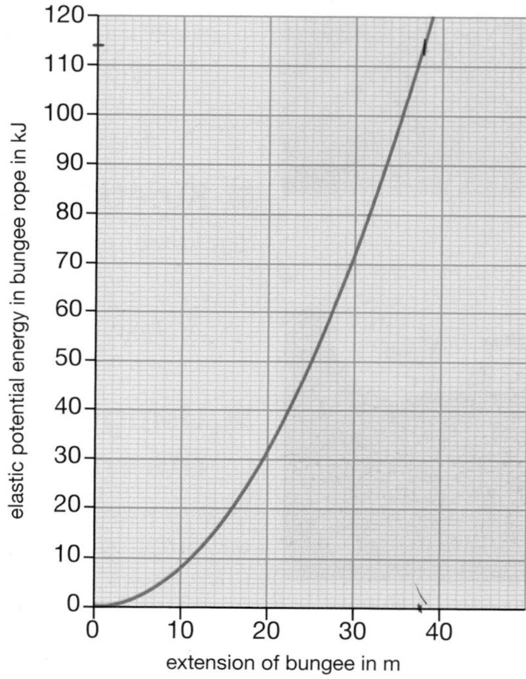

Figure 5.5

We can use this data to calculate the maximum speed of the jumper and to check that she falls a total of only 120 m and so avoids falling into the river.

She reaches her greatest speed after she has fallen 90 m, just before the rope begins to slow her down. So her potential energy store has been transferred to a kinetic energy store.

$$mgh = \tfrac{1}{2}mv^2$$
$$60 \times 10 \times 90 = \tfrac{1}{2} \times 60 \times v^2$$
$$v^2 = 2 \times 900$$
$$v^2 = 1800$$
$$v = 42 \text{ m/s}$$

When Chloe falls a total of 120 m the gravitational potential energy transferred is:

$$\text{GPE transferred} = mgh$$
$$= 60 \times 10 \times 120$$
$$= 72\,000\text{ J}$$

You can see from the graph (Figure 5.5) that when the rope is stretched by 30 m, it stores 72 000 J of elastic potential energy, so she stops there, having fallen a total of 90 m + 30 m, which is 120 m, 8 m above the river.

In the examples above we used the idea that the conservation of energy links gravitational potential energy, kinetic energy and work done.

- Example 1: the initial store of kinetic energy is transferred to gravitational potential energy.
- Example 2: the reduction in the car's kinetic energy store is transferred to the thermal store of the car's brakes and the surroundings. However, the transfer of energy is due to the work done by the car's brakes, so we can use this link to calculate the force exerted by the brakes.
- Example 3: the bungee jumper's initial store of gravitational potential energy is transferred to a kinetic energy store, and then to a store of elastic potential energy.

Power

Often when we want to do a job of work, we want to do it quickly. A crane that lifts a crate more quickly than another crane lifting the same crate is more **powerful**.

Power is defined as the rate at which energy is transferred or the rate at which work is done. Power can be calculated by using these equations.

Figure 5.6 When this train travels at 60 m/s, its engines run at a power of 2 MW.

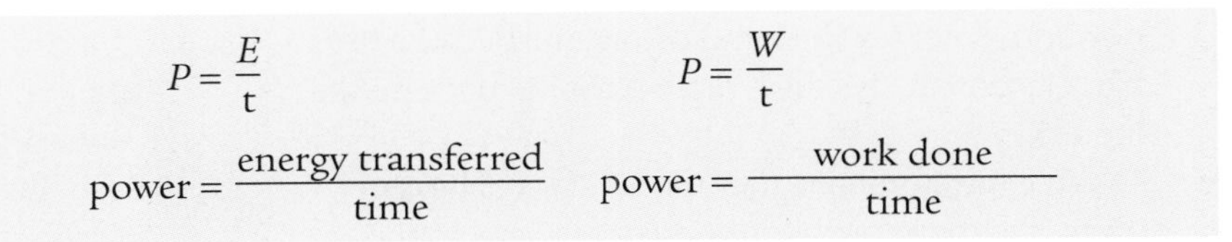

$$P = \frac{E}{t} \qquad P = \frac{W}{t}$$

$$\text{power} = \frac{\text{energy transferred}}{\text{time}} \qquad \text{power} = \frac{\text{work done}}{\text{time}}$$

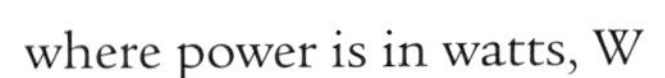

where power is in watts, W — energy transferred is in joules, J

time is in seconds, s — work done is in joules, J

An energy transfer of 1 joule per second is equal to a power of 1 watt.

Large powers are also measured in kilowatts, kW, and megawatts, MW.

1 kW = 1 000 W (10^3 W) 1 MW = 1 00 000 W (10^6 W)

Example. A weight lifter lifts a mass of 140 kg a height of 1.2 m in 0.6 s. Calculate the power developed by the weight lifter.

The potential energy transferred = mgh

$$E_p = 140 \times 10 \times 1.2 = 1680\text{ J}$$

$$\text{power} = \frac{\text{energy transferred}}{\text{time}}$$

$$= \frac{1680}{0.6} = 2800\text{ W or } 2.8\text{ kW (to 2 significant figures)}$$

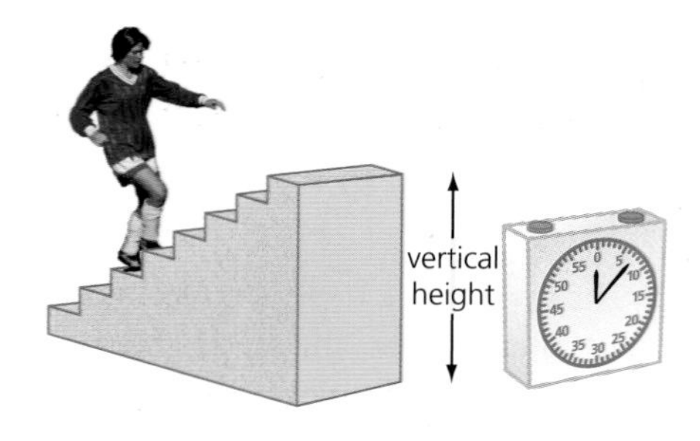

Figure 5.7 When you go up a flight of stairs, you are lifting your body weight and doing work against the force of gravity. The faster you run up the stairs, the greater the power.

Practical

Calculate your personal power by running up a flight of steps.
You need to know your mass, the time it takes you to run up the stairs and the vertical height of the stairs.

1 Record your time and calculate the increase in your gravitational potential energy store.
2 Now calculate your power.
3 Explain why you need the vertical height of the staircase and not the length along the staircase.

STUDY QUESTIONS

1 What is the unit of power?
2 State the connection between the energy transferred and power.
3 A charging rhinoceros moves at a speed of 15 m/s and its mass is 750 kg. Calculate its kinetic energy.
4 A car of mass 750 kg slows down from 30 m/s to 15 m/s over a distance of 50 m.
  a) Calculate its change in kinetic energy.
  b) Calculate the average braking force that acts on it.
5 A crane lifts a weight of 12 000 N through a height of 30 m in 90 s. Calculate the power output of the crane in kW.
6 Two students have an argument about who is more powerful. Peter says he is more powerful because he is bigger. Hannah says she is more powerful because she is quicker.
  To settle the argument, they run up stairs of height 4.5 m. Use the information about their weights and times to settle the argument.

Table 1

	Weight in N	Fastest time in s
Peter	760	3.80
Hannah	608	3.04

7 A catapult stores 10 J of elastic potential energy when it is fully stretched. It is used to launch a marble of mass 0.02 kg straight up into the air. Assume all the energy stored in the catapult is transferred to the marble's kinetic energy store.
  a) Calculate how high the marble rises.
  b) Calculate the speed of the marble when it is 30 m above the ground. (Ignore any affects due to air resistance.)
8 a) Gregg has a mass of 90 kg.
    i) Calculate the gravitational potential energy transferred if he falls 128 m.
    ii) Use the graph in Figure 5.5 to calculate the elastic potential energy stored when the rope has extended 38 m.
    iii) Discuss whether the rope is safe for Gregg to use in the bungee jump.
  b) Discuss the change or changes you would make to ensure a safe jump for someone of Gregg's mass.
9 When an express train travels at a speed of 80 m/s, the resistive forces acting against it add up to 150 kN.
  a) Calculate the work done against the resistive forces in 1 s.
  b) Calculate the power output of the train, travelling at 80 m/s.
10 Read this extract from a magazine article: *The hind legs of a locust are extremely powerful. The insect takes off with a speed of 3 m/s. The jump is fast and occurs in a time of 25 milliseconds. The locust's mass is about 2.5 g.*
  Use the information to answer the problems below.
  a) Calculate the locust's average acceleration during take off. (25 milliseconds = 0.025 s)
  b) Calculate the average force exerted by the locust's hind legs (remember to turn the mass into kg).
  c) The locust's legs extend by 5 cm (0.05 m) during its jump. Calculate the work done by its legs.
  d) Calculate the power the locust develops in its muscles.

4.6 Energy resources and electricity generation

Electricity generation and energy transfers

Figure 6.1 illustrates a simple way in which we can generate electricity in the laboratory. A magnet is moving towards a solenoid. The changing magnetic field induces a voltage in the solenoid, which then drives a current through the light emitting diode (LED). The direction of the current is such that the left hand side of the solenoid behaves like a north pole, so there is a repulsive force on the magnet that acts in the opposite direction to the magnet's movement. (You can read more about this in chapter 6.6.)

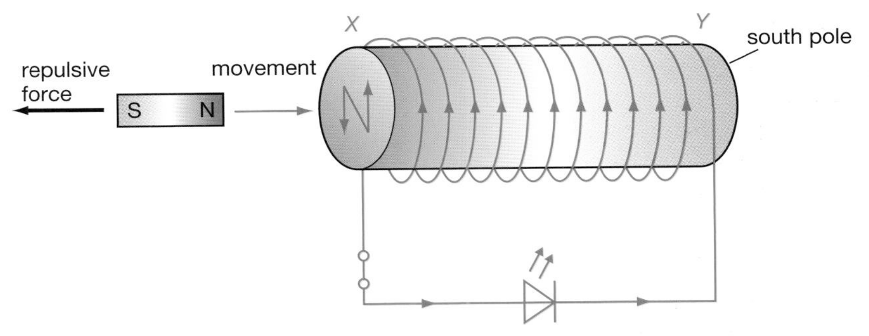

Figure 6.1

In the process shown in Figure 6.1 the following energy transfers occur.

- First, chemical energy stored in someone's muscles is transferred to kinetic energy stored in the moving magnet.
- As the magnet moves, work is done against the repulsive force from the magnet. There must be a continuous transfer of energy from the muscles' chemical store to do this work.
- The magnet exerts a force on the solenoid (this is an example of Newton's Third Law of motion.) Now electrical work is done by the induced voltage as it pushes the current through the LED.
- Finally, energy is transferred to the thermal store of the LED and the surroundings.

Energy transfers occur when electricity is generated by any means. Energy can be generated using the following resources:

- fossil fuels
- nuclear power
- wind
- water
- geothermal resources
- solar cells
- solar heating systems.

Fossil fuels

Fossil fuels include coal, oil and gas. In 2016, the UK generated just over half its electricity using fossil fuels: 30% was generated using gas, 20% using coal and 2% with oil. Figure 6.2 shows the principle of a power station using a fossil fuel; you will not be expected to remember this for IGCSE, but it is helpful to see the diagram to understand the energy transfers. In the UK the average power consumption of electricity is about 35 GW (a GW is a gigawatt or 10^9 W.)

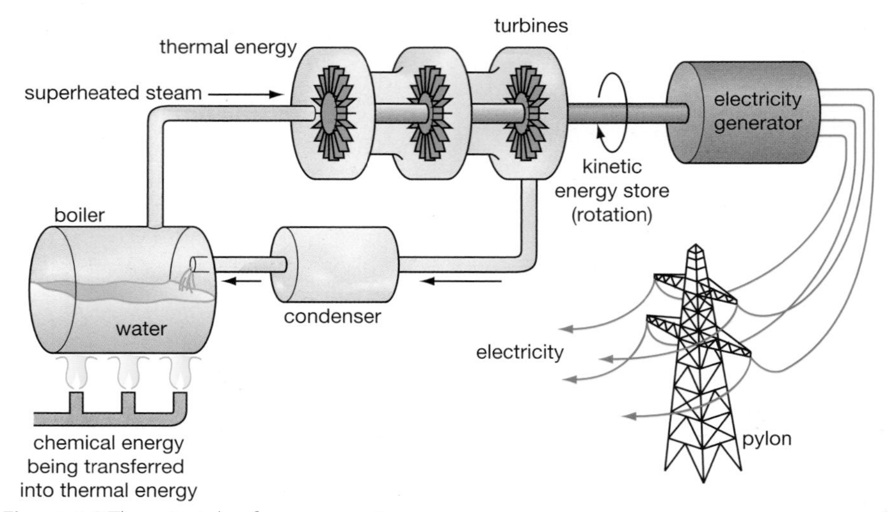

Figure 6.2 The principle of a power station.

Chemical energy stored in the fuel is transferred into thermal energy stored in water and steam. Superheated steam, at high pressure, transfers energy into the kinetic store of the turbines. Then the turbines do work to drive the generators. The end point of the energy transfer occurs in factories and houses, when electrical devices and machines transfer energy in to the thermal store of the surroundings. (This is the same endpoint for all electrical generation, so will not be mentioned in other discussions about energy transfers.)

- The advantages of using fossil fuels are that they are readily available and provide a reliable source of power. Gas fired power stations can be turned on and off at short notice, thereby allowing electricity companies to respond to changes in demand throughout the day.
- The disadvantages of using fossil fuels are that they are non-renewable and that they produce gases, which contribute to global warming and cause pollution. A **non-renewable** energy resource is one that runs out, we have limited reserves of all the fossil fuels, so we need to look for **renewable** energy sources, which will not run out. Burning a fossil fuel produces carbon dioxide, which is a 'greenhouse' gas that warms the Earth by trapping energy in the atmosphere's thermal store. Coal, in particular causes pollution, which can damage our health. The UK government is committed to phasing out coal burning power stations by 2025.

TIP

A **renewable** energy resource is one that never runs out: wind, water and solar power for example.
A **non-renewable** energy resource is one that has a limited supply: oil, gas, coal and nuclear.

Nuclear power

The principle of a nuclear power station is discussed in greater detail in chapter 7.7. The initial source of energy is the nuclear energy store of the nuclei, which undergo nuclear fission. The kinetic energy store of the nuclei is transferred into the thermal energy store of water and steam, which then drive turbines to generate electricity. In 2016 the UK generated 22% of its electricity using nuclear power; this percentage might increase in the future.

- Advantages. There is plenty of nuclear fuel so the resources will last for a long time; there is a great amount of energy released for each kilogram of nuclear fuel – this allows relatively small amounts of fuel to be transported. Nuclear power does not produce gases which contribute to global warming.
- Disadvantages. Nuclear power stations produce radioactive waste, which has a long half-life; waste needs to be stored safely for many years. Nuclear power is expensive to produce. There have been some serious accidents at nuclear power stations, which have contaminated the environment.

turbine blade
transmission shaft
tower
generator

Figure 6.3 a)

Figure 6.3 b)

Wind power

Figure 6.3 (a) shows the principle of a wind turbine. The source of energy is the kinetic energy stored in the moving wind. The wind turns a blade, which is attached to a generator.

- Advantages. Wind power is clean and has no waste products. It is a renewable source that will never run out.
- Disadvantages. Wind power is unreliable. If the wind is too light no power is generated, and sometimes when the wind is too strong the turbines can overheat and have to be turned off. Wind turbines can make a noise, that disturbs people living close to them, and some people complain that they are unsightly.

In the UK in 2016, about 11% of the country's electricity was generated using wind power. Wind power works well provided there are other sources of energy: when the wind blows electricity is generated using wind turbines, but gas can be used when the wind turbines are not in use. But, over a year, using wind power reduces the overall use of fossil fuels.

Water power (hydroelectric power)

The most common type of hydroelectric power plant uses a dam on a river to store water in a reservoir. Water released from the reservoir flows through a turbine, spinning it, which then activates a generator to produce electricity. Here the gravitational potential energy stored in the water, is transferred to a kinetic energy store to drive the turbines. This is illustrated in Figure 6.4.

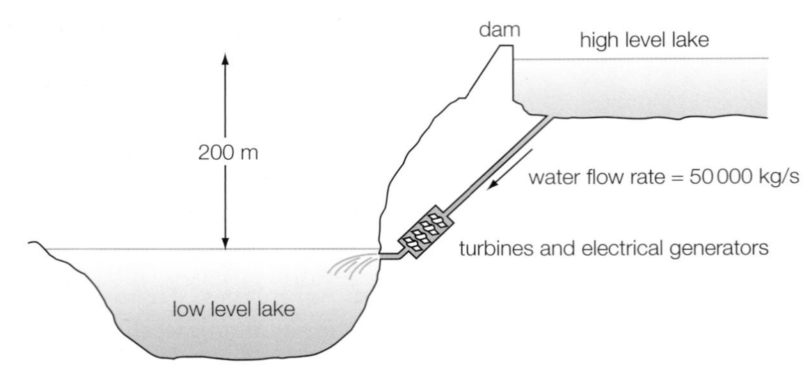

Figure 6.4 A pumped storage power station.

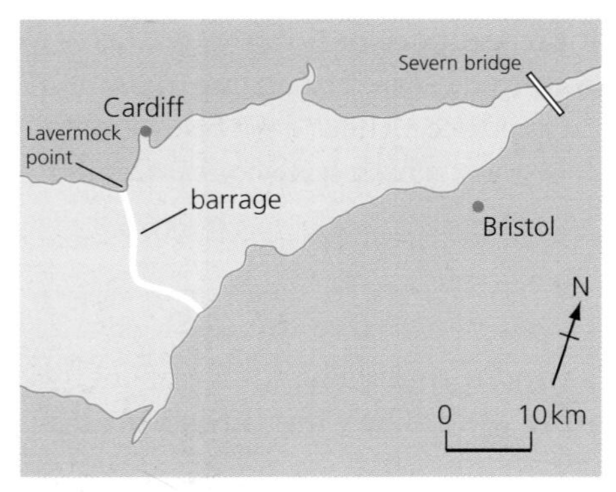

Figure 6.5

Hydroelectric power contributes to 2% of the UK requirement. In other countries hydroelectric power is the main source of electricity; for example, Norway 99%, Brazil 75% and Kenya 70%.

Tidal power is also used to generate electricity, but on a relatively small scale. Every day tides rise and fall. Massive amounts of water move in and out of river estuaries. It is estimated that tides could generate about 20% of the UK's electricity. It has been proposed to build a barrage across the Severn Estuary as shown in Figure 6.5. The barrage would have underwater gates that open as the tide comes in, and then close at high tide. (Figure 6.6). When the tide goes out a second set of gates opens. Water flows out of these gates to drive turbines, which are connected to generators. The stored gravitational potential energy in the water is transferred to a kinetic energy store to drive the turbines.

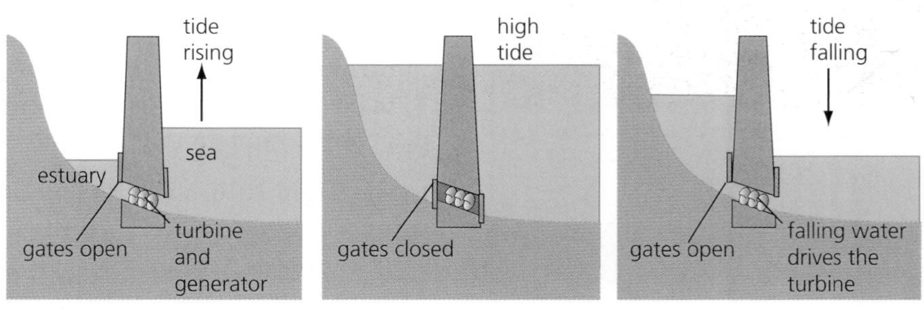

Figure 6.6

- Advantages. Hydroelectric power and tidal power are reliable and have no waste products. Hydroelectric power is reliable because rivers usually flow: in Norway there is no shortage of rain and snow to fill the rivers. Hydroelectric power generated from rivers is relatively cheap to provide – the water is free and does not need mining or transporting. High tides come and go reliably twice a day.
- Disadvantages. Many hydroelectric power schemes have environmental impacts, because building a dam causes a new lake to be formed. In the process of building: forests have been cut down; farmland is lost; wildlife habitats have been destroyed; people lose their homes. Opponents to the building of the Severn Barrage point out that birds and other animals would lose their habitat. Tidal power can only be used when the tides are going out.

Figure 6.7 The Three Gorges Dam in China can generate 21 GW of power, which is over half of the UK requirement.

Geothermal power

Iceland is able to generate all of electricity using clean renewable energy resources; 70% of the country's electricity is generated using hydroelectric power and 30% using geothermal power (Figure 6.8). Iceland is a volcanic country and there are regions where the ground is very hot. In some places steam is used directly to warm houses. In other areas water is pumped down into the ground where it is heated. Steam then returns to drive turbines that are attached to generators. In this case thermal energy builds up a pressure in the steam (which is a type of potential energy store), and the kinetic energy store of the escaping steam does work to drive the turbines.

Figure 6.8 A geothermal power station.

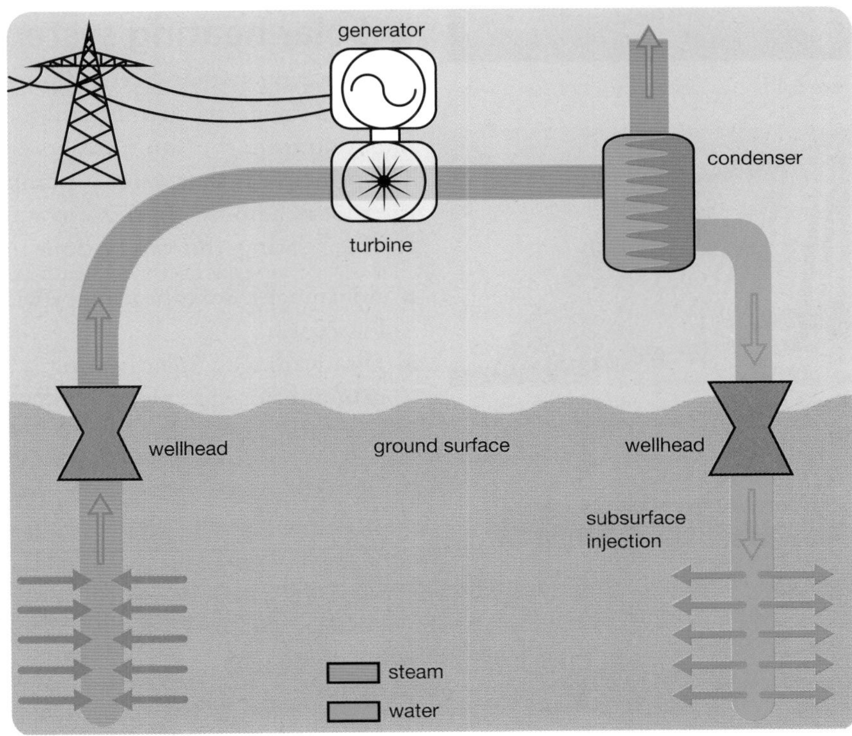

Figure 6.9

- Advantages. Geothermal power is renewable and reliable; it never runs out. There are no waste products. It is cheap to provide; Iceland uses more electricity per head of population than any other country.
- Disadvantages. Geothermal power is only available in some regions of geological activity. In Iceland there appear to be few disadvantages of using geothermal power. However, there are risks associated with setting up a geothermal power station: it requires great expertise, and some poisonous gases could be emitted, and need to be contained.

Solar power

Solar cells use energy directly from the Sun to generate electricity. Solar cells usually generate electricity on a small scale; solar panels can be used to power traffic signs, and we see solar panels on the roofs of houses. The panels on houses can put electricity into the national grid. In some places there are large scale solar 'farms'. In 2016 the UK generated about 4% of its electricity using solar power.

Figure 6.10 This traffic slowing sign is powered by solar cells.

- Advantages. Once the panels are installed solar power is free at source. The panels require little maintenance, so solar power is a good long-term investment. Solar power is indefinitely renewable, and has no waste products.
- Disadvantages. There is an initial high investment. African countries (where there is a great potential to use solar power) cannot afford the initial investment. In countries such as the UK, solar power is unreliable, due to the weather; solar power cannot operate at night, and is less effective in winter. But, used alongside other technologies, solar power makes a useful contribution to a country's electricity supply.

Figure 6.11 Solar panels on a house.

Figure 6.12 Explain why the heating pipes on this roof are black.

Solar heating systems

Solar heating systems do not directly generate electricity, but they reduce the need for electricity (or fossil fuels) by using the Sun to heat water; radiation from the Sun transfers energy to the thermal store of the water. The principle is that water is pumped through pipes, which are placed on the roof of a house. In this way water is warmed; then if the water requires further heating, this can be done using electricity or gas.

- Advantages. Solar heating provides a clean, renewable source of energy at low cost.
- Disadvantages. Solar heating is unreliable; it depends on the weather and cannot be used at night. However, it reduces the need to heat a house or hot water from other sources.

STUDY QUESTIONS

1 **a)** Explain what is meant by a non-renewable energy resource. Give two examples.
  b) Explain what is meant by a renewable energy resource. Give two examples.

2 Name a common fuel used in a nuclear power station. (You can refer to chapter 7.6). State whether the fuel is renewable or non-renewable.

3 **a)** Discuss the advantages and disadvantages of generating electricity using coal.
  b) Explain the energy transfers, which occur in the generation of electricity using coal.

4 **a)** Discuss the advantages and disadvantages of generating electricity using tidal power.
  b) Explain the energy transfers, which occur in the generation of electricity using tidal power.

5 **a)** Explain why tides provide a more reliable way of generating electricity than using wind power.
  b) The UK has 12 000 wind turbines spread from the north to the south of the country. Explain why spreading the turbines over a wide area improves the reliability of wind power.

6 A wind turbine is designed to produce a maximum electrical power of 4 MW. However, due to variations in wind speed, the generator only produces this power 10% of the time.
Calculate how many wind turbines are required to replace a coal fired power station, which produces a power of 2 000 MW all the time.

7 This question refers to the hydroelectric power station shown in Figure 6.4.
  a) Explain the energy transfers that occur to generate electricity.
  b) Use the information in Figure 6.4 to calculate the gravitational potential energy transferred per second when the generators are working.
  c) The generators are 80% efficient. Calculate the power output of the station in MW.

Summary

I am confident that:

✓ **I can recall these facts about energy**

- The energy stores are: chemical, kinetic, gravitational, elastic, thermal, magnetic, electrostatic, nuclear.
- Energy can be transferred by: mechanical work, electrical work, heating and radiation (electromagnetic and mechanical e.g. sound).
- Energy can be transferred from one store of energy to another.
- The principle of conservation of energy: energy cannot be created or destroyed.
- $\text{Efficiency} = \dfrac{\text{useful energy}}{\text{energy put in}}$
- Efficiency is expressed as a fraction or a percentage.
- Thermal energy can be transferred by conduction. Metals are good conductors of heat. Trapped air is a poor conductor (or insulator).
- Thermal energy can be transferred by convection. Warm air is less dense than cold air, so it rises. This difference of density causes convection currents.
- Thermal transfers may be reduced by insulation.
- Thermal energy can be transferred by infrared radiation, which travels at the speed of light. Black surfaces are good emitters and absorbers of radiation; shiny surfaces are poor emitters and absorbers of radiation.
- Work = force × distance moved in direction of force
- 1 joule = 1 newton × 1 metre
- Gravitational potential energy =
  mass × gravitational field strength × height

  GPE = mgh
- Kinetic energy = ½ mass × velocity^2

  KE = $\frac{1}{2} mv^2$
- $\text{Power} = \dfrac{\text{energy transferred}}{\text{time}}$

  1 watt = 1 joule/second
- There are advantages and disadvantages of the generation of electricity using renewable and non-renewable energy resources.
- Electricity can be generated on a large scale.
- Energy transfers occur during the generation of electricity.

Sample answers and expert's comments

1 A commercial hovercraft service runs in the UK between Portsmouth and the Isle of Wight.
The hovercraft has a mass of 60 000 kg and a maximum speed of 13 m/s.

a) The hovercraft approaches the beach and slows down.
   i) State the equation linking kinetic energy, mass and velocity. (1)
   ii) Calculate the change in kinetic energy, in J, when the hovercraft slows down from 13 m/s to 10 m/s. (3)

b) The hovercraft moves along the beach as shown in the diagram.
   i) State the equation which links potential energy, mass, gravitational field strength and height. (1)
   ii) Calculate the increase in potential energy stored when the hovercraft has moved 50 m up the beach. The gravitational field strength = 10 N/kg. (2)

c) Explain how the slope on the beach allows the hovercraft to save fuel and time during its regular journeys. Include ideas about energy in your answers. (4)

(Total for question = 11 marks)

Student response Total 7/11	Expert comments and tips for success
a) i) $KE = ^1/_2 mv^2$ ✔	Correct equation.
ii) The change in speed is 3 so the change in kinetic energy is $^1/_2 \times 60\,000 \times 3^2$ $= 270\,000$ J ✔	The calculation is wrong, but a mark is gained for correct calculation with the numbers used. To find the change in the kinetic energy stored you need to work out the kinetic energy stored at 13 m/s and the kinetic energy stored at 10 m/s, and then find the difference. So it should be: change in kinetic energy stored $= ½ \times 60\,000 \times 13^2 - ½ \times 60\,000 \times 10^2$ $= 5\,070\,000 - 3\,000\,000$ $= 2\,070\,000$ J = 2.0 MJ (to 2 significant figures)
b) $GPE = mgh$ ✔ $60\,000 \times 10 \times 5$ ✔ $= 3000\,000$ J ✔	The formula has been applied correctly. The important point to realise is that the hovercraft increases its potential energy store through the vertical height, not the distance moved along the beach.
c) Energy is stored at the top of the slope, ✔ so the hovercraft can start quickly. ✔	Potential energy is stored at the top, so this gains a mark, but this has not been linked to saving fuel. The potential energy store does allow the hovercraft to start quickly, as it is transferred to the kinetic energy store. To get the last 2 marks, the answer needs to show an understanding of the stores of energy and transfer from one store to another.

2 In Sweden a small hydroelectric power station has been built to provide electricity for the local community.
The graph shows how the demand for electrical power varies though the year.

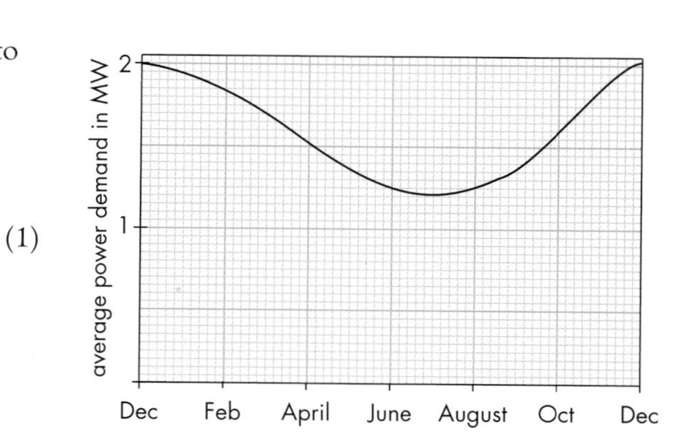

a) Explain why the demand for electrical power varies throughout the year. (1)

The second graph shows how the flow rate of the river, which provides water for the power station, changes during the year.

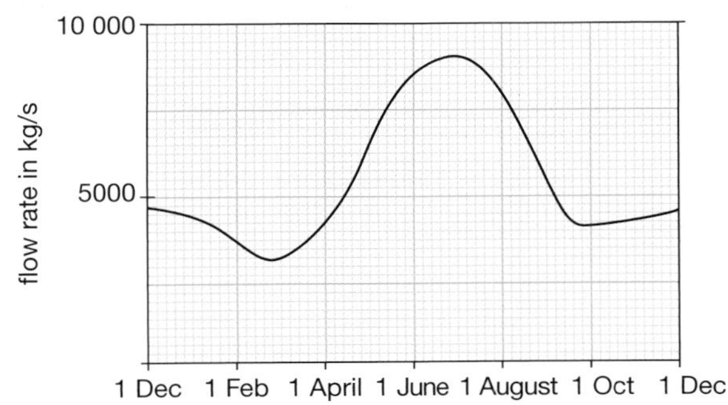

The flow of water is greatest in summer months because the winter snow melts from the mountains to fill the rivers.

b) Discuss the energy transfers that occur when electricity is generated in this power station. (3)

c) The water entering the power station is directed through pipes so that the water falls through a height of 80 m before reaching the generators.

i) State the equation which links gravitational potential energy to mass, gravitational field strength and height. (1)

ii) Calculate the gravitational potential energy transferred by 1 kg of water which falls through a height of 80 m. (3)

The gravitational field strength is 10 N/kg.

d) Use values from the second graph to show that the maximum power available from the river in July is 7.2 MW. (4)

e) The turbines have an efficiency of 60% in transferring the potential energy from the water into electrical energy. Calculate the maximum electrical power output from the power station in July. (2)

(Total for question = 14 marks)

Student response Total 9/14	**Expert comments and tips for success**
a) It is cold in winter and people turn the heating up. ✓	Correct
b) Potential energy is transferred to electricity. ○	The student needs to mention that energy is initially stored in the gravitational potential energy store, which is transferred to the kinetic energy store of the water. Then the moving water does work to drive the turbines and generator.
c) i) $E_P = mgh$ ✓	Correct
ii) $E = 1 \times 10 \times 80$ ✓ $= 800$ ✓	Correct substitution, but the unit has been forgotten. You must include the unit (J).
d) $P = \frac{E}{t}$ ✓ $=$ kg/s $\times 800$ ✓ $= 9000 \times 800$ ✓ $= 7200\,000$ ✓	This student gets the marks, but it could be explained more clearly. P = mass flow per second × 800 J as each kilogram transfers 800 J $P = 9000 \times 800$ $= 7.2$ MW As the unit was given, the student did not lose a mark this time – but it should have been included.
e) efficiency $= \frac{\text{useful power}}{\text{power input}}$ $0.6 = \frac{\text{useful power}}{7.2\text{ MW}}$ ✓ useful power $= 0.08$ MW	The student made a careless slip and loses a mark. But the substitution is correct and gains a mark. Put numbers into the equation first, as you will be given credit for correct substitutions. Useful power $= 0.6 \times 7.2$ MW $= 4.3$ MW

Exam-style questions

1 Which of the following is not a unit of power?

A W **B** J **C** J/s **D** MW [1]

2 Energy is best radiated from a surface which is:

A shiny metal

B matt white

C shiny black

D matt black [1]

3 Energy is transferred from the Sun to Earth by:

A conduction

B radiation

C convection and radiation

D conduction and radiation [1]

4 Energy is transferred through the air by convection currents. These occur when air is heated because:

A the molecules moved more quickly

B the molecules expand

C air become less dense

D the molecules become less dense [1]

5 A 2 kg mass is lifted through a height of 15 m. The energy transferred to the gravitational store of the mass is:

A 30 J **B** 30 N **C** 300 J **D** 3 J [1]

6 Which of the following is not an energy store?

A chemical **B** electrical **C** elastic **D** magnetic

7 Which of the following is a renewable source of energy?

A coal **B** nuclear **C** wood **D** oil [1]

8 A force of 250 N moves a box along the floor a distance of 200 cm. The work done by the force is:

A 500 000 W **B** 50 000 J **C** 0.8 J **D** 500 J [1]

9 A ball of mass 100 g is moving with a speed of 8 m/s. Its store of kinetic energy is:

A 320 J **B** 80 J **C** 6.4 J **D** 3.2 J [1]

10 A car engine has an efficiency of 0.28. When the car travels 1 km at 20 m/s, the engine does 420 000 J of work against resistive forces. How much energy must the fuel must transfer as the car travels?

A 1.5 MJ **B** 120 KJ **C** 8.4 MJ **D** 84 KJ [1]

11 People can take various actions to reduce thermal energy transfers from their homes. Describe one thing that can be done to reduce thermal energy transfers through:

a) the gaps around the doors [1]

b) the glass in the windows [1]

c) the outside walls [1]

d) the roof. [1]

e) The diagram shows three pieces of aluminium, each with a different surface. They each have a drop of water placed on them, then they are heated by a radiant heater placed about 30 cm above them. Explain the order in which the water drop on each piece of aluminium will dry.

A Shiny white paint

B Matt black paint

C Shiny unpainted metal [3]

12 A conveyor belt is used to lift bags of cement on a building site.

a) A 30 kg bag of cement is lifted from the ground to the top of the building. Calculate the energy transferred to the gravitational store. (1 kg has a weight of 10 N.) [3]

b) The machine lifts five bags per minute to the top of the building. Calculate the useful work done by the machine each second. [2]

c) The machine is 20% efficient. Calculate the input power to the machine while it is lifting the bags. [3]

13 A car and its passengers have a combined mass of 1500 kg. It is travelling at a speed of 15 m/s. It increases its speed to 25 m/s.

a) Calculate the energy transferred to its kinetic store as a result of this increase in speed. [3]

b) Explain why the increase in the kinetic energy store is a much greater fractional increase than the increase in speed. [1]

14 The diagram shows a Pirate Boat theme park ride, which swings from A to B to C and back.

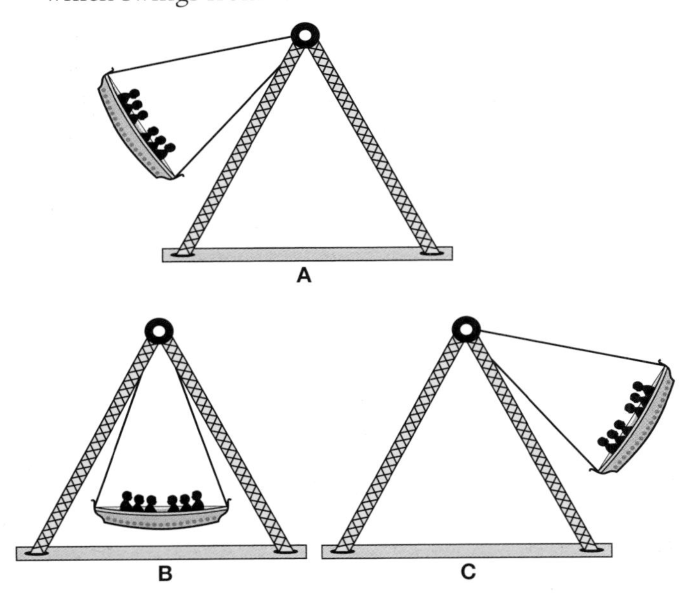

a) As the boat swings from A to B a child's kinetic energy store increases by 10 830 J. The child has a mass of 60 kg and sits in the centre of the boat. Calculate the speed of the child as the boat passes through B. [3]

b) Sketch a graph to show how the child's gravitational energy store changes as the boat swings from A to B to C. [3]

c) By using the idea that the store of gravitational potential energy is transferred to a kinetic energy store as the child falls, calculate the height change of the ride. [3]

15 In the following cases explain the energy transfers that occur.

a) You throw a ball vertically upwards. [3]

b) You accelerate a car, then bring it to rest using the brakes. [3]

c) A firework rocket takes off and explodes. [3]

16 A hot liquid transfers thermal energy by conduction, convection, radiation and evaporation. The vacuum flask shown below is designed to prevent hot liquids becoming cold.

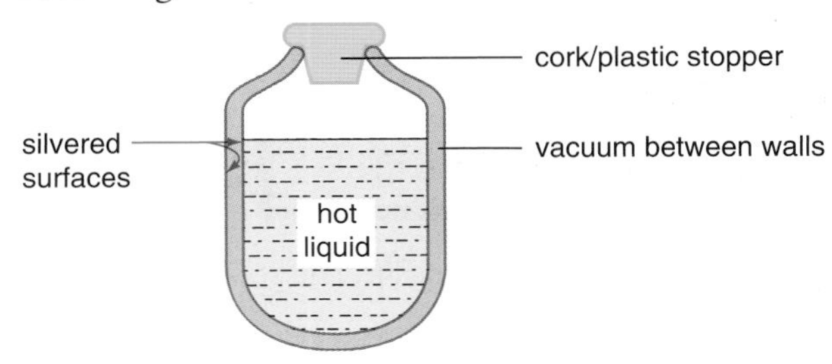

a) Explain how the vacuum between the walls reduces the thermal energy transfer. [2]

b) Explain how and why the cork/plastic stopper reduces the thermal energy transfer. [2]

c) Explain how the silvered surfaces reduce thermal energy transfer. [2]

17 a) i) Name a renewable and a non-renewable energy resource. [2]

ii) Explain the difference between renewable and non-renewable resources. [2]

b) i) State and explain the energy transfers, which occur in a gas fired power station when electricity is generated and then used to warm peoples' homes. [4]

ii) Explain the advantages and disadvantages of generating electricity using gas as the energy resource. [4]

18 The next diagram shows how energy from the Sun can be used to heat a house.

Water from the storage tank is pumped through the solar panel.

In the solar panel, the water passes through copper pipes that are painted black.

The water is then returned to the storage tank.

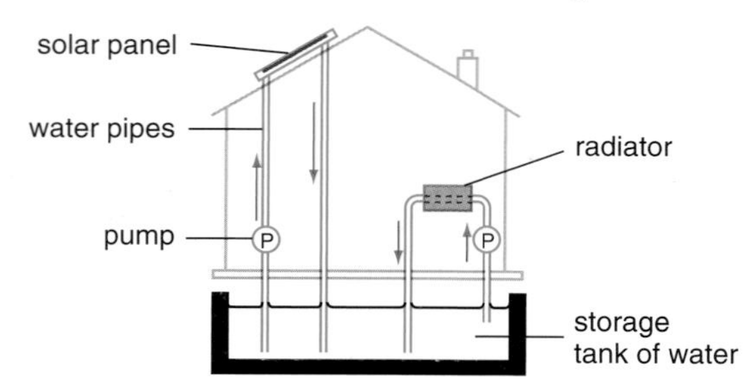

a) i) Explain why the pipes in the solar panel are painted black. [1]

ii) Plastic pipes are cheaper than copper pipes. Explain why copper pipes are used in the panel rather than plastic ones. [1]

b) Write down **one** advantage and **one** disadvantage of using solar panels rather than gas to heat a house. [2]

c) The solar panel takes water from the bottom of the storage tank. The radiator takes water from the top of the storage tank. Explain why. [2]

d) The house is heated by pumping warm water from the storage tank through radiators. The diagram below shows the position of a radiator in a room. Describe how the radiator heats the room. Copy the diagram and add to it to illustrate your answer. [2]

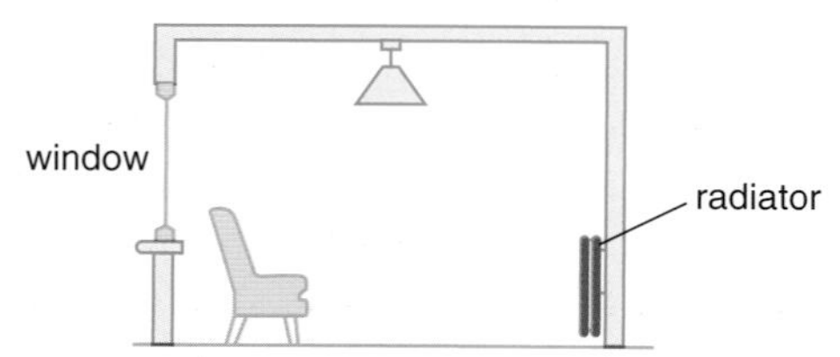

e) i) State two ways in which thermal energy transfers may be reduced from a house. [2]

ii) For each of your examples, explain why thermal energy transfers are reduced. [4]

19 The diagram shows a section of track near a railway station. The track at the station is slightly higher than the rest of the track. What advantage does this give? Explain your answer in terms of energy transfers. [3]

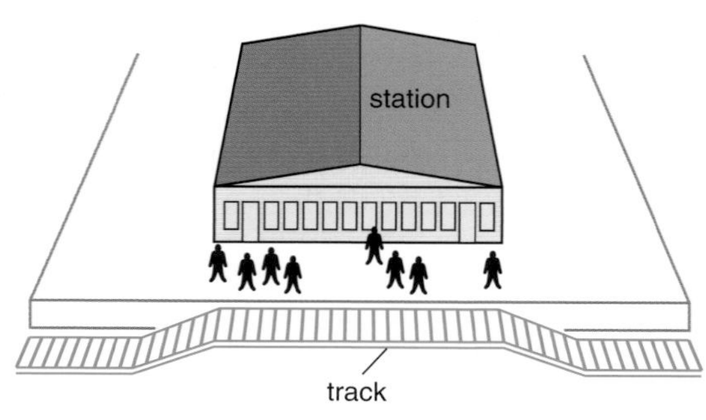

20 The Eiffel Tower in Paris is 300 m high and can be climbed using its 1792 steps. Jacques decided to climb the tower; he took 15 minutes to do it and he has a mass of 60 kg.

a) How much work did he do climbing the steps? [3]

b) Calculate his average power output during the climb. [3]

c) A croissant provides Jacques with 400 kJ of chemical energy when digested. How many croissants should Jacques eat for breakfast if his body is 20% efficient at transferring this energy into useful work? [3]

d) Discuss the energy transfers that occur as Jacques climbs the tower. [2]

EXTEND AND CHALLENGE

1 **a)** The diagram (below) shows a wind turbine with a rotor 100 m across, used to generate electricity.

i) Estimate the total area of the rotor blades.

ii) The wind has a velocity of 10 m/s. The force it can cause on a surface of 1 m^2 is 90 N. Calculate the work the wind can do on 1 m^2 in 1 s.
(work = force × distance)

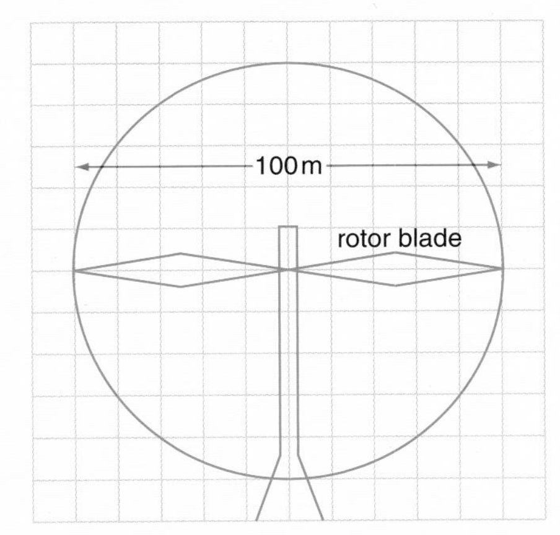

iii) Wind is used to turn the rotor blade. The effective work done on 1 m^2 of the blade is 50% of that calculated in **ii)**. State how much energy is transferred to the rotor in 1 s.

iv) Calculate the maximum electrical power, in watts, that would be generated by this wind turbine.

v) A conventional power station generates about 900 MW of electrical power (1 MW = 1 000 000 W). Calculate how many of these wind turbines would be needed to replace a power station.

b) In Chapter 4.6 you saw where a barrage could be built across the estuary of the River Severn. This would make a lake with a surface area of about 200 km^2 (200 million m^2).
The diagram below shows that the sea level could change by 9 m between low and high tide; but the level in the lake would only change by 5 m.

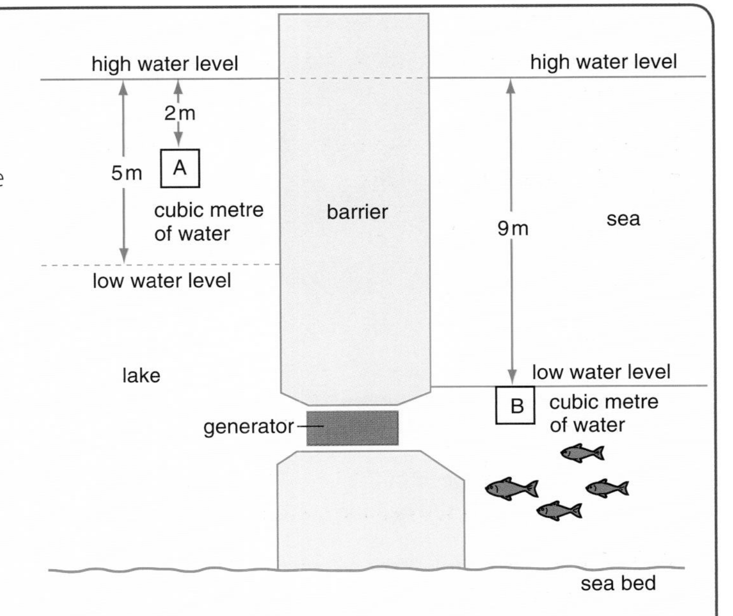

i) State the energy transfer that occurs when a cubic metre of water falls from position A to position B.

ii) A cubic metre of water has a mass of 1000 kg. The gravitational field strength is 10 N/kg. Calculate the force of gravity on a cubic metre of water.
(force = mass × acceleration)
Calculate the work that this can do as it falls from A to B.
(work = force × distance)

iii) Calculate the volume of water that could flow out of the lake between high and low tide.

iv) Use your answers to parts **ii)** and **iii)** to determine how much energy could be obtained from the tide. Assume that position A is the average position of a cubic metre of water between high and low levels in the lake.

v) The time between high and low tide is approximately 20 000 seconds (about 6 hours). Use this figure to estimate the power available from the dam. Give your answer in megawatts.

c) Explain the advantages and disadvantages of the wind turbine and the Severn Barrage as sources of power.

5 Solids, liquids and gases

A hot air balloon works on the principle that the density of hot air is less than the density of cold air. So, when hot air is put into the balloon from the burners, the cold air surrounding the balloon causes the balloon to rise.

CONSIDER • CALCULATE

1. Describe how the speed of the air molecules inside the balloon compares with that of the air molecules outside the balloon.
2. Describe how the spacing of the air molecules inside the balloon compares with that of the air molecules outside the balloon.
3. Do some research to find out:
   a) the average temperature of the air in a hot air balloon
   b) the volume of a balloon that lifts a load of 3000 kg – the typical mass of five passengers, the basket, the burners and the balloon itself.

By the end of this section you should:

- know and be able to use the relationship between density, mass and volume
- investigate density using direct measurements of mass and volume
- know and be able to use the relationship between pressure, force and area
- understand how the pressure in a gas or liquid acts equally in all directions
- know and be able to use the relationship between pressure difference, height, density and gravitational field strength
- be able to explain that heating a system will change the energy stored in that system
- be able to describe the changes that occur when a solid melts or a liquid boils or evaporates
- be able to describe the arrangements of atoms in solids, liquids and gases
- be able to obtain a temperature-time graph to show a constant temperature during a change of state
- know that the specific heat capacity of a substance is the energy required to change the temperature of 1 kg of a substance by 1°C
- be able to use the equation relating the change in thermal energy to mass, specific heat capacity and temperature change
- be able to investigate the specific heat capacity of materials including water
- explain how molecules in a gas have random motion and exert a pressure on the sides of their container
- understand why there is an absolute zero at – 273 °C
- be able to describe the Kelvin scale of temperature and be able to convert between the Kelvin and Celsius scales
- understand why an increase in temperature results in an increase in average speed of molecules
- know that kelvin temperatures of a gas are directly proportional to the average kinetic energies of its molecules
- be able to explain, for a fixed mass of gas, the qualitative relationship between: pressure and volume at constant temperature; pressure and kelvin temperature at constant volume
- be able to use the relationship between the pressure and kelvin temperature of a fixed mass of gas at constant volume
- be able to use the relationship between the pressure and volume of a fixed mass of gas at constant temperature.

5.1 Density

Figure 1.1 Eurofighter Typhoon F1.

The Eurofighter Typhoon is made from low-density materials to keep its weight to a minimum and allow it to climb or change direction quickly. It is made up mainly of the following materials: carbon-fibre composites such as Kevlar; lightweight alloys such as aluminium/lithium alloy; glass-reinforced plastics. Find out how much more expensive these materials are than aluminium and steel. Why would steel be a very poor choice for this purpose?

What is density?

A tree obviously weighs more than a nail. Sometimes, you hear people say 'steel is heavier than wood'. What they mean is this: a piece of steel is heavier than a piece of wood with the same volume.

To compare the heaviness of materials we use the idea of **density**. Density can be calculated using this equation:

$$\text{density} = \frac{\text{mass}}{\text{volume}} \quad \text{or} \quad \rho = \frac{m}{V}$$

The Greek letter ρ (rho) is used to represent density. Density is usually measured in units of kg/m^3. Some typical values of density are in Table 1.

Table 1

Material	Density in kg/m^3
gold	19 300
lead	11 400
steel	8000
aluminium	2700
glass	2500
Kevlar	1440
water	1000
lithium	500
cork	240
air hydrogen	1.3 0.09 } at standard temperature and pressure

TIP

Use the correct vocabulary – say 'steel is denser than water' rather than 'steel is heavier than water'.

Experiment: measuring density of a regular solid

You can calculate the density of an object by measuring its mass and calculating its volume. Figure 1.2 shows a regularly shaped piece of material. Its dimensions have been measured carefully with an accurate ruler. Its mass has been measured on an electronic balance and found to be 173.2 g.

Calculate the density in kg/m^3.

$$\begin{aligned}\text{volume} &= 0.101\text{ m} \times 0.048\text{ m} \times 0.013\text{ m} \\ &= 0.000\,063\text{ m}^3\end{aligned}$$

$$\begin{aligned}\text{density} &= \frac{\text{mass}}{\text{volume}} \\ &= \frac{0.1732\text{ kg}}{0.000\,063\text{ m}^3} \\ &= 2750\text{ kg/m}^3\end{aligned}$$

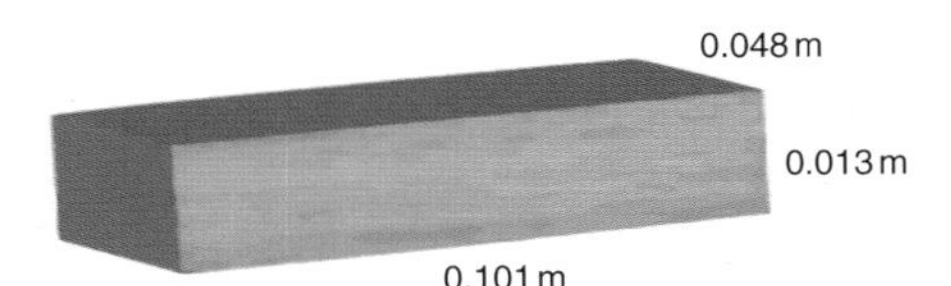

Figure 1.2

This method has limitations:

- A ruler can only measure to an accuracy of 0.1 cm, so the volume may not be calculated exactly.
- The shape of the block may not be regular.

Follow this procedure for yourself and determine the density of a block of wood or metal.

PRACTICAL

Make sure you can describe how to take appropriate measurements so that you can calculate the density of an object.

The volume of an irregularly shaped object can be calculated by immersing it in water. This is illustrated in Study Question 5 on page 161.

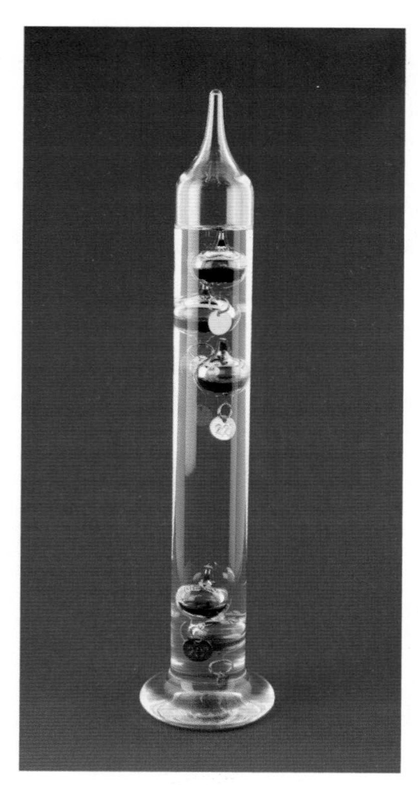

Figure 1.3 Galileo's thermometer (see Study Question 2).

Table 2

	Steel	Fibreglass
relative strength	40 000	50 000
density in kg/m^3	8 000	2 000
$\frac{\text{strength}}{\text{density}}$	5	25

TIP

School beakers are often marked in ml or millilitres. 1 ml = 1 cm^3

Using density in engineering

Knowing the density of materials is very important to an engineer. This allows the mass of building materials to be calculated.

Example. Calculate the mass of a steel girder, which is 9.0 m long, 0.10 m high and 0.10 m wide.

$$\text{volume of girder} = 9.0\,\text{m} \times 0.10\,\text{m} \times 0.10\,\text{m}$$

$$= 0.090\,\text{m}^3$$

$$\rho = \frac{m}{V}$$

$$\text{So } m = \rho \times V$$

$$= 8000\,\text{kg/m}^3 \times 0.090\,\text{m}^3$$

$$= 720\,\text{kg}$$

Steel is a very common building material because it is so strong. In aircraft construction, low-density and high-strength materials are used such as aluminium and carbon-fibre reinforced plastic. These materials are far too expensive to use for building houses or bridges.

Fibreglass is an important modern material. It is made by strengthening plastic with glass fibres. Table 2 allows us to compare steel and fibreglass. Fibreglass is actually a little stronger than 'mild' steel. This means a larger force is needed to break it. Fibreglass has a much lower density than steel. This makes it ideal for building small boats. Fibreglass cannot be used for large boats because it bends too much.

Experiment: measuring the density of a liquid

Method

1 Place an empty 100 ml measuring cylinder on to an electronic balance. Record the mass of the cylinder.

2 Now pour 20 ml of the liquid provided (e.g. water or paraffin) into the cylinder. Measure the mass of the liquid and cylinder.

3 Determine the mass of the liquid, by subtracting the mass of the empty cylinder.

4 Make a suitable table to record the volume and mass of the liquid.

5 Add 20 ml more of the liquid, and repeat the process above. Continue adding 20 ml at a time until you have 5 pairs of values of volume and mass.

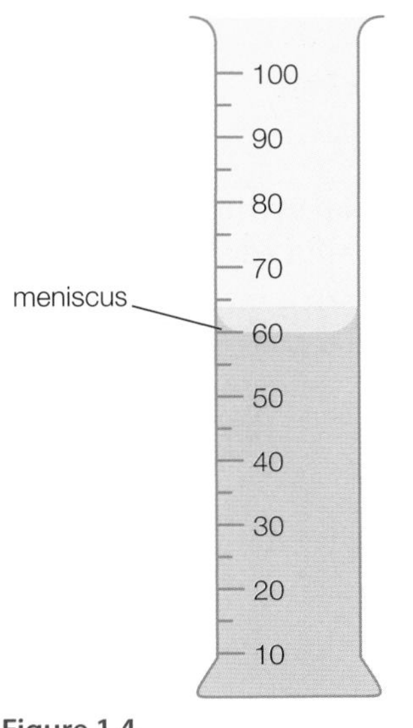

Figure 1.4

Calculating the density

1 Use each set of readings to calculate the density of the liquid in g/cm^3. Remember that 1 ml is the same as 1 cm^3.

2 Which of your five readings is likely to be the most accurate? Explain your answer.

3 Calculate the mean of your five results.

4 You can calculate your density in kg/m^3 by multiplying your answer by 1000. Explain where the factor of 1000 comes from.

STUDY QUESTIONS

1 A student wrote the following sentence in an exam paper; read it and correct any mistakes you see.
'A cork floats in a pond because it is lighter than water; a stone sinks because it is too heavy to float in water.'

2 Look at the photograph of Galileo's thermometer in Figure 1.3. The density of water reduces as it warms up. Each float has a slightly different mass. As the temperature rises, do the floats rise or sink? Explain your answer.

3 **a)** Describe a method to determine the density of a regular object. Explain what measurements you would take.
b) Evaluate your method and comment on its accuracy.

4 Copy Table 3 and fill in the gaps.

Table 3

Material	Volume in m^3	Mass in kg	Density in kg/m^3
osmium	0.02	450	
potassium	0.000 02	0.017	
titanium	0.5		4500
water		3000	1000
alcohol		3200	800
radium	0.35		5000

5 Carole is a geologist. She wants to work out the density of a rock. First she measures the mass of the rock, then she puts it into a beaker of water to work out its volume.
a) Use the diagrams to calculate the rock's volume. Give your answer in m^3.
b) Now calculate the rock's density in kg/m^3.

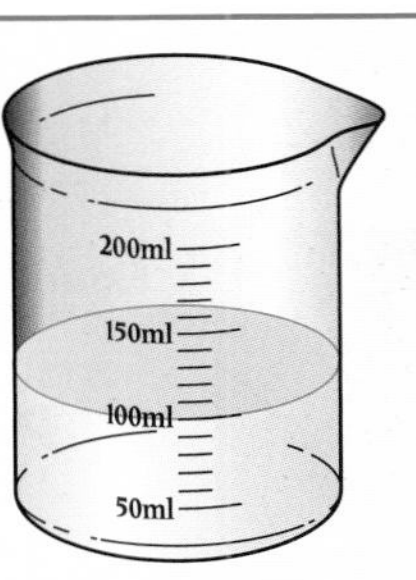

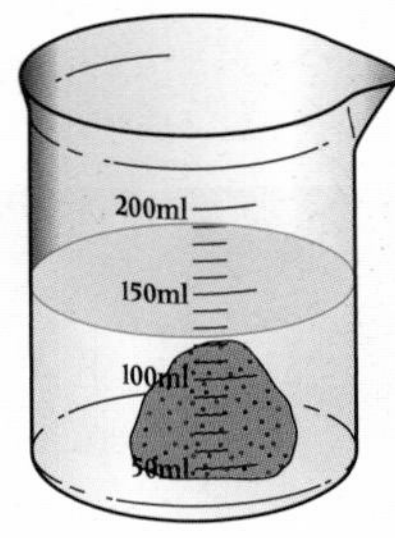

Figure 1.5

- Mass of rock = 0.09 kg
- $1 \text{ ml} = \frac{1}{1\,000\,000} \text{ m}^3$

6 Use Table 1 to calculate the volume of:
a) 1000 kg of aluminium
b) 100 kg of cork.

7 White dwarf stars are extremely dense. They have a density of about 100 million kg/m^3. If you had a matchbox full of material from a white dwarf, determine what its mass would be. (Hint: A matchbox has a volume of about 0.000 05 m^3.)

8 An aluminium/lithium alloy is stronger than aluminium. What other advantages does this alloy have for building aircraft? Use Table 1 to help you with your answer.

9 **a)** Explain why aluminium and carbon-fibre reinforced plastic are used to build aeroplanes.
b) In Table 2, the last row is headed 'strength/density'. Explain why this is an important ratio in aircraft construction.

5.2 Pressure

You always choose a sharp knife when you want to chop up meat or vegetables ready for cooking. A sharp knife has a very thin edge to the blade. This means that the force which you apply is concentrated into a very small area. We say that the pressure under the blade is large.

Figure 2.2 Why should this woman take her stiletto heels off before walking on a wooden floor?

Figure 2.1

Figure 2.3 Why does the jagged edge of the saw make it easier to cut through wood?

TIP

Use the correct vocabulary – pressure, weight and area – to answer the questions in the photo captions.

Pressure points

$$\text{pressure} = \frac{\text{force}}{\text{area}} \quad \text{or} \quad P = \frac{F}{A}$$

The unit of pressure is **N/m²** or **pascal (Pa)**; 1 N/m² = 1 Pa.

Figure 2.4 shows a physics teacher lying on a bed of nails. How can he lie there without hurting himself? You all know that nails are sharp and will make a hole in you if you tread on one. The teacher has spread his body out, so that it is supported by a lot of nails. The area of nails supporting him is large enough for it not to hurt (too much!).

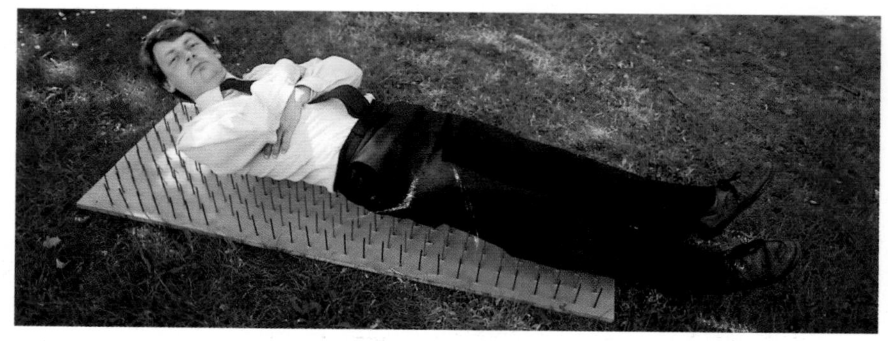

Figure 2.4 Why does this bed of nails not cause pain? (Don't try this at home!)

Figure 2.5 Why do these snow shoes stop the man sinking into the snow?

When an engineer designs the foundations of a bridge, he or she must think about pressure. In the example shown in Figure 2.6, the bridge will sink into the soil if it causes a pressure greater than 80 kN/m² (80 000 N/m²). Calculate the minimum area the foundations must have to stop this happening.

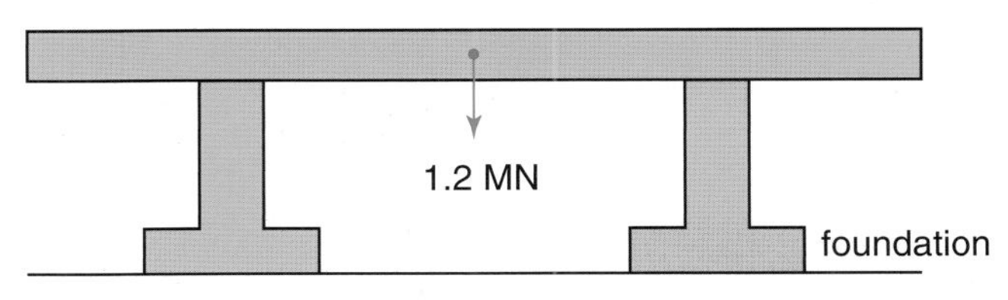

Figure 2.6

The bridge has a weight of 1.2 MN, so each pillar will support 0.60 MN (600 000 N).

$$\text{pressure} = \frac{\text{force}}{\text{area}}$$

$$\text{So area} = \frac{\text{force}}{\text{pressure}}$$

$$= \frac{0.60\ \text{MN}}{80\ \text{kN/m}^2}$$

$$= \frac{600\ 000\ \text{N}}{80\ 000\ \text{N/m}^2}$$

$$= 7.5\ \text{m}^2$$

STUDY QUESTIONS

1 a) Describe how a drawing pin takes advantage of high pressure at one end and low pressure at the other end.
b) Give two examples in sport or other activities in which something is used to reduce pressure.
c) Give two examples of a device which is used to increase pressure.

2 Five girls stand on one heel of their shoes. The weights of each girl and the area of their heel is shown in the table. Which girls put:
a) the greatest,
b) the least pressure on the floor?

	Weight of girl in N	Area of heel in cm^2
Amanda	800	4.0
Becky	600	2.0
Chloe	600	4.0
Debbie	500	2.5
Eli	400	1.0

3 a) When a pressure of 2.3 N/mm^2 is applied to your skin it hurts. Each nail in the bed in Figure 2.4 has a point with an area of 1.5 mm^2. Determine how much weight the teacher can put on each nail without being hurt.
b) The teacher's weight is 690 N. Calculate how many nails (at least) he should put in the bed.
c) Explain why he must be very careful getting on and off the bed.

4 Use the idea of pressure to explain why:
a) a saw has a jagged edge,
b) a knife cuts better when sharpened,
c) you should not walk over a wooden floor wearing stiletto heels,
d) Canadians wear snow shoes to get about in winter,
e) a tank has caterpillar tracks on it.

5 The diagram shows three boxes, A, B and C. Calculate which box would you use and how would you position it to exert:
i) the greatest pressure on the ground
ii) the smallest pressure on the ground.

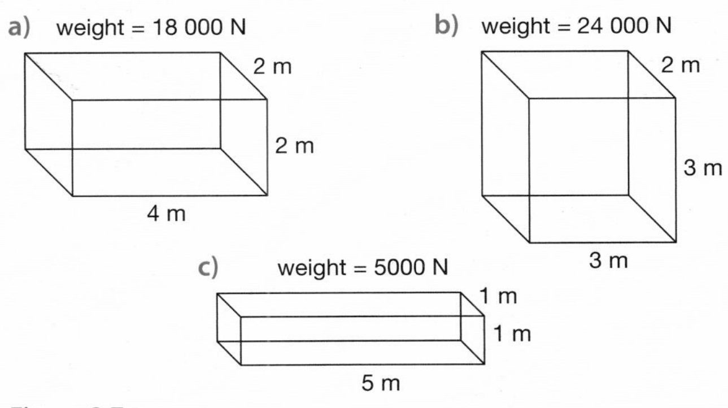

Figure 2.7

5.3 Pressure in liquids

Figure 3.1 This diver is 20 m below the surface of the sea.

This scuba diver is enjoying warm tropical waters, where she can see brightly coloured fish in the coral reefs. Although the water looks clear and safe, there are hidden dangers. One danger is the pressure of the water itself. For every 10 m below the surface of the sea, there is an increase in pressure equal to 1 atmosphere pressure. When the pressure increases on your body gases dissolve in your blood, which can be very painful and dangerous to your health. You have to take care to change depth slowly, as rapid changes of pressure can cause injuries to your ears or lungs.

Increase of pressure with depth

The pressure that acts on a diver depends on the weight of water above her. As she goes deeper, the height and weight of the water above her increase. This increases the pressure on the diver.

For every 10 m below the surface of the sea, there is an increase of about 1 atmosphere pressure. The diver in Figure 3.1 is experiencing a pressure of 3 atmospheres at a depth of 20 m.

The diver feels a bigger pressure under 10 m of sea water than under 10 m of fresh water. This is because sea water has a higher density than fresh water. So a column of sea water would have a greater mass and therefore greater weight than an identical column of fresh water.

The pressure at the bottom of a column of liquid depends on three things:

- height of the column, h in metres, m
- density of the liquid, ρ in kg/m^3
- gravitational field strength, g (g = 10 N/kg).

The pressure due to a column of liquid can be calculated using the equation:

$$P = h \rho g$$

pressure = height of the column × density of the liquid × gravitational field strength where pressure is in pascals, Pa, or N/m^2.

Measuring atmospheric pressure

Figure 3.2 shows how we can measure atmospheric pressure using a mercury barometer. There is a vacuum above the mercury, so atmospheric pressure is large enough to support a column of mercury 76 cm long. The pressure at **x** under the column of mercury is the same as the pressure at **y** due the atmosphere. The density of mercury is 13 600 kg/m^3. We can now calculate atmospheric pressure:

$$\begin{aligned} P &= h\rho g \\ &= 0.76 \times 13\,600 \times 10 \\ &= 103\,000\ N/m^2 \\ &= 100\ kN/m^2 \end{aligned}$$

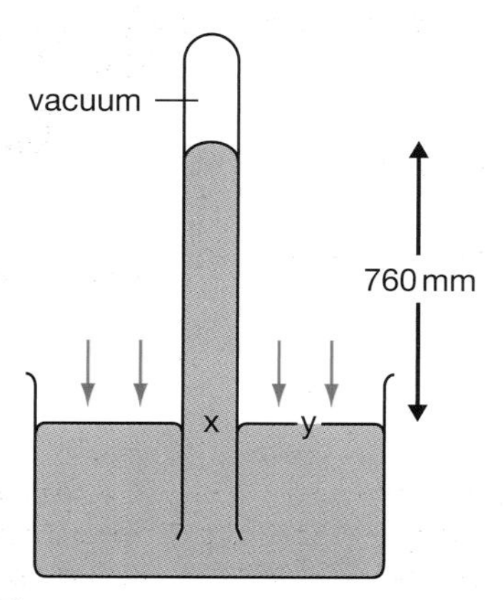

Figure 3.2 A mercury barometer

This is a large pressure, but we tend not to notice this as the pressures inside and outside out bodies are the same. However, we do notice pressure changes, when we drive up or down a hill for example.

TIP

The answer above was rounded to 2 significant figures to make the lowest number of significant figures in the data provided.

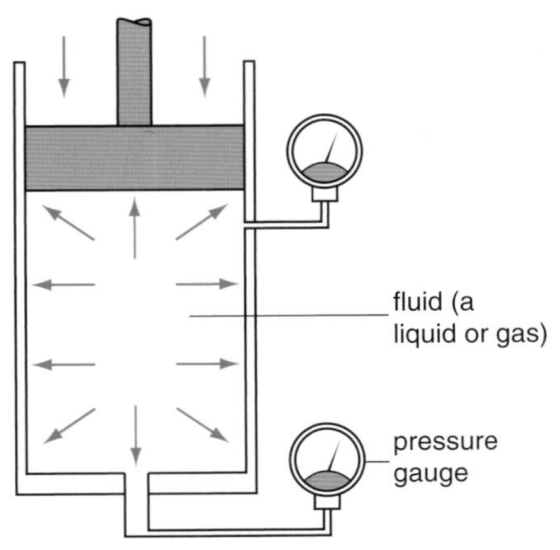

Figure 3.3 The pressure in a fluid acts equally in all directions

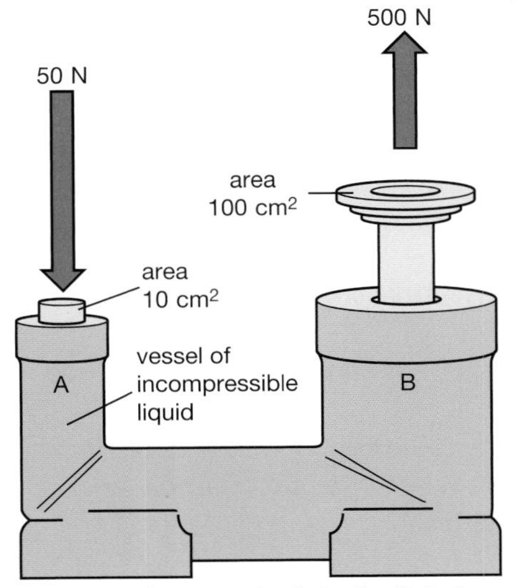

Figure 3.4 The principle of a hydraulic jack.

Transmitting pressures

When you hit a nail with a hammer the pressure is transmitted downwards to the point. This happens only because the nail is rigid.

When a ruler is used to push a lot of marbles lying on a table, they do not all move along the direction of the push. Some of the marbles give others a sideways push. The marbles are behaving like a fluid.

In Figure 3.3 you can see a cylinder of fluid that has been squashed by pushing a piston down. The pressure increases everywhere in the fluid, not just next to the piston. The fluid is made up of lots of tiny particles, which act rather like the marbles to transmit the pressure to all points.

Hydraulic machines

We often use liquids to transmit pressures. Liquids can change shape but they hardly change their volume when compressed. Figure 3.4 shows how a hydraulic jack works.

A force of 50 N presses down on the surface above A. The extra pressure that this force produces in the liquid is:

$$P = \frac{F}{A}$$
$$= \frac{50\,\text{N}}{10\,\text{cm}^2}$$
$$= 5.0\,\text{N/cm}^2$$

The same pressure is passed through the liquid to B. So the upwards force that the surface B can provide is:

$$F = P \times A$$
$$= 5.0\,\text{N/cm}^2 \times 100\,\text{cm}^2$$
$$= 500\,\text{N}$$

With the hydraulic jack you can lift a load of 500 N by applying a force of only 50 N. Figure 3.5 shows another use of hydraulics.

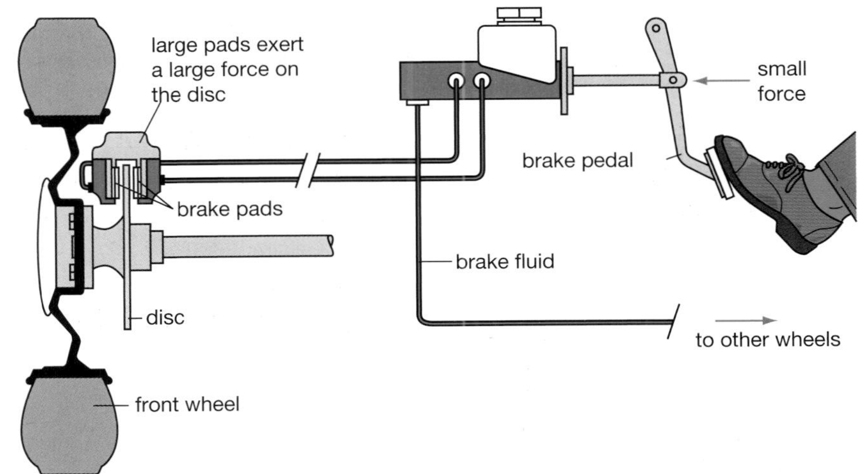

Figure 3.5 Cars use a hydraulic braking system. The foot exerts a small force on the brake pedal. The pressure created by this force is transmitted by the brake fluid to the brake pads. The brake pads have a large area and exert a large force on the wheel disc. The same pressure can be transmitted to all four wheels.

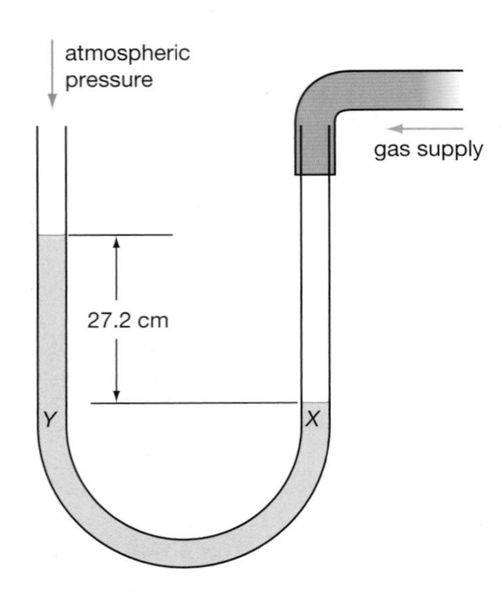

Figure 3.6 A manometer.

The manometer

Figure 3.6 shows how you can use a **manometer** to measure the pressure of a gas supply. The two points *X* and *Y* are at the same level in the liquid. This means that the pressures at *X* and *Y* are the same.

Pressure at X = gas supply pressure.

Pressure at Y = atmospheric pressure + pressure due to 27.2 cm of water.

So the gas supply pressure is greater than atmospheric pressure, by an amount equal to the pressure due to a column of water 27.2 cm high.

STUDY QUESTIONS

1 a) Name the three factors that affect the pressure at a point under the sea.
   b) Explain why the pressure is greater under a depth of 5 m of sea water, than it is under a depth of 5 m of fresh water.
   c) Explain why atmospheric pressure is lower at the top of a mountain, than it is at sea level.

2 Explain why a liquid exerts the same pressure in all directions.

3 Explain why the pressure on the diver in Figure 3.1 is equivalent to three atmosphere pressures.

4 This question refers to the manometer in Figure 3.6.
   a) On another day the manometer shows a level difference of 29.8 cm. Explain what has happened to the gas pressure.
   b) Calculate the gas pressure for the supply shown in the diagram. [The density of water is 1000 kg/m^3].
   c) A second manometer has oil in it, with a density of 800 kg/m^3. It is attached to the gas supply (with the pressure shown in Figure 3.6). Calculate the level differences between the two columns.

5 a) Use the equation $P = h\rho g$ to calculate the pressure acting on a diver at depths of **i)** 10 m, **ii)** 30 m. The density of the water is 1000 kg/m^3. You will need to include the pressure of the atmosphere at the surface of the sea, which is 100 000 N/m^2.
   b) Calculate the force exerted by the pressure on the diver's mask at a depth of 30 m. His mask has an area of 0.02 m^2.
   c) Explain why the diver's mask is in no danger of breaking under this pressure.

6 Figure 3.7 shows the principle of a hydraulic jack. Two cylinders are connected by a reservoir of oil.
A 200 N weight resting on piston A can be used to lift a larger load on piston B.
   a) Calculate the extra pressure at x, due to the 200 N weight.
   b) Explain why the pressure at y is the same as the pressure at x.
   c) Why would the jack not work if the oil were replaced by a gas?
   d) Calculate the size of the load, *W*, which can be lifted.
   e) If you need to lift *W* by 0.05 m, calculate how far you need to move piston A.

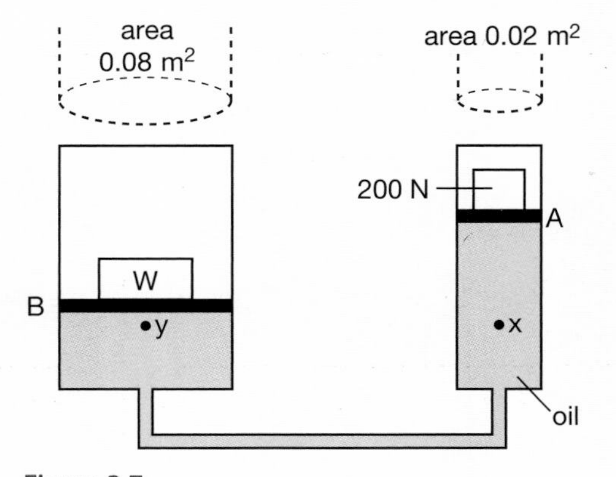

Figure 3.7

5.4 Solids, liquids and gases

Figure 4.1

Ice, water and steam are three different **states** of the same substance. We call these three states solid, liquid and gas.

- **Solid:** In a solid, the atoms (or molecules) are packed in a regular structure. The atoms cannot move out of their fixed position, but they can vibrate. The atoms are held close together by strong forces, so it is difficult to change the shape of a solid.
- **Liquid:** The atoms (or molecules) in a liquid are also close together. The forces between the atoms keep them in contact, but atoms can move from one place to another. A liquid can flow and change shape to fit any container. Because the atoms are close together, it is very difficult to compress a liquid.
- **Gas:** In a gas, the atoms (or molecules) are separated by relatively large distances. The forces between the atoms are very small. The atoms in a gas are in constant, random motion. A gas can expand to fill any volume, and can be compressed.

a)

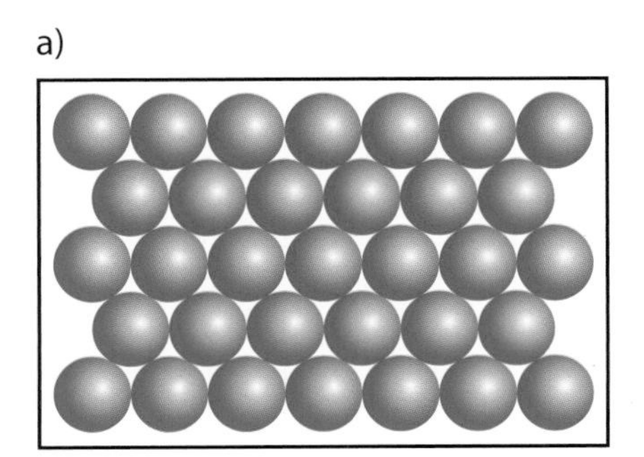

b)

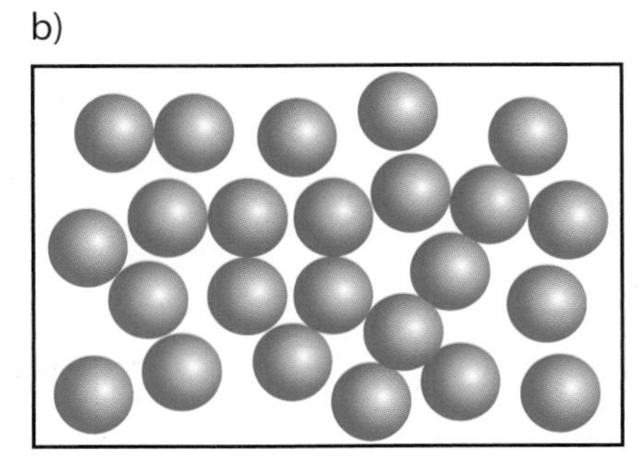

c)

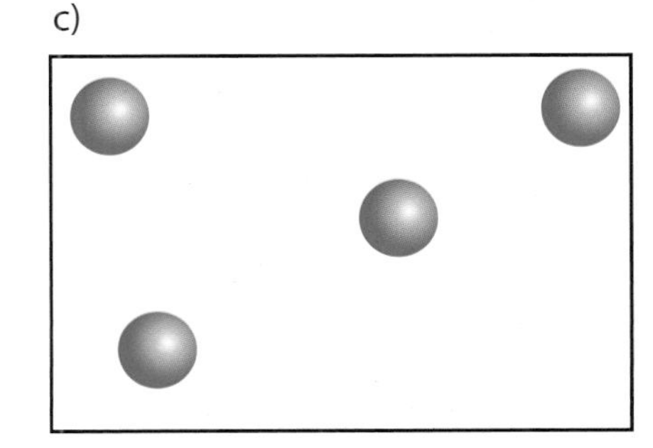

Figure 4.2 The particle arrangement in a) a solid, b) a liquid and c) a gas.

Table 1 The densities of some solids, liquids and gases. (The gases are at room temperature and pressure.)

Material	Density in kg/m³
Lead (solid)	11 400
Glass (solid)	2500
Water (liquid)	1000
Lithium (solid)	500
Cork (solid)	240
Air (gas)	1.3
Hydrogen (gas)	0.09

Density of solids, liquids and gases

The densities of solids and liquids are usually much higher than the density of gases. In solids and liquids, the atoms are closely packed together, so there is a lot of mass in a small volume. In gases the atoms are much further apart, so there is less mass in the same small volume.

- Lead has a much higher density than lithium because the atoms of lead have a much greater mass than lithium atoms. (The lead atoms are only slightly larger than lithium atoms.)
- The density of air is larger than the density of hydrogen, because the nitrogen and oxygen molecules in the air have a greater mass than hydrogen molecules.

Internal energy

Energy is stored inside a system by the particles (atoms or molecules) that make up that system. This is called **internal energy**. Internal energy is the total kinetic and potential energy of all the particles that make up the system.

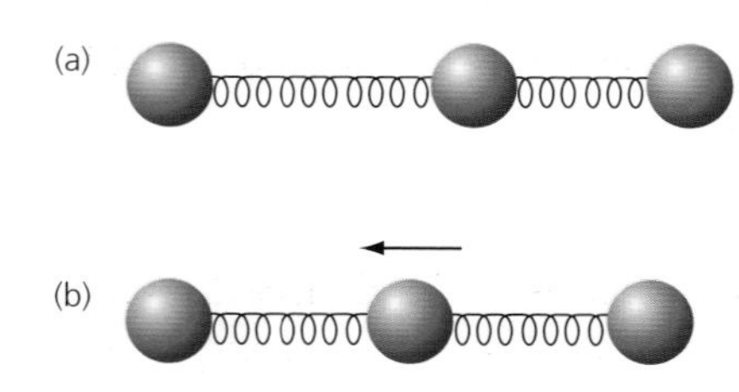

Figure 4.3

We can use a model of several balls and springs to help us understand the nature of internal energy in a solid. The balls represent the atoms and the springs represent the forces or 'bonds' that keep the atom in place.

In Figure 4.3(a), the middle atom has been displaced to the right. Now potential energy is stored in the stretched bond. In Figure 4.3(b), the atom has kinetic energy as it moves to the left.

Heating

Heating changes the energy stored within a system by increasing the energy store of the particles that make up the system.

- Heating can increase the temperature of a system. For example, when a gas is heated, the atoms (or molecules) move faster and the kinetic energy store of the atoms increases.
- Heating a system can also cause a change of state, for example when a solid melts to become a liquid. Usually when a solid melts, there is a small increase in volume as the solid turns to liquid. The atoms increase their separation and there is an increase in the potential energy stored, so the internal energy increases.

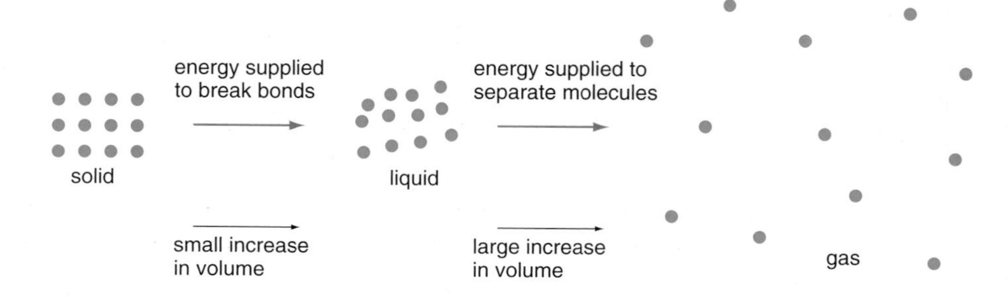

Figure 4.4 Change of state as thermal energy is supplied.

Changes of state

We use these terms to describe changes of state.

- **Melting** occurs when a solid turns to a liquid. The internal energy of the system increases.
- **Freezing** occurs when a liquid turns to a solid. The internal energy of the system decreases.
- **Boiling** or **evaporation** occurs when a liquid turns to a gas. The internal energy of the system increases.
- **Condensation** occurs when a gas turns to a liquid. The internal energy of the system decreases.
- **Sublimation** occurs when a solid turns directly into a gas. The internal energy of the system increases. Sublimation is rare. An example is carbon dioxide (CO_2). Solid CO_2 (dry ice) turns directly into the gas CO_2 missing out the liquid state at normal atmospheric pressure.
- **Deposition** is the reverse process of sublimation - a gas turns to a solid. The internal energy of the system decreases.

A change of state of a substance is a **physical change**. The change does not produce a new substance, and the process can be reversed. For example, a cube of ice from the freezer can be allowed to melt into water. The water can be put back into its container and then into the freezer. The water will freeze back into ice. No matter what its state, water or ice, the mass is the same. So when a substance changes state, the mass is conserved. This is because the total number of particles (atoms or molecules) stays the same. The internal energy of the system decreases.

PRACTICAL

Make sure you are familiar with this practical.

A cooling curve temperature against time

When you leave a cup of coffee to cool, the temperature drops gradually, as shown in Figure 4.5. When something is hot, its temperature drops quickly. When water is close to room temperature, its temperature drops more slowly.

In this practical, you investigate what happens when some stearic acid cools and solidifies.

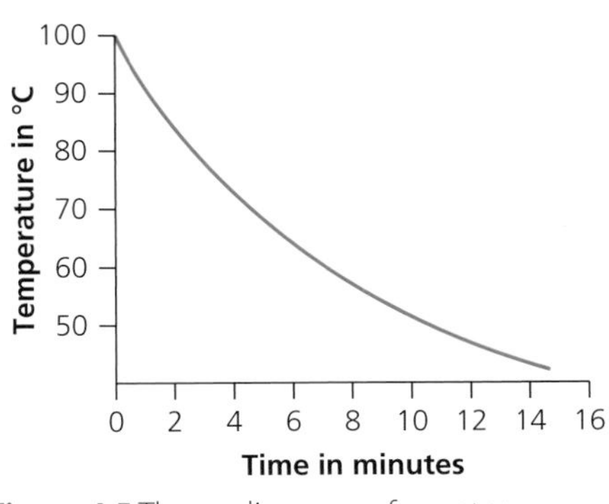

Figure 4.5 The cooling curve for water.

Method

1. Put on your safety glasses.
2. Set up a tripod and gauze. Fill a beaker of water with 150 ml of water.
3. Put your test tube of stearic acid into a clamp supported by a retort stand. Include a cotton wool or mineral fibre plug in the tube neck to reduce fumes.
4. Construct a table to record pairs of readings of the temperature of the stearic acid and the time (recorded in minutes).
5. Light the Bunsen burner and bring the water to the boil. Place your stearic acid into the water. Do not heat the test tube directly in the Bunsen flame.
6. Put your thermometer into the stearic acid and make sure it reaches a temperature of 100 °C.
7. Get a stopwatch ready. Start the stopwatch and quickly move the stearic acid back to the retort stand.

8 Record the temperature every minute until the temperature reaches 50 °C.

9 Plot a graph of the temperature against time. You should produce a graph similar to that shown in Figure 4.7.

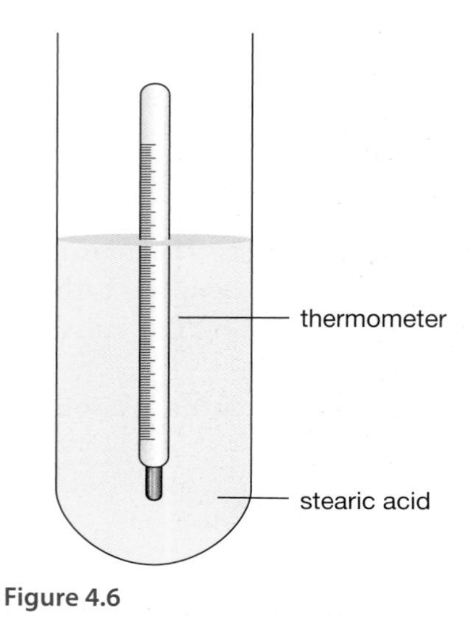

Figure 4.6

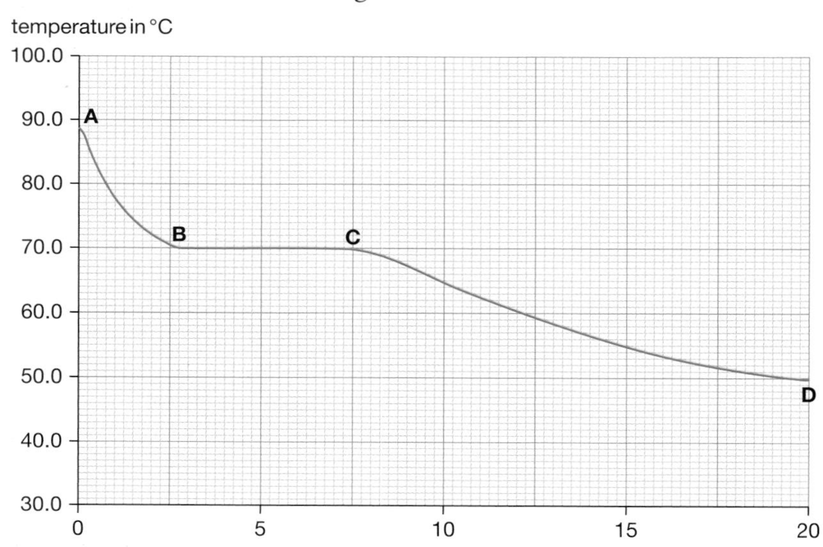

Figure 4.7 Cooling curve for stearic acid. The horizontal line indicates a change of state.

Questions

1 Explain why the stearic acid cools more quickly over the region AB than it does over the region CD, as shown in Figure 4.7.

2 Over the region BC, the stearic acid is transferring thermal energy to the surroundings. Explain why the temperature remains constant.

STUDY QUESTIONS

1 Draw diagrams to show the arrangement of the particles in each of the three states of matter.

2 Explain why gases are less dense than liquids and solids.

3 **a)** State what is meant by a change of state of a substance.
b) Give two examples of changes of state.

4 Which of the following changes are physical changes?
- melting snow
- burning a matchstick
- breaking a matchstick
- boiling an egg
- mixing salt and sugar together.

5 **a)** What happens to the internal energy of a system when the system is heated?
b) How is it possible to heat a system without the temperature of the system increasing?

6 **a)** When we exercise, we sweat. The sweat evaporates from our skin. Why does sweating help us stay cool?
b) Even on a warm day, having wet skin can soon make you feel cold. Explain why.

7 A solid is heated at a constant rate until it becomes a gas. Figure 4.8 shows how the temperature of the substance increases with time.
a) Explain why the temperature of the substance remains constant over the periods:
i) B to C
ii) D to E
b) Determine the temperature at which the substance melts.

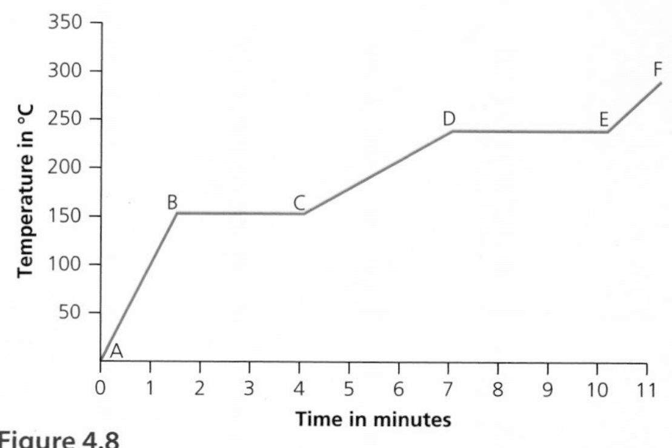

Figure 4.8

5.5 Specific heat capacity

Figure 5.1

When we exercise, energy is transferred from our chemical store to a thermal store. If we get too hot, our bodies cannot cope and we become ill. We are kept safe by two properties of water. First, we sweat, and evaporating water cools us. Secondly, water has a very high specific heat capacity – this means that water requires a lot of thermal energy to warm it up.

When the temperature of a system is increased by supplying energy to it, the increase in temperature depends on:

- the mass of the substance heated
- what the substance is made of
- the energy put into the system.

Water needs much more energy to increase its temperature by 1 °C than the same mass of concrete. This also means that when water cools by 1 °C, it gives out more energy than the same mass of concrete cooling by 1 °C.

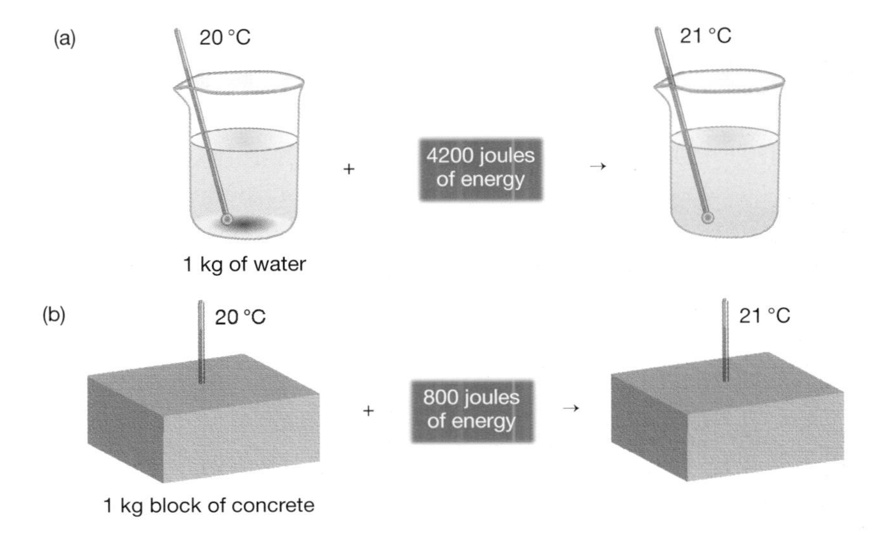

Figure 5.2 (a) 4200 joules of energy are needed to increase the temperature of 1 kg of water by 1 °C. **(b)** 800 joules of energy are needed to increase the temperature of 1 kg of concrete by 1 °C.

We say that the specific heat capacity of water is 4200 joules per kilogram per degree Celsius (J/kg °C).

The specific heat capacity of a substance is the amount of energy required to raise the temperature of one kilogram of the substance by one degree Celsius.

To calculate the change in thermal energy in a substance we use the equation:

$$\Delta Q = mc\,\Delta T$$

change in thermal energy = mass × specific heat capacity × temperature change

Where change in thermal energy is in joules, J

mass is in kilograms, kg

specific heat capacity is in joules per kilogram per degree Celsius, J/kg °C

temperature change is in degrees Celsius, °C.

Example. A domestic hot water tank contains 200 kg of water. Calculate how much energy is required to warm the tank from 15 °C to 45 °C.

Temperature rise = 45 °C − 15 °C = 30 °C

Energy supplied = increase in thermal energy of the water

$$\begin{aligned}\Delta Q &= mc\,\Delta T\\ &= 200 \times 4200 \times 30\\ &= 25\,200\,000\ \text{J}\\ &= 25\ \text{MJ}\end{aligned}$$

TIP

Note that the final answer to the calculation has been rounded to 2 significant figures to match the accuracy of the data.

The table gives some examples of specific heat capacities for various substances at 20 °C.

Table 1 Examples of specific heat capacities *(c)*

Substance	Specific heat capacity J/kg °C
Water	4200
Alcohol	2400
Ice	2100
Dry air	1000
Aluminium	880
Concrete	800
Glass	630
Steel	450
Copper	380

The specific heat capacity of water

Water has a very high specific heat capacity. This means that 1 kg of water requires a lot of energy to heat it up, and a lot of energy must be transferred from the water when it cools down. This high specific heat capacity is very important.

- We are made mostly of water. The high specific heat capacity of water means that our body temperature does not increase too much when we exercise, or cool too quickly when we go outside on a cold day.
- Water is used for keeping many homes warm. A house central heating system pumps hot water around the house. The hot water transfers a lot of energy as it flows through radiators. If water had a low specific heat capacity, it would cool down before it got to some of the radiators in your house.

PRACTICAL

Make sure you are familiar with this practical.

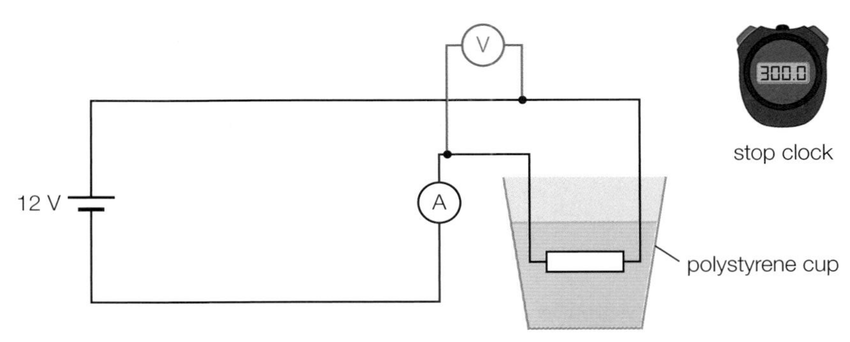

Figure 5.3

Method

TIP

Most school beakers are marked in ml. 1 ml is the same volume as 1 cm^3.

1 Measure 100 ml of water with a beaker and pour it into an insulated cup (such as a polystyrene one). Remember that 1 ml of water has a mass of 1 g.

2 Set up the electrical circuit shown and place the heater in the water. Do not switch the circuit on.

3 Stir the water with the thermometer and record the temperature. This is T_1.

4 Switch on the heater for 5 minutes, then switch it off. Stir the water and record the new temperature. This is T_2.

5 Record the voltmeter and ammeter readings during the heating.

Analysing the results

1 Calculate the temperature rise during the heating: $\Delta T = T_2 - T_1$.

2 Use the following equation to determine the energy transferred to the water by the heater:

$$E = VIt$$

3 Calculate the specific heat capacity in J/kg °C using the equation:

$$\Delta Q = mc\Delta t$$

TIP

In the two equations, E and ΔQ are the same size.

Questions

1 Explain why it is important to use an insulated beaker.

2 Discuss the errors that could occur in this experiment. Does the heater require energy to warm it up?

TIP

An unsmoothed d.c. current varies with time. A d.c. ammeter is calibrated to measure a constant current.

Experimental note

One error that can easily occur here is the measurement of voltage and current. If you use d.c. from a power pack, the current is likely to be unsmoothed. This means that the meters (which are designed to measure smooth d.c. from a cell) will be inaccurate.

If the school has a.c. meters, you could use a.c. If your school had calibrated heaters, you could read off their power e.g. 36 W.

STUDY QUESTIONS

1 **a)** Define the term specific heat capacity.
b) State the unit of specific heat capacity.

In these questions you will need to refer to the information in the table on page 172 (specific heat capacities table).

2 **a)** A night storage heater contains 60 kg of concrete. A heater embedded in the concrete heats the concrete up from 12 °C to 37 °C. How much energy is transferred to the concrete?
b) A heater supplies 4180 J to a block of copper of mass 0.5 kg. Calculate the temperature rise of the block.
c) A heater supplies 21 120 J to a block of aluminium. The temperature of the block rises from 18 °C to 34 °C. Calculate the mass of the block.

3 **a)** In Figure 5.4, a block of tin is heated from a temperature of 20 °C to 65 °C. The mass of the block is 1.2 kg. Use the reading on the joulemeter to calculate the specific heat capacity of tin.
b) Give two reasons why the specific heat capacity calculated in part **a)** is likely to be inaccurate.

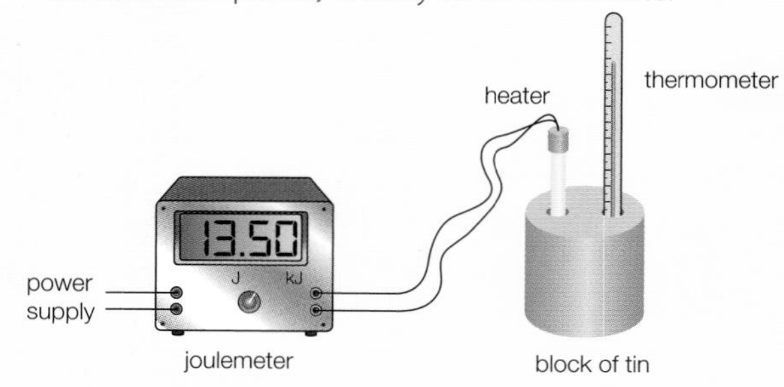

Figure 5.4

4 An electric kettle has a power rating of 2.0 kW. The kettle is filled with 0.75 kg of water.
a) Calculate the energy required to warm the water from 20 °C to 100 °C.
b) Calculate how long it takes the kettle to bring the water to the boil at 100 °C from 20 °C.

5.6 Ideal gas molecules

When we walk past pipes in a factory or refinery we assume that they are all perfectly safe. But someone will have calculated that they are. What factors need to be taken into account when checking the safety of these pipes?

Figure 6.1 What factors affect the pressure of the gases in these pipes?

The particle model of gases

As a result of studying the behaviour of gases, we have built up a model (or theory) to help us understand, explain and predict the properties of gases. This is called the particle model or kinetic theory of gases. The main points of the model are listed here.

- The particles in a gas (molecules or atoms) are in a constant state of random motion.
- The particles in a gas collide with each other and the walls of their container without transferring any of their kinetic energy.
- The temperature of the gas is related to the average kinetic energy of the molecules.
- As the kinetic energy of the molecules increases, the temperature of the gas increases.

Gas pressure

When the particles of a gas collide with a wall of their container, the particles exert a force on the wall. Figure 6.2 shows three particles bouncing off a container wall. Each particle exerts a force on the wall at right angles to it.

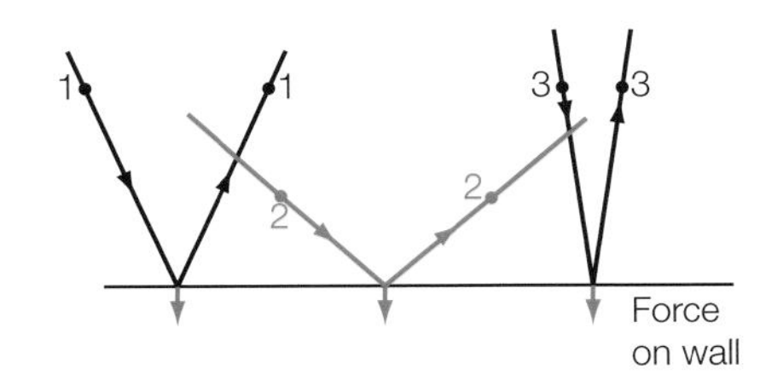

Figure 6.2

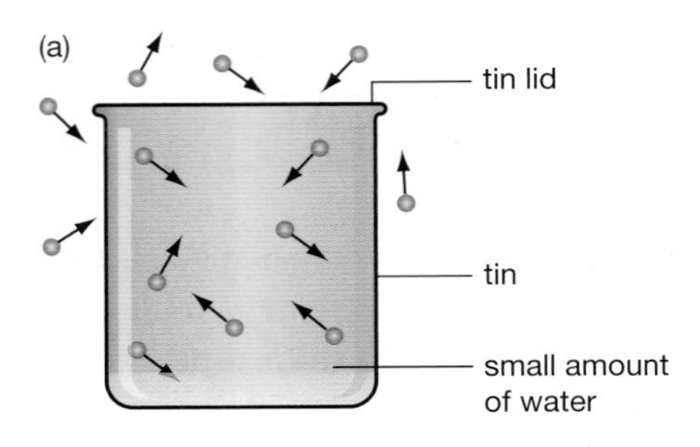

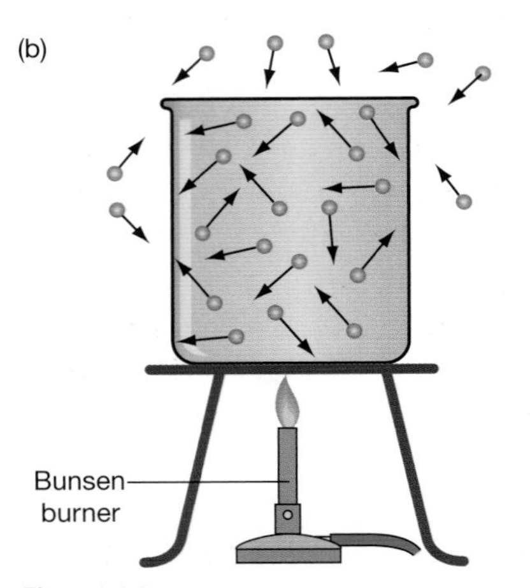

Figure 6.3

The pressure inside a container of gas with a fixed volume is increased when the temperature of the gas is increased. When this happens, the average speed of the particles in the gas increases. This means that the particles hit the walls of the container with a greater force, and the particles hit the walls more frequently, so the pressure increases.

Demonstrating gas pressure

Your teacher might demonstrate the effect of gas pressure as follows:

- Take a tin with a close fitting lid and put a small amount of water in it. Press the lid firmly in place.
- Then put the tin on a tripod and heat with a Bunsen burner. After a while, the lid flies off.

So why does the lid fly off?

- In Figure 6.3(a), the molecules inside the tin move at the same speed as the molecules outside the tin. There is no resultant force on the tin lid.
- In Figure 6.3(b), two things have happened. As the tin warms up, some water evaporates, so the number of molecules of gas inside the tin increases. The molecules travel faster as the temperature rises (shown with longer arrows on the molecules). The molecules inside the tin exert a force on the lid large enough to blow it off.

Linking the pressure of a gas to its temperature

Figure 6.5 shows apparatus that can be used to investigate how the pressure of a gas changes as it is heated up. To make it a fair test, the volume of the gas is kept constant. The pressure of the gas is measured at several different temperatures with a pressure gauge as the temperature of the water bath is raised.

The results of such an experiment are shown in Figure 6.4. As the temperature is raised from 0 °C to 100 °C the pressure rises steadily from about 100 kPa to about 137 kPa. We are able to draw a straight line through these results. If this line is extended below 0 °C you can predict what will happen to the pressure at lower temperatures. You can see that the pressure will be about 50 kPa at −136 °C; this is half of the pressure at 0 °C. At a temperature of −273 °C the gas pressure has reduced to zero.

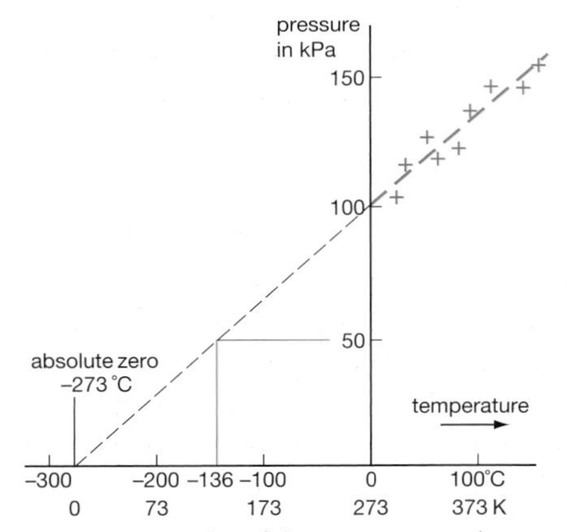

Figure 6.4 Results of the experiment shown in Figure 6.5.

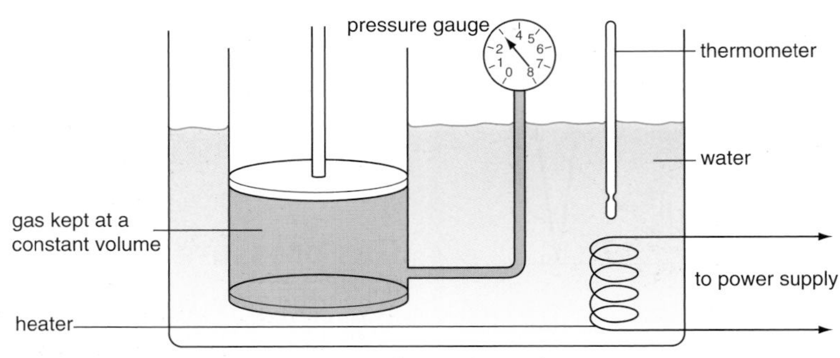

Figure 6.5 Investigating how the pressure of a gas depends on temperature.

We can begin to make sense of these results when we think about the particle model of gases described above. The theory tells us that molecules are constantly on the move and exert a pressure by hitting the walls of their container. We can explain the results as follows.

- At 0 °C moving molecules exert a pressure of 100 kPa because they hit the walls of their container.
- At 100 °C the molecules are moving faster, so they hit the walls of the container harder and more often; the pressure rises to 137 kPa.
- At −136 °C the molecules have slowed down, so the collisions with the container walls are less frequent and less forceful; the pressure drops to 50 kPa.
- At −273 °C the graph predicts that the pressure will be zero. The conclusion we draw from this is that at this temperature the molecules stop moving altogether. If the molecules have stopped moving, they do not hit the walls of their container at all, so no pressure is exerted.

This means that −273 °C is the lowest possible temperature. The temperature of an object depends on the random kinetic energy of its molecules. At −273 °C molecules stop moving, so we cannot cool them any further below that. We call −273 °C **absolute zero**.

PRACTICAL

Make sure you understand how to investigate how the pressure of a gas depends on its temperature. What are the independent, dependent and constant variables in this investigation?

The Kelvin scale

We use an **absolute temperature scale** to help us predict how the pressure of a gas changes with temperature. Absolute temperatures are measured in **kelvin** or **K**. Absolute zero is 0 K; 0 °C is 273 K. The size of a kelvin is the same size as 1 °C. Note that we do not use a degree sign with kelvin: we write 77 K, *not* 77 °K.

We convert degrees Celsius to kelvin by adding 273; we convert kelvin to degrees Celsius by subtracting 273.

For example:

$$150\,\text{K} = (150 - 273)\,°\text{C}$$
$$= -123\,°\text{C}$$
$$200\,°\text{C} = (200 + 273)\,\text{K}$$
$$= 473\,\text{K}$$

The Kelvin scale is helpful is enabling us to see the pressure change more clearly. If the kelvin temperature doubles then we know that the pressure of the gas also doubles (provided the volume of the gas remains constant).

Boyle's law

Figure 6.6 shows a very similar apparatus to that used in the previous experiment. But now the piston is moved to allow the volume of the gas to change; the cylinder is leak-tight so that no air can escape or enter. To make it a fair test, the piston is moved slowly to ensure the temperature of the gas remains the same throughout the experiment. When the volume of the air in the cylinder is changed, by moving the piston, the pressure in the cylinder also changes. Figure 6.7 shows in graph form how the changes between pressure and volume are linked.

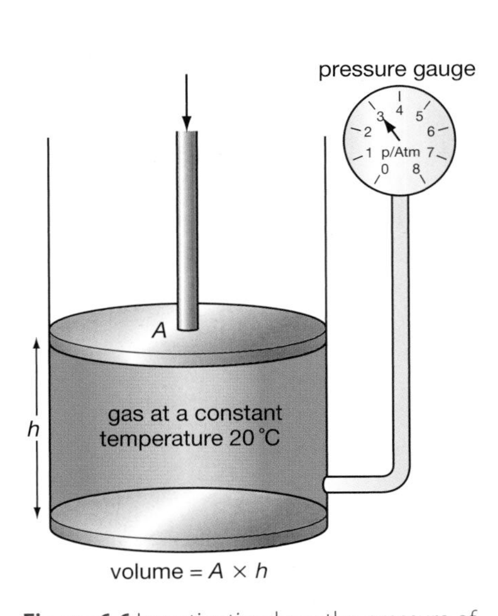

Figure 6.6 Investigating how the pressure of a gas depends on its volume.

PRACTICAL

Make sure you understand how to investigate the relationship between the pressure and volume of a gas at constant temperature.

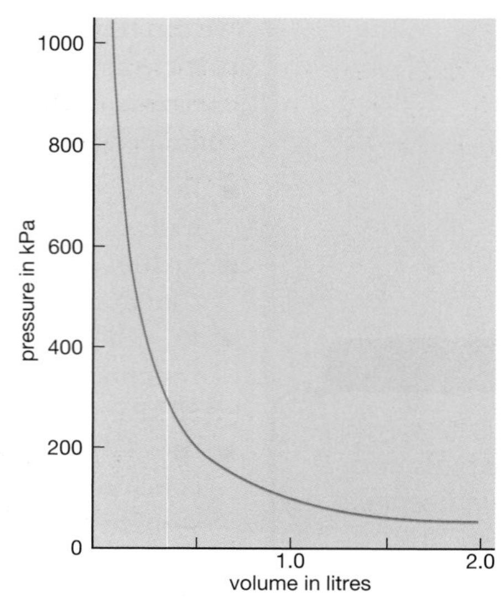

Figure 6.7 As the volume of the gas is halved, the pressure doubles.

Table 1

Pressure in kPa	Volume of air in l
50	2.0
100	1.0
200	0.5
400	0.25
1000	0.1

Table 1 shows the sort of results you can expect to get if you do the experiment; when the gas is compressed to half its volume the pressure doubles. Note also from the values in the table that if you multiply the pressure and volume together, you always get the same number.

So, for a constant mass of gas, at a constant temperature:

$$P_1V_1 = P_2V_2$$

This is known as **Boyle's law**.

We can also make sense of this in terms of molecular motion. When the volume of a gas is reduced, the pressure increases because the molecules hit the walls of the container more often.

Calculating the pressure

Here we look again at the results of the experiment using the apparatus in Figure 6.5. The results are plotted again, using only the Kelvin scale, in Figure 6.8. We can now see more clearly the relationship between the pressure of the gas and the Kelvin scale. The straight line drawn through the data passes through the origin. So we can say that the pressure of a fixed mass of gas at a constant volume is directly proportional to the kelvin temperature. This means that the ratio of pressure to temperature (in K) for the gas is always the same.

We can use this equation to calculate the pressure or temperature:

$$\frac{P_1}{P_2} = \frac{T_1}{T_2}$$

Remember this is for a change at constant volume.

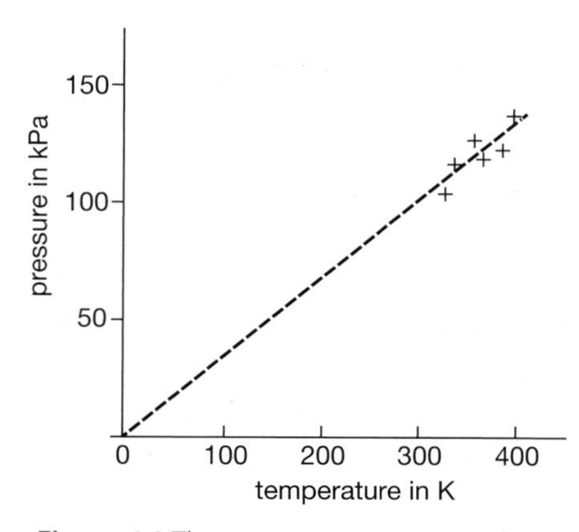

Figure 6.8 The pressure is proportional to the kelvin temperature.

TIP

When calculating the pressure of a gas, make sure you have changed the temperature into kelvin.

For example, the graph in Figure 6.8 shows us that the pressure is 110 kPa at a temperature of 300 K. How can we calculate the pressure when the container is heated to a temperature of 600 K? Using T_1 as 600 K, T_2 as 300 K and P_2 as 110 kPa, we can substitute those numbers into the equation to give:

$$\frac{P_1}{110} = \frac{600}{300}$$

$$P_1 = 2 \times 110 \text{ kPa}$$

$$= 220 \text{ kPa}$$

Kelvin temperature and kinetic energy

A further important point that comes from the Kelvin scale is that the kinetic energy of the molecules in a gas is directly proportional to the kelvin temperature. We can put this the other way round: the kelvin temperature of a gas is a measure of the average random kinetic energy of its molecules.

STUDY QUESTIONS

1 Describe how the molecules of a gas move.

2 Explain, in terms of molecular motion, why a gas exerts a pressure on its container.

3 a) Explain what happens to the speed of the molecules in a gas as the temperature increases.
  b) Explain, in terms of molecules, why the pressure of a gas increases as the temperature rises.

4 This question refers to the experiment shown in Figure 6.6. Use Table 1 or Figure 6.7 to determine the volume of the air when the pressure was:
  a) 300 kPa
  b) 800 kPa

5 a) Explain what is meant by absolute zero.
  b) Why is it not possible to have a temperature of −300 °C?

6 a) Convert these temperatures to kelvin: 100 °C, 327 °C, −173 °C, −50 °C.
  b) Convert these temperatures to °C: 10 K, 150 K, 350 K, 400 K.

7 This question refers to the experiment described in Figure 6.5.
  Use the graph in Figure 6.4 to predict the pressure in the cylinder at:
  a) a temperature of −200 °C
  b) a temperature of 173 K.

8 Figure 6.9 shows a gas holder. It is filled with gas from the grid pipeline where the pressure is 800 kPa. In the gas holder, the volume of gas is 160 000 m^3 and the pressure is 100 kPa. Calculate the volume of gas when it was in the pipeline.

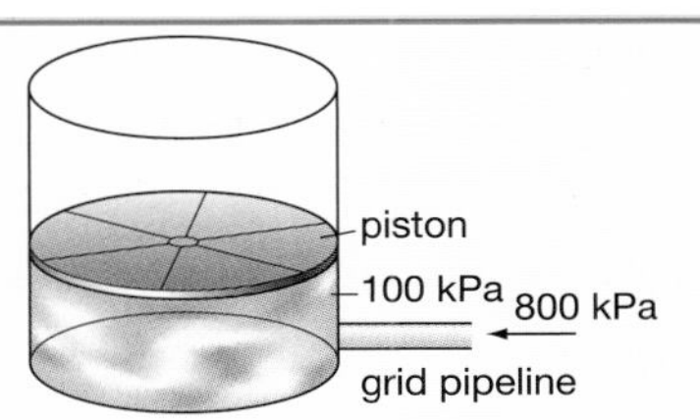

Figure 6.9

9 A gas in a container of a fixed volume has a pressure of 200 kPa at a temperature of 200 K. Calculate the pressure at:
  a) 100 K b) 600 K c) 350 K

10 A cylinder of gas has a pressure of 450 kPa at a temperature of 27 °C.
  a) Express 27 °C in kelvin.
  b) Now calculate what the pressure in the gas will be at:
    i) 177 °C ii) −73 °C.

11 An airtight cylinder of gas has a volume of 0.4 m^3 and a pressure of 250 kPa at 20 °C.
  a) The cylinder expands slightly but the temperature of the gas does not change. State what happens to the pressure of the gas.
  b) The cylinder is then warmed to 100 °C without changing its volume. Calculate the pressure after it has been heated.

12 a) Explain what happens to the temperature of a gas in kelvin if the average kinetic energy of the molecules doubles.
  b) Explain what happens to the temperature of gas in kelvin if the average speed of its molecules is doubled.

Summary

I am confident that:

✓ **I can recall these facts and concepts about density and pressure**

- I can recall and use the equation: density $= \frac{\text{mass}}{\text{volume}}$
- I know that the unit of density is kg/m^3
- I know and can describe how to measure density
- I can recall and use the equation: pressure $= \frac{\text{force}}{\text{area}}$
- I know that the unit of pressure is N/m^2 or pascal (Pa)
- I can recall and use the relationship:

pressure difference = height × density × gravitational field strength

$$P = h \times \rho \times g$$

✓ **I can recall these facts and concepts about change of state**

- I understand that a substance can change state from solid to liquid by the process of melting.
- I understand that a substance can change state from a liquid to a gas by the process of evaporating or boiling.
- I know that particles in a liquid have a random motion within a close-packed irregular structure.
- I know that particles in a solid vibrate about fixed positions within a close-packed regular structure.
- know that the particles in a gas are widely separated, and that the particles move quickly and randomly.

✓ **I can recall these facts about specific heat capacity**

- Specific heat capacity is the energy required to raise the temperature of an object by one degree Celsius per kilogram of mass.
- Specific heat capacity has the unit J/kg °C
- I can use the equation $\Delta Q = mc\Delta T$

change in thermal energy = mass × specific heat capacity × temperature change

✓ **I can recall these facts and concepts about ideal gas molecules**

- I know that molecules exert a pressure by hitting the walls of their container.
- I understand that absolute zero is −273 °C.
- I can convert temperatures between the Kelvin and Celsius scales.
- I understand that the kelvin temperature of the gas is proportional to the average kinetic energy of its molecules.
- I can use the relationship between the pressure and kelvin temperature for a fixed mass of gas at a constant volume:

$$\frac{P_1}{T_1} = \frac{P_2}{T_2}$$

- I can use the relationship between the pressure and volume of a fixed mass of a gas at a constant temperature: $P_1V_1 = P_2V_2$

Sample answers and expert's comments

1 The diagram shows how steel piles are put into a river bed to make the foundation for building a bridge. A large downward force is applied to push them into the river bed.
The second diagram shows two types of pile, A and B, which can be used for building the bridge.

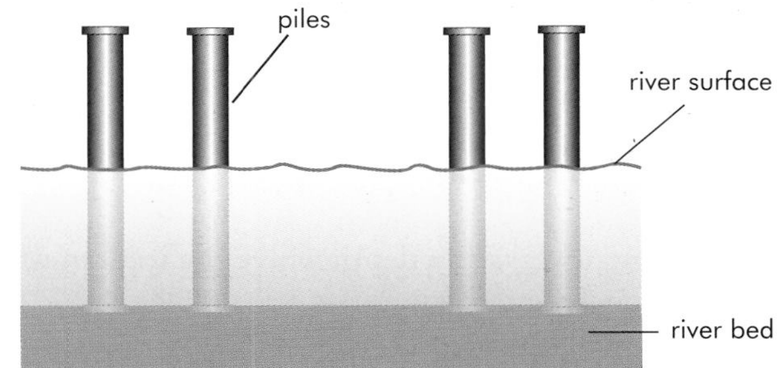

a) The density of steel is 9000 kg/m^3. Explain why the steel piles sink when they are put into the river. (1)

b) Calculate the volume of pile A. (2)

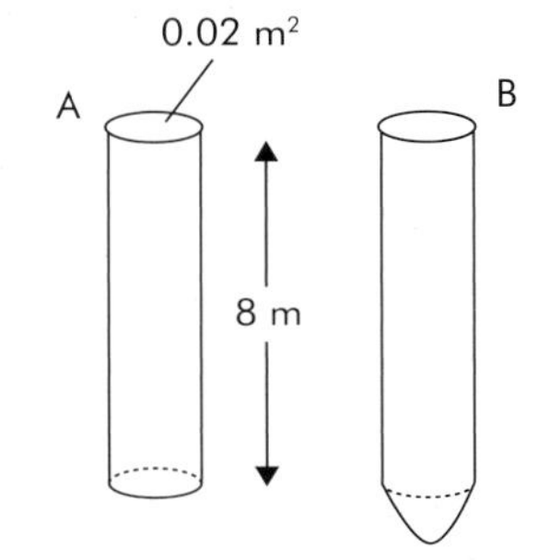

c) Write an equation, which connects density, mass and volume. Calculate the mass of pile A. (3)

d) An engineer suggests that pile B would be better to use than pile A. Explain why she thinks this. (2)

(Total for question = 8 marks)

Student response Total 6/8	Expert comments and tips for success
a) *Steel has a greater density than water.* ✔	This is a correct answer. The correct scientific term has been used so this gets the mark.
b) *volume = area × height* *= 0.02 × 8* ✔ *= 1.6 m^3* O	This is well set out, but a careless mistake has been made in the calculation. One mark is lost. Always check your work carefully. The correct answer is 0.16 m^3
c) $\rho = \frac{m}{V}$ ✔ $9000 = \frac{m}{1.6}$ ✔ $m = 9000 \times 1.6 = 14\,440$ kg ✔	The calculation is correct, so full marks are earned, even though an incorrect value for the volume was carried over from part b). The correct answer is: 9000 × 0.16 = 1440 kg or 1400 kg to 2 significant figures.
d) *The smaller pile concentrates the force.* ✔	This answer is too vague to gain full marks. Remember to use the correct vocabulary when you are asked to 'explain'. Here the key word to use is *pressure* – there is the same force acting on a smaller area.

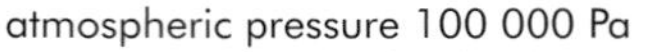

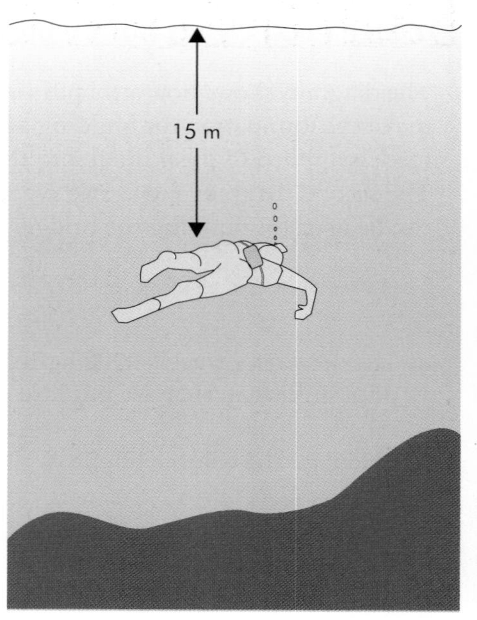

2 a) The diagram below shows a diver. He is swimming at a depth of 15 m below the surface of a lake.

The atmospheric pressure above him is 100 000 Pa.

i) State the equation which links pressure difference, height, density and gravitational field strength. (1)

ii) Calculate the total pressure on the diver. (3)

The density of the water is 1000 kg/m^3. The gravitational field strength is 10 N/kg.

b) A second diver is at a depth where the total pressure on him, from the atmosphere and the water, is 400 000 Pa.

The face mask of the diver has an area of 0.012 m^2.

i) State the equation that links pressure, force and area. (1)

ii) Calculate the force that acts on the outside of the mask of the diver (3)

c) A diver inflates a balloon with 1.6 litres of air, at the surface of the lake, where atmospheric pressure is 100 000 Pa.

i) Calculate the volume of the balloon when the diver dives to a depth where the total pressure on the balloon is 400 000 Pa. (3)

ii) Explain why a diver must breathe air out of his lungs when he swims back to the surface of the lake, after diving at a depth where the pressure is high. (2)

(Total for question = 13 marks)

Student response Total 9/13	Expert comments and tips for success
a) i) $P = h\rho g$ ✔	
ii) $P = h\rho g$ $= 15 \times 1000 \times 10$ ✔ $= 150\ 000$ Pa ✔	A correct calculation, but the student forgot to add 100 000 Pa for atmospheric pressure. So the pressure is 250 000 Pa.
b) i) $P = \frac{F}{A}$ ✔	
ii) $P = \frac{F}{A}$ $400\ 000 = \frac{F}{0.012}$ ✔ $F = 4800$ ✔	Note the student substituted the numbers into the equation first – this always earns a mark. The answer was calculated correctly, but the unit was left out – which loses a mark.
c) i) $P_1V_1 = P_2V_2$ ✔ $100\ 000 \times 1.6 = 400\ 000 \times V_2$ ✔ $V_2 = 0.4$ litres ✔	Again the student substitutes the numbers first. Then it is easy to see the answer.
ii) Without breathing out his lungs would explode. O	What is written is probably true, but it is a bit of a guess with no physics. At depth where the pressure is 400 kPa, the pressure inside the lung is the same outside – so the forces on the lung balance. If you go to the surface without breathing out, the pressure inside the lung remains at 400 Pa, but it is only 100 kPa outside. This large pressure difference will cause the lung to expand, thus doing great damage to it.

Exam-style questions

1 The correct unit for specific heat capacity is:

A J/kg °C **B** J **C** J/kg **D** J kg/°C [1]

2 The side of a cube of wood has a length of 4.0 cm. The mass of the cube is 32.0 g. The density of the wood is:

A 8 g/cm^3 **B** 4 g/cm^3 **C** 2 g/cm^3 **D** 0.5 g/cm^3 [1]

3 A sheet of steel has a mass of 360 kg. The density of steel is 9000 kg/m^3. The volume of the steel sheet is:

A 3240 m^3 **B** 25 m^3 **C** 0.04 m^3 **D** 0.02 m^3 [1]

4 The density of sea water is 1030 kg/m^3. Atmospheric pressure is 101 000 Pa.

The pressure at a depth of 30 m is:

A 310 000 Pa **B** 410 000 Pa **C** 404 000 Pa
D 41 000 Pa [1]

5 A gas has a temperature of 130 °C.

This temperature in kelvin is:

A −143 K **B** 27 K **C** 403 K **D** 1300 K [1]

6 A gas in a cylinder of volume 0.2 m^3 has a pressure of 2 MPa. The volume of the gas is increased to 0.5 m^3. The pressure changes to:

A 1 MPa **B** 0.8 MPa **C** 0.5 MPa **D** 0.2 MPa [1]

7 A gas is in a cylinder of volume 0.5 m^3. The pressure in the cylinder is 0.2 MPa, and its temperature is 270 K. The cylinder is heated until the pressure rises to 0.6 MPa, and the volume remains at 0.5 m^3. The new temperature of the gas is:

A 90 K **B** 180 K **C** 540 K **D** 810 K [1]

8 A block of metal has a mass of 2 kg and a specific heat capacity of 400 J/kg °C. 48 000 J of thermal energy are transferred to the block. Its temperature rises by:

A 6 °C **B** 30 °C **C** 60 °C **D** 120 °C [1]

9 A heater transfers 80 kJ of thermal energy to a liquid of mass 0.4 kg. The temperature rises by 50 °C. The specific heat capacity of the liquid is:

A 500 J/kg °C **B** 1000 J/kg °C **C** 4000 J/kg °C
D 15 000 J/kg °C [1]

10 A cube of metal has a side length of 10 cm. The weight of the metal is 45 N. When the cube rests on the ground the pressure it exerts on the ground is:

A 4500 Pa **B** 45 000 Pa **C** 450 000 Pa **D** 4.5 MPa [1]

11 **a)** Write down the formula that connects density, mass and volume. [1]

b) Explain how you would use a measuring cylinder, electronic balance and some glass marbles to calculate the density of glass. [5]

12 When a drop of ether is placed on the skin, the skin feels cold. Explain why. [3]

13 A kettle is being heated on a camping stove. Thermal energy is transferred by burning gas from the cylinder.

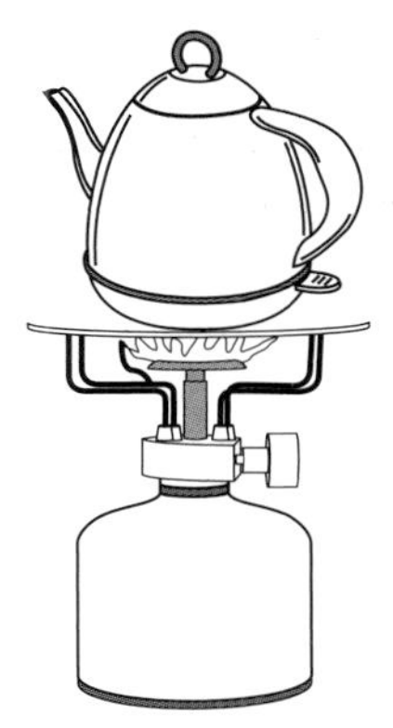

a) Water in the kettle boils at 100 °C and steam is produced. Convert this temperature to a temperature on the kelvin scale. [1]

b) Explain **one** difference between the molecules in steam and the molecules in water. [1]

c) Explain how the molecules in the steam exert a pressure on the kettle walls. [2]

d) The wind blows the flame out and 520 cm^3 of gas, at a pressure of 135 kPa, escapes from the cylinder. As the gas escapes, its pressure reduces to 100 kPa. Calculate the volume of the escaped gas at a pressure of 100 kPa. [3]

e) After the tap is closed to prevent further loss of gas, the cylinder is left in the sunshine and it warms up. Explain what happened to the gas in the cylinder. [2]

14 The apparatus in the figure below is used to heat up a block of metal of mass 2 kg. When the heater is turned on, the temperature of the block of metal increases as shown in the graph.

a) Use the graph to determine the temperature rise of the metal in the first 10 minutes of heating. [1]

b) During 10 minutes of heating, 48 000 J of thermal energy is transferred to the block. Calculate the specific heat capacity of the block. [4]

c) Use the information in part (b) and information from the graph to show that the power of the heater is 80 W. [3]

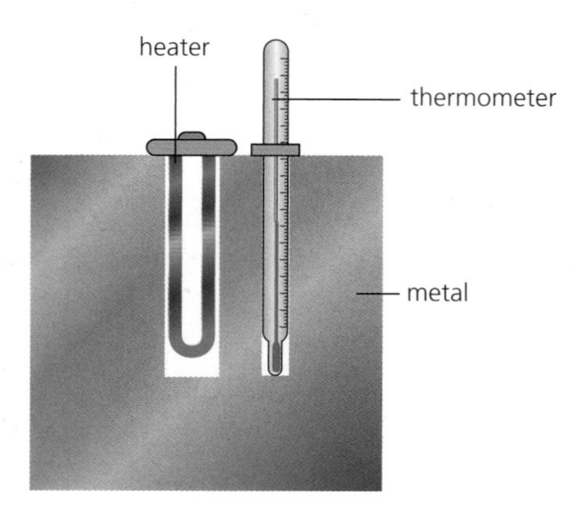

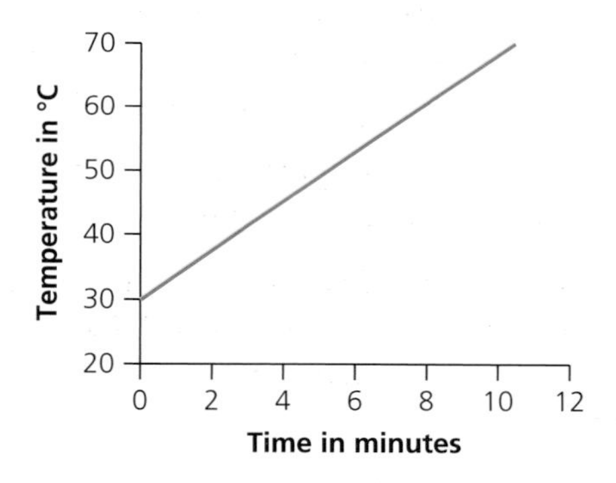

15 The figure below shows a heater at the bottom of a boiling tube of solid wax. The heater is then connected to a power supply. A joulemeter measures the energy supplied to the heater as it melts the wax. The graph shows how the temperature of the wax changes with the energy supplied.

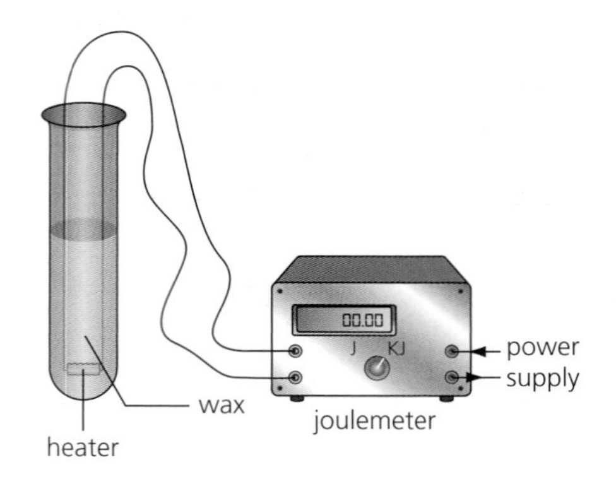

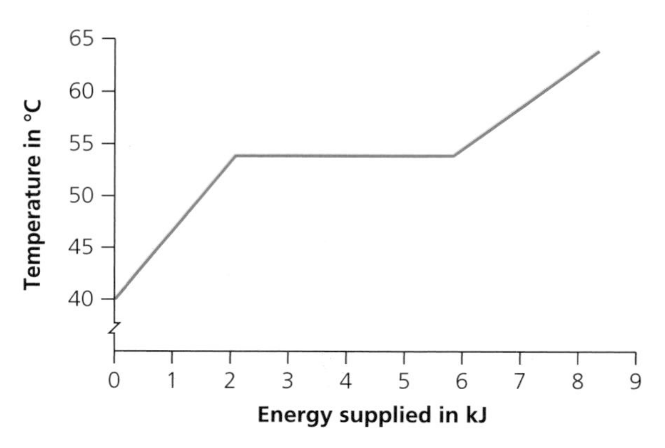

a) State the melting temperature of wax. [1]

b) The temperature of the wax remains constant as the wax melts. Explain why. [2]

c) i) State what is meant by: specific heat capacity. [1]

ii) Use the graph to show that the specific heat capacity of liquid wax is greater than the specific heat capacity of solid wax. [3]

16 These questions are about a scuba diver and the pressure in his body. You will need the following information to help you answer them:

The density of water is 1000 kg/m^3; the pull of gravity is 10 N/kg; on the day of the dive, the atmospheric pressure is 100 kPa.

a) The diver descends to a depth of 25 m.

i) Use the formula $P = h \times \rho \times g$ to calculate the extra pressure on his body caused by the water.

ii) Calculate the total pressure on the diver's body. [3]

b) When the diver is at a depth of 25 m his lungs can hold 5 litres of air.

i) Determine the volume the air would occupy at normal atmospheric pressure.

ii) Explain why divers must breathe out as they rise to the surface. [4]

17 At the beginning of the compression stroke in the cylinder of a diesel engine, the air is at a temperature of 127 °C and a pressure of 120 kPa.

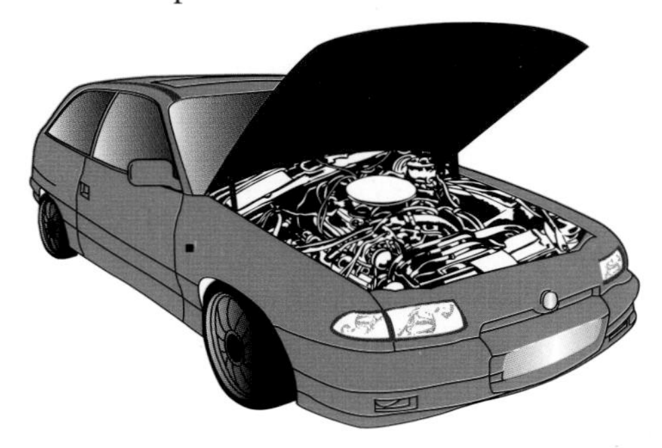

a) Convert this temperature to kelvin. [1]

b) During the compression stroke the volume of the air is squashed to 1/20th of its original volume.

i) Show that the pressure in the air after it has been compressed to 1/20th of its volume is 2400 kPa. [2]

ii) The volume of the cylinder remains constant and is now heated to 727 °C. Calculate the pressure of the gas now. [4]

18 Cylinders of gas that are caught in the flames of a burning building can be a significant hazard. At a temperature of 20 °C the pressure in a cylinder of gas is 300 kPa – this is three times normal atmospheric pressure. The temperature of a fire might reach 1200 °C.

a) Explain why the pressure rises in the cylinder, in terms of molecular motion. [2]

b) i) Calculate the pressure in the cylinder at a temperature of 1200 °C.

ii) Explain why this is a dangerous pressure. [4]

19 A balloon seller has a cylinder of helium gas which he uses to blow up his balloons.

The volume of the cylinder is 0.10 m^3.
It contains helium gas at a pressure of 1.0×10^7 Pa.
The balloon seller fills each balloon to a volume of 1.0×10^{-2} m^3 and a pressure of 1.2×10^5 Pa.

a) Explain, in terms of particles, how the helium in the cylinder produces a pressure. [2]

b) Calculate the total volume that the helium gas will occupy at a pressure of 1.2×10^5 Pa. You can assume that there is no change in the temperature of the helium gas. [3]

c) Calculate the number of balloons of volume 1.0×10^{-2} m^3 that the balloon seller can fill using the gas. [2]

EXTEND AND CHALLENGE

1 Below you can see Boris, Vladimir and Leonid. Boris is twice as tall, twice as long and twice as wide as Vladimir, but they have the same density.

a) How many times more massive than Vladimir is Boris?

b) The area of Boris's paw is greater than the area of Vladimir's paw. By how many times is it bigger?

c) Is the pressure bigger under Boris's paws or under Vladimir's paws? By what factor does the pressure differ under the two cats' paws?

d) Leonid the lion is a distant cousin to Boris. Lions, as you know, are much bigger than domestic cats. Can you use the result of part **c)** to explain why a lion's legs are proportionately thicker than a cat's?

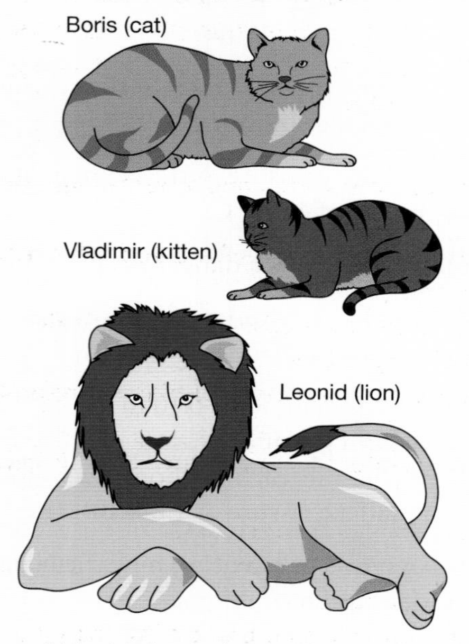

2 A metal block is heated using the arrangements shown in the diagram. It is heated for about 16 minutes then allowed to cool down. The temperature is plotted every minute, as shown in the graph.

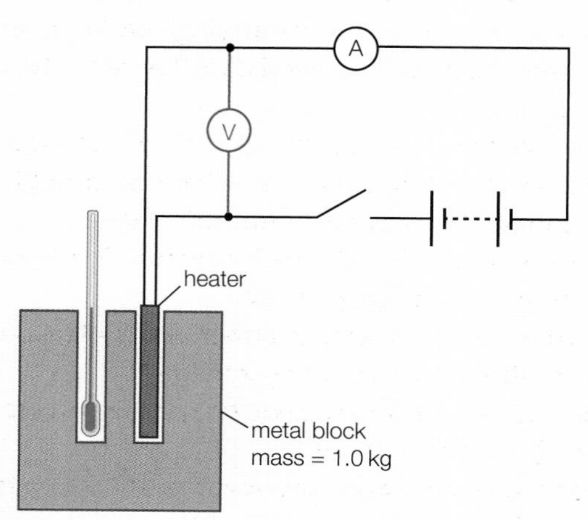

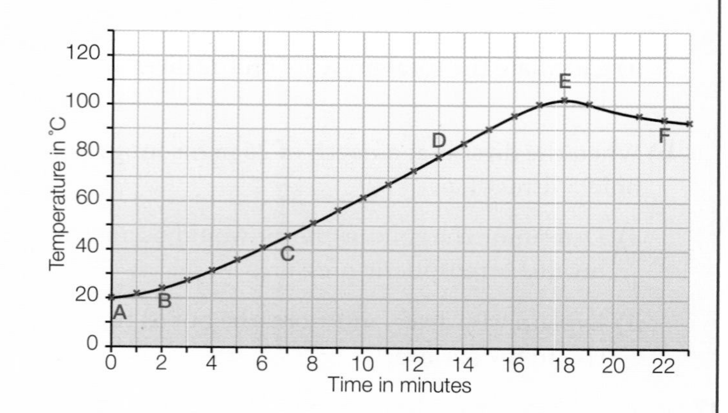

a) i) Explain the shape of the graph over the region AB.

ii) Although the heater was switched off after 16 minutes, the temperature continued to rise. Explain why.

b) Determine the gradient of the graph over the regions:
i) BC **ii)** DE **iii)** EF

c) Comment on the connection between the three gradients.

d) When the heater is switched on the voltmeter reads 12.0 V and the ammeter reads 4.5 A. Calculate the specific heat capacity of the metal. Explain any assumptions you make in this calculation.

e) i) A student measures the specific heat capacities of several metals and notices that the metals with higher atomic masses have lower specific heat capacities. She wonders what the specific heat capacity is for each metal per mole. So she designs the table below. Complete the table.

Metal	Specific heat capacity in J/kg °C (at 0 °C)	Atomic mass	No of moles in 1 kg	Specific heat capacity in J/mol °C
Lithium	3480	6.9	145	24.0
Alumini-um	880	27.0		
Copper	379	63.5		
Gold	128	197.0		

ii) Most pure metals have a very similar arrangement of atoms. Use this idea to help you make a comment about the molar specific heat capacities of metals.

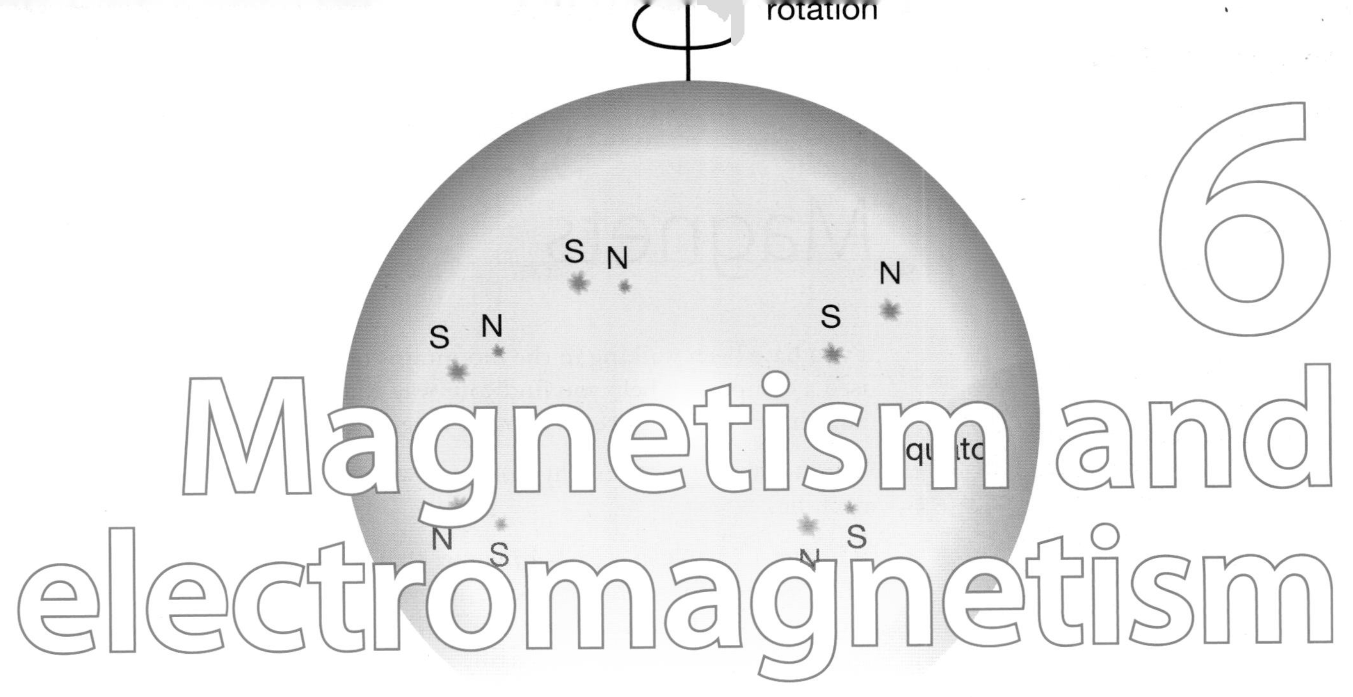

6 Magnetism and electromagnetism

CONSIDER • CALCULATE

Use the internet to help you to answer the following questions:

1 State the average surface temperature of the Sun.
2 State the surface temperature of a sunspot.
3 How can you tell from the photograph that sunspots are colder than the rest of the Sun's surface?

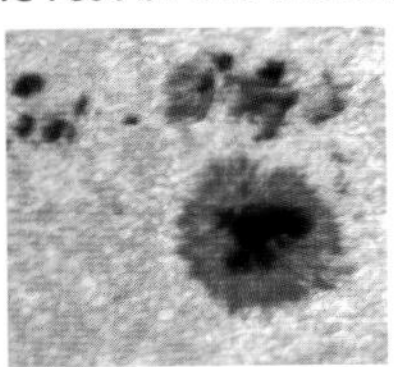

4 How long is the sunspot cycle?
5 Use the information in the photograph and text on this page to sketch the magnetic field lines close to a pair of sunspots. The diagram at the top of the page shows how sunspots are paired.

The Sun has a strong magnetic field. This is produced by the movement of charged particles (a plasma) caused by convection currents in the Sun's core. At the surface, the magnetic field emerges from pairs of sunspots, one of which is a north pole and the other a south pole.

By the end of this section you should:

- **know that magnets repel and attract each other and attract magnetic substances**
- **be able to describe the properties of magnetically hard and soft materials**
- **understand the term magnetic field line**
- **know that a magnetic field can induce magnetism in some materials**
- **be able to investigate magnetic field patterns**
- **be able to describe how to use two bar magnets to produce a uniform magnetic field**
- **know that an electric current in a conductor produces a magnetic field around it**
- **be able to describe the construction of electromagnets**
- **be able to draw magnetic field patterns for a straight wire, flat circular coil and solenoid when each is carrying a current**
- **know that there is a force on a moving charged particle in a magnetic field**
- **understand why a force is exerted on a current-carrying wire in a magnetic field, and how this effect is applied in dc motors and loudspeakers**
- **be able to use the left hand rule to predict the direction of the resulting force when a wire carries a current perpendicular to a magnetic field**
- **be able to describe how the force on a current-carrying conductor in a magnetic field changes with the magnitude and direction of the field and current**
- **know that a voltage is induced in a conductor or coil when it moves through a magnetic field or when a magnetic field changes through it and describe the factors that affect the size of the induced voltage**
- **be able to describe the generation of electricity by the rotation of a coil within a magnetic field or rotating magnet in a coil**
- **be able to describe the construction of a transformer, and understand that a transformer changes the size of an alternating voltage**
- **be able to explain the use of step-up and step-down transformers in the large scale transmission of electrical energy**
- **know the relationship between input and output voltages and the turns ratio for a transformer**
- **know and be able to use the relationship input power = output power**

6.1 Magnets

Figure 1.1 This is a simple everyday device, but can you explain how it works?

If you have been walking in the mountains or in the forest, you might have used a compass to help you find your way. Magnets are very useful to us, as a magnet can exert a force on another magnetic object at a distance.

Write a list of at least six ways in which we use magnets in the home or in machines.

In Figure 1.2 you can see a bar magnet that is hanging from a fine thread. When it is left for a while, one end always points north. This end of the magnet is called the **north-seeking pole**. The other end of the magnet is the **south-seeking pole**. We usually refer to these poles as the north (N) and south (S) poles of the magnet.

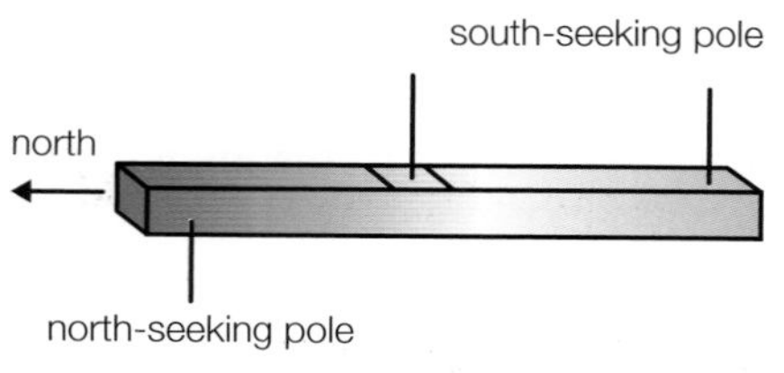

Figure 1.2

Poles

Some metals, for example iron, cobalt and nickel, are **magnetic**. A magnet will attract them. If you drop a lot of pins on to the floor the easiest way to pick them all up again is to use a magnet.

The forces on pins, iron filings and other magnetic objects are always greatest when they are near the poles of a magnet. Every magnet has two poles that are equally strong.

When you hold two magnets together you find that two north poles (or two south poles) repel each other, but a south pole attracts a north pole (Figure 1.3).

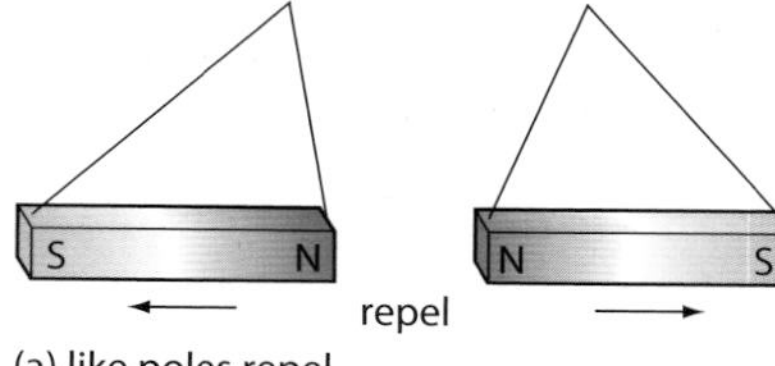

(a) like poles repel

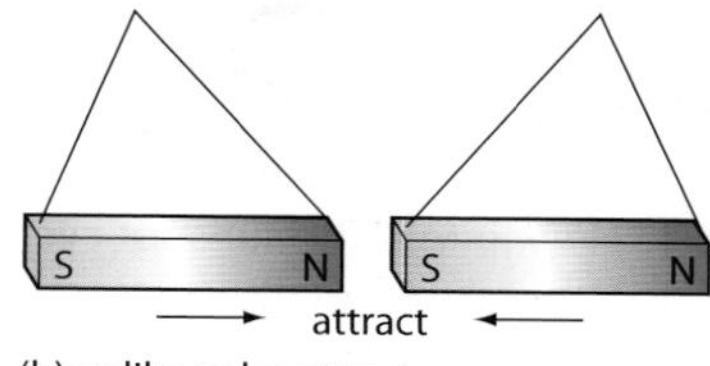

(b) unlike poles attract

Figure 1.3

Investigating magnetic field patterns

PRACTICAL

Make sure you understand how to trace field lines near to a magnet.

Method 1

1 Figure 1.4 shows a simple way to plot a magnetic field line. On a piece of paper, place a small plotting compass near to the north pole of a magnet. Mark each end of the compass needle with pencil dots.

2 Place a second compass in contact with the first and mark the position of the second needle.

3 Repeat the process with more compasses so that a field line is traced from the north pole to the south pole of the magnet.

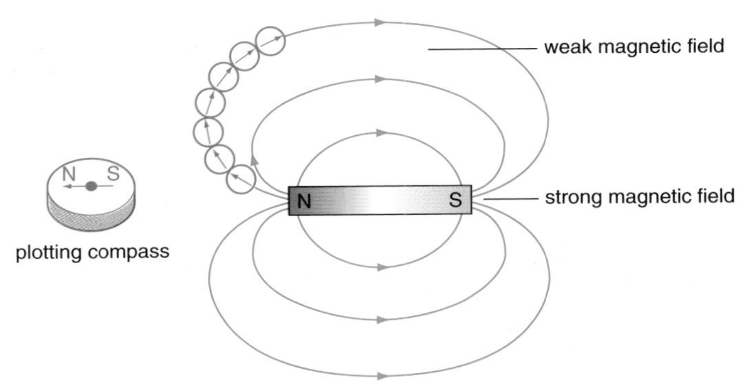

Figure 1.4 Plotting magnetic field lines around a bar magnet.

Method 2

1 Place a piece of paper over the top of the magnet.

2 Sprinkle some iron filings on top of the paper and tap the paper.

3 Sketch the pattern of the iron filings – this shows the shape of the magnetic field. (See Figure 1.5)

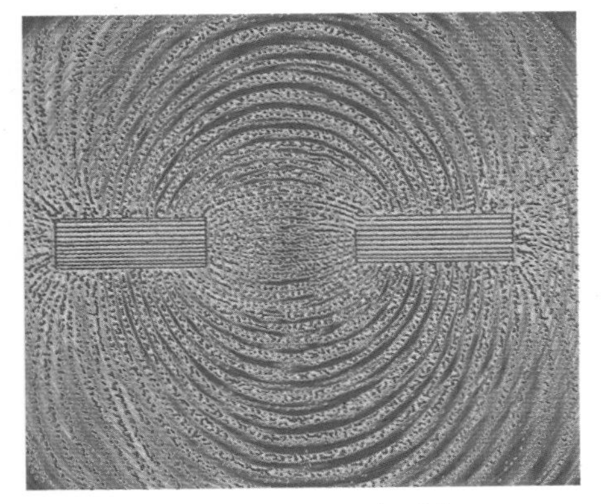

Figure 1.5 Computer artwork of the magnetic field around two magnets, as revealed by iron filings.

Taking it further

Use either or both of the methods above to investigate the magnetic field pattern between:

a) two bar magnets which attract each other, (see Figure 1.7).

b) two bar magnets which repel each other, (see Figure 1.6)

Questions

1 Explain how the magnetic field lines in Figure 1.4 show strong and weak areas of magnetic field.

2 Explain why two field lines can never cross each other. Hint: if two field lines crossed, in which direction would a compass point?

TIP

Magnetic field lines are not real, but they are a useful idea that helps us to understand magnetic fields. Magnetic field lines never cross.

Magnetic fields

There is a magnetic field in the area around a magnet. In this area there is a force on a magnetic object. If the field is strong the force is big. In a weak field the force is small.

We use magnetic field lines to represent a magnetic field. Magnetic field lines always start at a north pole and finish on a south pole. Where the field lines are drawn close together, the field is strong. The further apart the lines are, the weaker the field is.

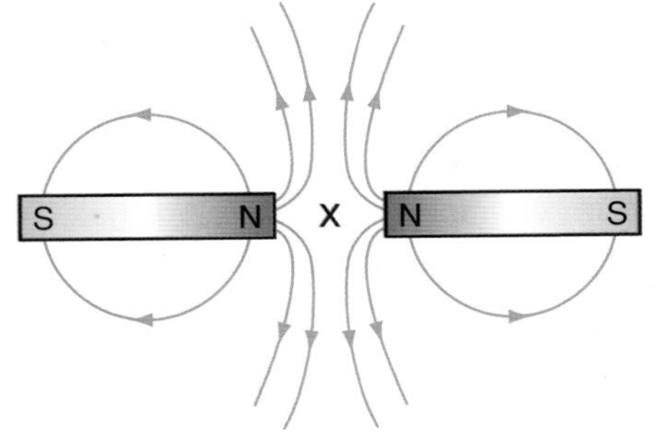

Figure 1.6 X is a neutral point; a compass placed here can point in any direction.

Combining magnetic fields

The pattern of magnetic field lines close to two (or more) magnets becomes complicated. The magnetic fields from the two magnets combine. The field lines from the two magnets never cross. (If field lines did cross, it would mean that a compass would have to point in two directions at once.)

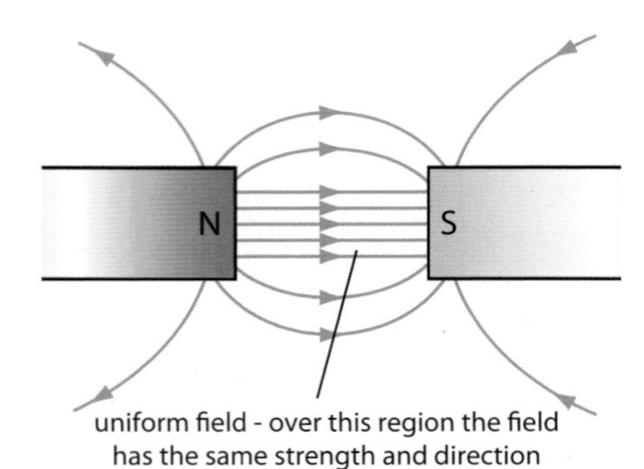

Figure 1.7

TIP

There is a magnetic south pole near the geographic North Pole.

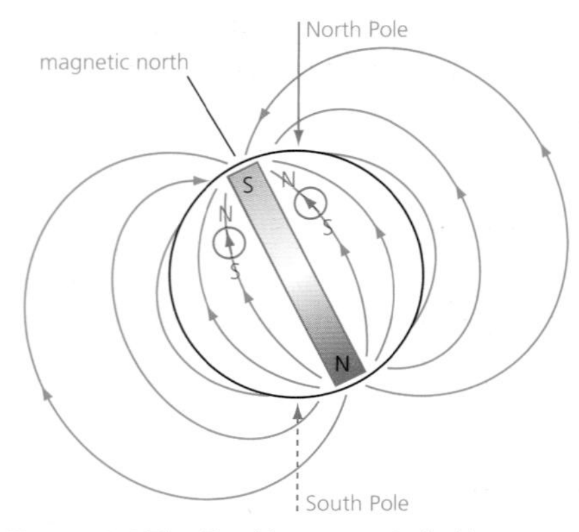

Figure 1.8 The Earth's magnetic field.

Figure 1.6 shows the sort of pattern when two north poles are near each other. The field lines repel. At the point X there are no field lines. The two magnetic fields cancel out. X is called a neutral point.

Figure 1.7 shows the pattern produced by a north and south pole. Notice that there is an area where the field lines are equally spaced and all point in the same direction. This is called a uniform field.

The Earth's magnetic field

Figure 1.8 shows the shape of the Earth's magnetic field. The north (seeking) pole of a compass points north. Unlike poles attract. This means that there is actually a magnetic south pole near the north pole, at a place we call magnetic north.

However, magnetic north is not in the same place as the geographic North Pole. At the moment magnetic north is in the sea north of Canada. The position of magnetic north changes slowly over a period of centuries.

STUDY QUESTIONS

1 a) Figure 1.9 shows a bar magnet surrounded by four plotting compasses. Copy the diagram and mark in the direction of the compass needle for each of the cases B, C, D.

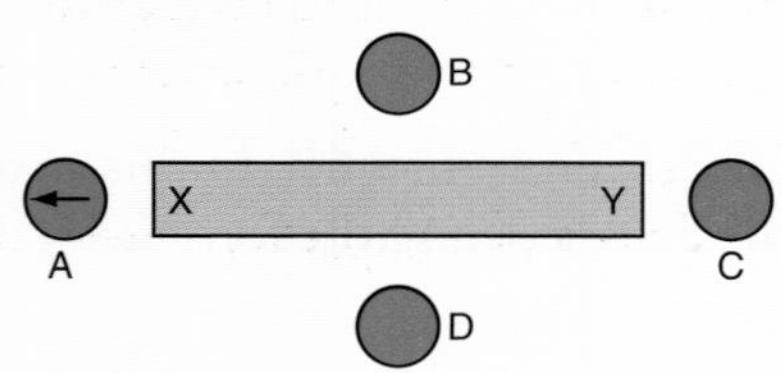

Figure 1.9

b) Which is a north pole, X or Y?

2 Draw carefully the shape of the magnetic field between and surrounding these two magnets. Mark in any neutral points.

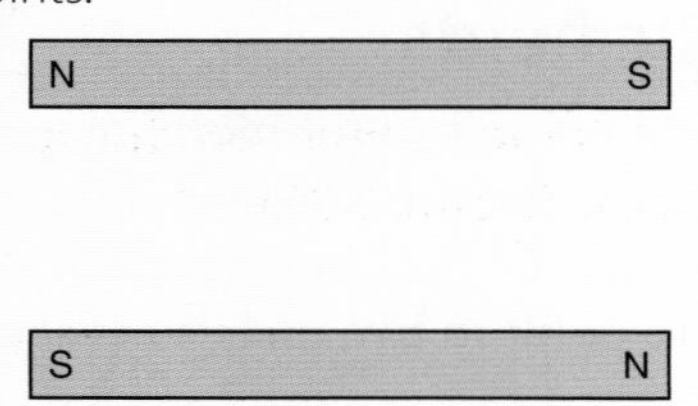

3 Two bar magnets have been hidden in a box. Use the information in Figure 1.10 to suggest how they have been placed inside the box.

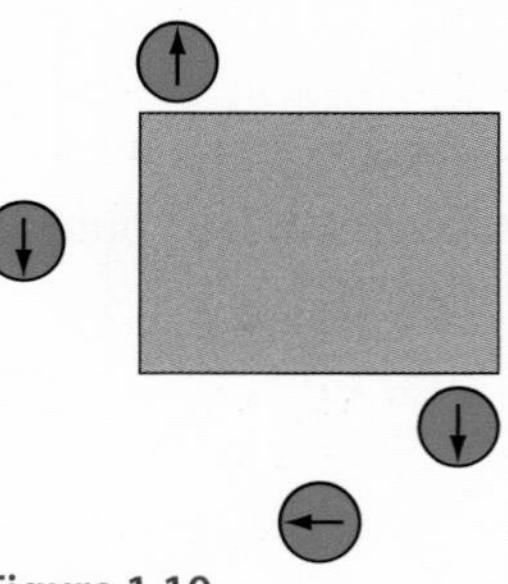

Figure 1.10

4 Refer to Figure 1.8 showing the Earth's magnetic field.
a) Compare the strength of the magnetic field at the Equator and at the geographic North Pole. Which is stronger? Give a reason for your answer.
b) Explain why a compass is very difficult to use near to the Earth's magnetic poles.

6.2 Magnetising

Magnetic domains

TIP

You will not be examined on domains. But knowing about domains helps to understand about induced magnets.

Steel is a permanent magnet; once a piece of steel has been magnetised it remains a magnet. Iron is an induced magnet; iron is only magnetised when in a magnetic field. This field can be provided by a magnet or by a coil of wire carrying a current, which also produces a magnetic field (see Chapter 6.3). A magnetically **hard** material becomes a permanent magnet when it is magnetised; a magnetically **soft** material will make an induced or temporary magnet.

The insides of magnetic materials are split up into small regions, which we call **domains**. Each domain acts like a very small magnet. In any iron or steel bar there are thousands of domains. The idea helps us to understand permanent and induced magnets.

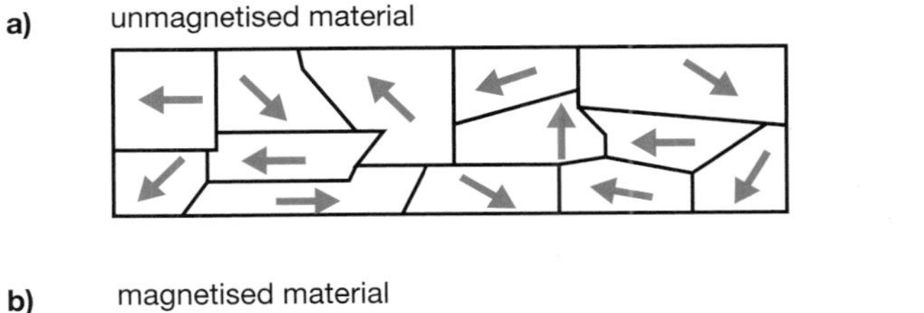

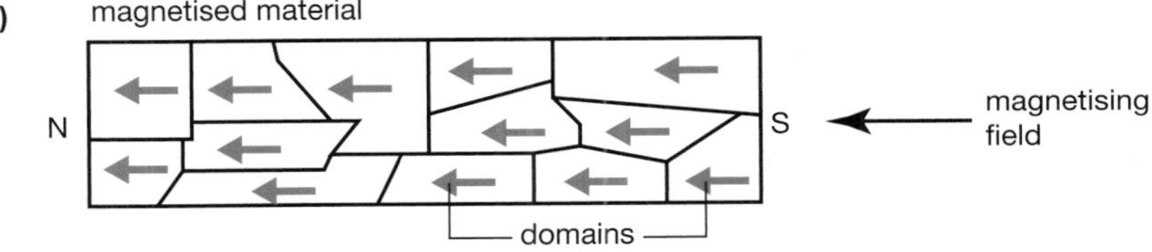

Figure 2.1 Domains in a magnetic material become aligned in a magnetic field.

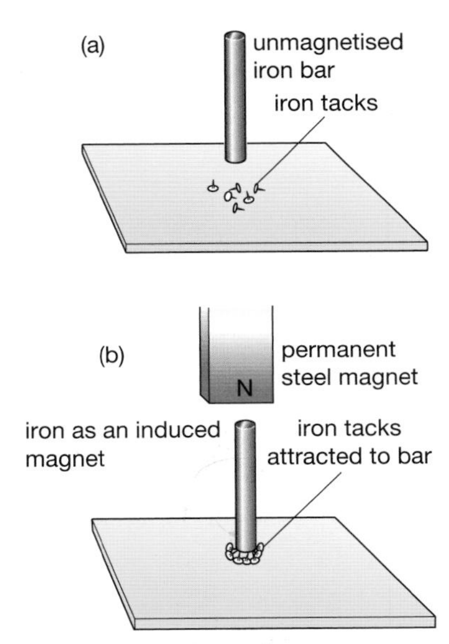

Figure 2.2 An iron bar will act as a magnet when there is a permanent magnet near it.

In an iron bar the domains are jumbled up when there is no magnet near (Figure 2.1(a)). But as soon as a magnet is put near to the bar the domains are made to line up (Figure 2.1(b)). So an iron bar will be attracted to either a north or south pole of a permanent magnet.

In a steel bar, once the domains have been lined up, they stay pointing in one direction.

Magnetically soft materials are induced magnets. Permanent magnets are magnetically hard materials.

An induced magnet is only magnetic as long as it is placed in a magnetic field.

Magnetising

One way to make a magnet from an unmagnetised steel bar is to stroke it repeatedly with a permanent magnet. The movement of the magnet along the steel bar is enough to make the domains line up (Figure 2.3(a)).

A magnet can also be made by putting a steel bar inside a solenoid. A short but very large pulse of current through the solenoid produces a strong magnetic field. This magnetises the bar (Figure 2.3(b)).

Figure 2.4 This photograph shows the design of a practical laboratory electromagnet. It is made strong by two coils with many turns of wire to increase the magnetising effect of the current. When the current is switched off the iron filings will fall off the magnet.

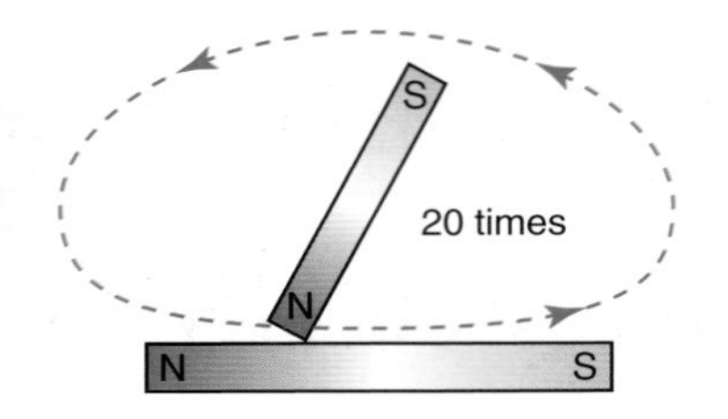

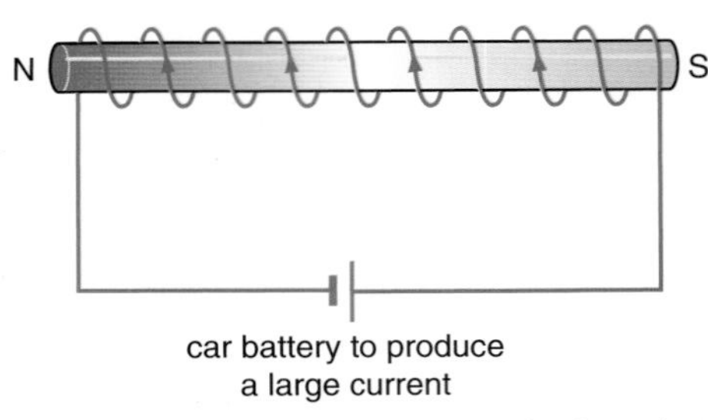

Figure 2.3 (a) Magnetisation of a steel bar by stroking. **(b)** Magnetisation using a solenoid.

The apparatus shown in Figure 2.3(b) can be used as an **electromagnet** if an iron bar is used instead of a steel bar in the solenoid. When a current is switched on the iron becomes magnetised; when the current is switched off the iron is no longer magnetised. The presence of iron greatly increases the strength of the solenoid's magnetic field. Electromagnets are widely used in industry.

STUDY QUESTIONS

1 Pins can be attracted to a magnet but they are not permanent magnets. Explain how you would demonstrate that something which is magnetic is a *permanent* magnet.

2 **a)** Tracey has magnetised a needle as shown below. She then cuts the needle into four smaller bits as shown. Copy Figure 2.5 and label the poles A–H.

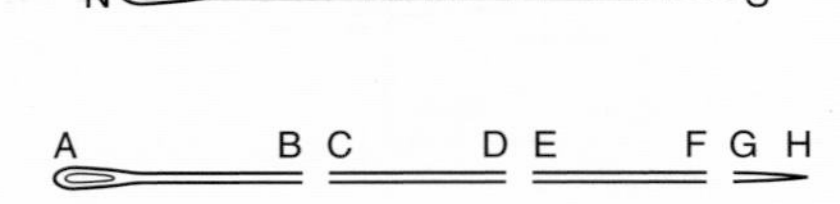

Figure 2.5

b) Tracey says 'This experiment helps to show that there are magnetic domains in the needle'. Comment on this observation.

3 Figure 2.6 below shows an electromagnet. As the current increases the magnet can lift a larger load because the domains in the iron core line up more and more.

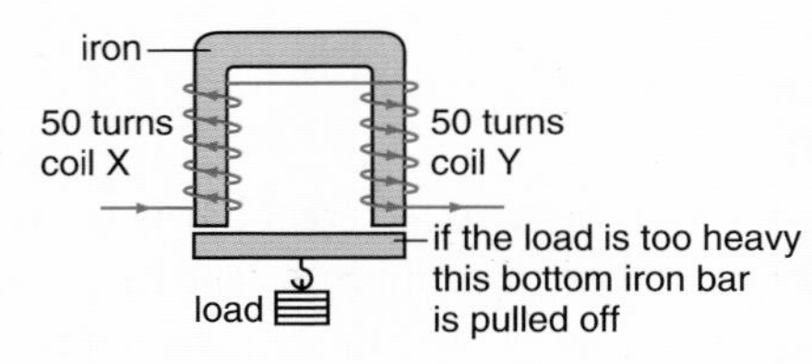

Figure 2.6

Maximum load in N	Current in A
0	0
5.0	0.5
8.5	1.0
12.0	2.0
14.0	3.0
14.8	4.0
15.0	5.0

a) Use the data provided to draw a graph of the load supported (*y*-axis) against current (*x*-axis).

b) Use the graph to predict the load supported when the current is:

i) 1.5 A **ii)** 6.0 A.

c) Sketch on your graph how the load varies with current when each of the coils X and Y has:

i) 25 turns **ii)** 100 turns.

d) Suggest what happens when you reverse the windings on coil X.

e) Make sketches to show how the domains are lined up when the current (in the original diagram) is:

i) 0 **ii)** 0.5 A **iii)** 5 A.

f) Use your answer to part

g) to explain why there is a maximum load that an electromagnet can lift.

6.3 Currents and magnetism

The magnetic field near a straight wire

An electric current in a conductor produces a magnetic field around it.

Figure 3.1 Iron filings are arranged in circles around this wire, which is carrying a current. What can you deduce from this? Can you identify any North and South poles?

In Figure 3.2, a long straight wire carrying an electric current is placed vertically so that it passes through a horizontal piece of hardboard. Iron filings have been sprinkled onto the board to show the shape of the field. If you do this experiment, take care – the wire may become very hot. Below are summarised the important points of the experiment:

- When the current is small, the field is weak. But when a large current is used the iron filings show a circular magnetic field pattern.
- The magnetic field gets weaker further away from the wire.
- The direction of the magnetic field can be found using a compass. If the current direction is reversed, the direction of the magnetic field is reversed.

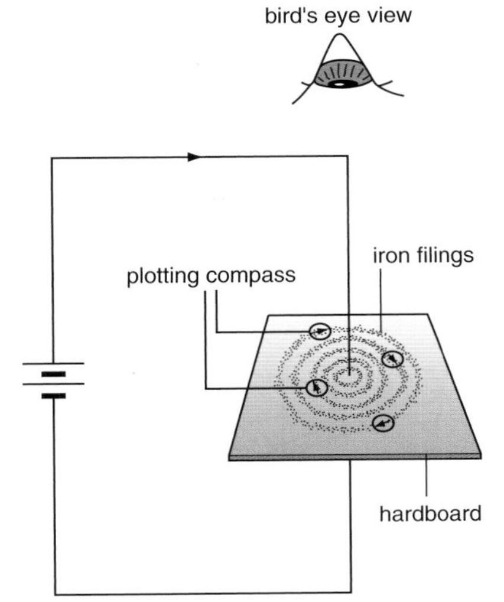

Figure 3.2 This experiment shows there is a magnetic field around a current-carrying wire.

Figure 3.3 shows the pattern of magnetic field lines surrounding a wire. When the current direction is into the paper (shown ⊗) the field lines point in a clockwise direction around the wire. When the current direction is out of the paper (shown ⊙) the field lines point anti-clockwise. **The right-hand grip rule** will help you to remember this (Figure 3.4). Put the thumb of your right hand along a wire in the direction of the current. Now your fingers point in the direction of the magnetic field.

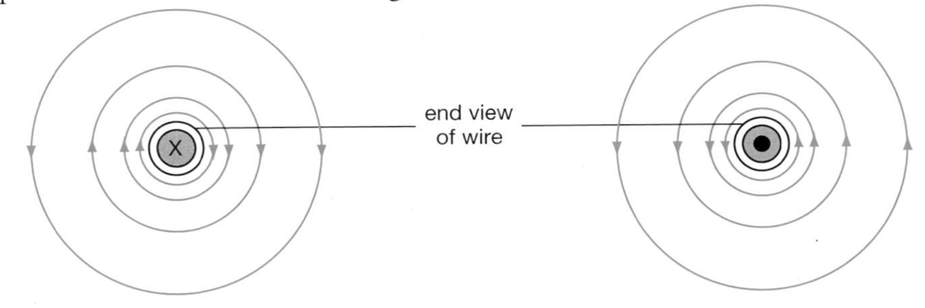

Figure 3.3 The magnetic field pattern around a current-carrying wire.

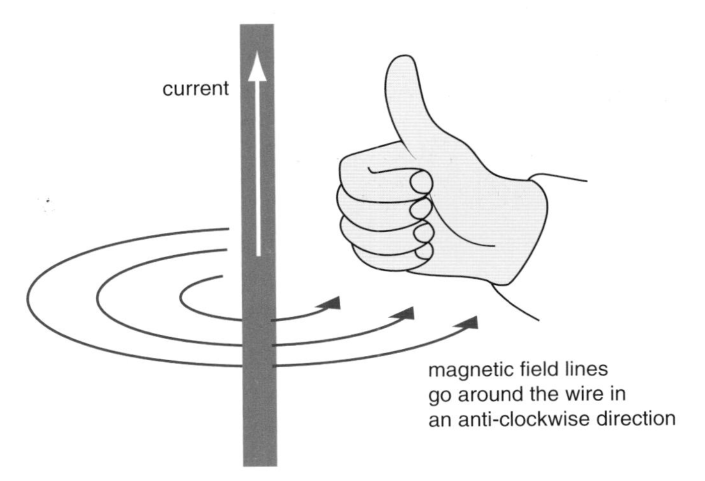

Figure 3.4 The right-hand grip rule.

PRACTICAL

Use the right-hand grip rule at A to check that the field lines go around the wire in an anti-clockwise direction. Take care – the wire may become very hot. Why do they go in the opposite direction at B?

The magnetic field near coils of wire

Figure 3.5 shows the magnetic field around a single loop of wire that carries a current. You can use the right-hand grip rule to work out the field near to each part of the loop. Near A the field lines point anti-clockwise as you look at them, and near B the lines point clockwise. In the middle, the fields from each part of the loop combine to produce a magnetic field running from right to left. This loop of wire is like a very short bar magnet. Magnetic field lines come out of the left-hand side (north pole) and go back into the right-hand side (south pole). Figure 3.6 shows the sort of magnetic field that is produced by a current flowing through a long coil or **solenoid**. The magnetic field from each loop of wire adds on to the next. The result is a magnetic field which is like a long bar magnet's field.

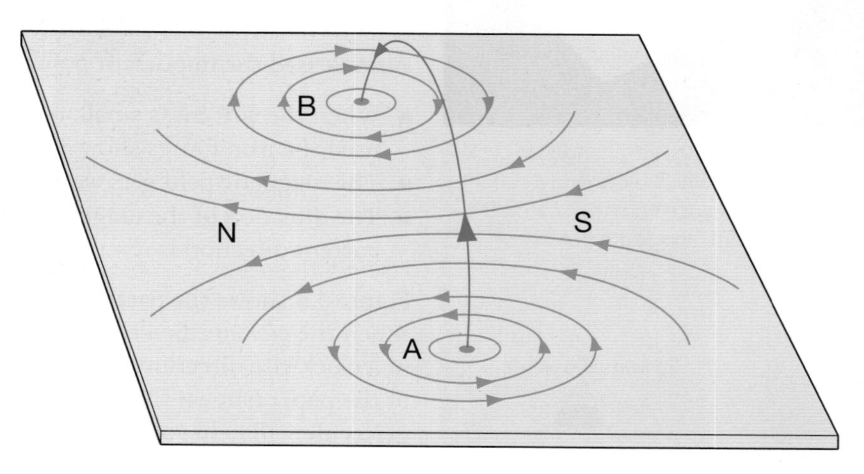

Figure 3.5 The magnetic field near a single loop of wire.

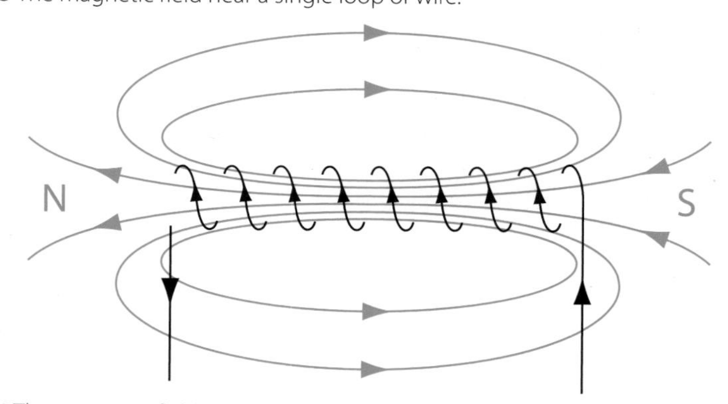

Figure 3.6 The magnetic field near a long solenoid.

TIP

You are expected to know the magnetic field patterns for the following when they carry a current: a long straight wire, a flat circular coil and a solenoid.

Producing large magnetic fields

The strength of the magnetic field produced by a solenoid can be increased by:

- using a larger current
- using more turns of wire
- putting some iron into the middle of the solenoid.

But there is a limit to how strong you can make the magnetic field. If the current is made too large the solenoid will get very hot and start to melt. For this reason large solenoids must be cooled by water. There is also a limit to the number of turns of wire you can put into a space.

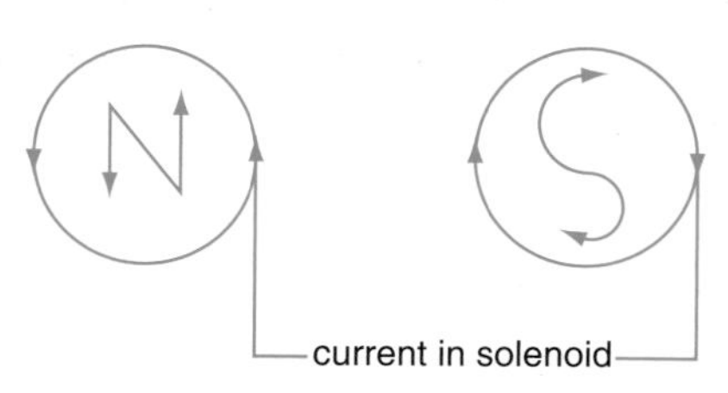

Figure 3.7 This is a good way to work out the polarity of the end of a solenoid.

TIP

A long solenoid and a bar magnet both produce a magnetic field of the same shape.

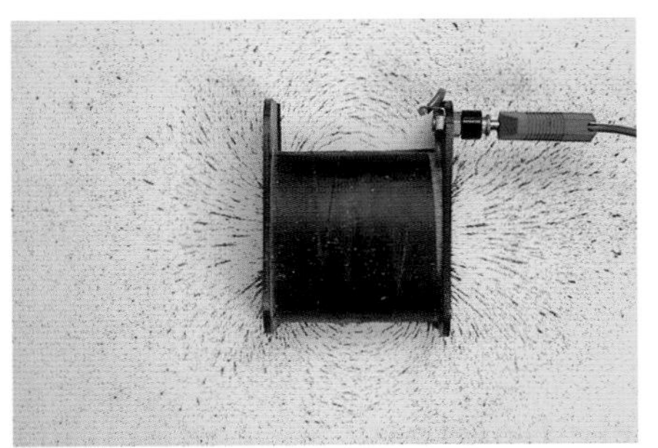

Figure 3.8 The iron filings show the shape of the magnetic field around a solenoid.

Nowadays, the world's strongest magnetic fields are produced by **superconducting magnets**. At very low temperatures (about 4 K) some materials, such as niobium and lead, become superconductors. A superconductor has no electrical resistance. This means that a current does not cause any heating effect. So a large magnetic field can be produced by an enormous current (5000 A) through a solenoid, which is kept cold in liquid helium.

Figure 3.9 Some superconducting magnets are so strong that they can lift the weight of a train. The Shanghai Airpot Maglev Train (derived from magnetic levitation) is lifted and driven by superconducting magnets. The magnetic levitation reduces friction; the train has a top speed of 430 km/h.

STUDY QUESTIONS

1 Figure 3.10 shows two plotting compasses, one above and one below a wire. Draw diagrams to show the position of the needles when:

a) there is no current

b) the current is very large (30 A)

c) the current is small (1 A). A current of 1 A produces a magnetic field near the wire, which is about the same size as the Earth's magnetic field.

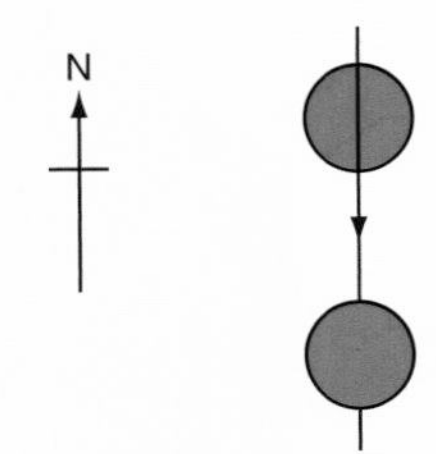

Figure 3.10

2 Figure 3.11 shows a long Perspex tube with wire wrapped around it to make a solenoid.

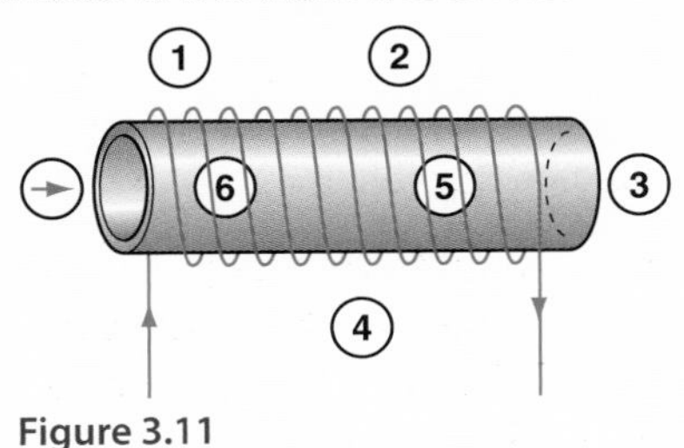

Figure 3.11

a) Copy the diagram and mark in the direction of the compass needles 1–6, when there is a current through the wire.

b) Which end of the solenoid acts as the south pole?

c) In which direction do the needles point when the current is reversed?

d) Copy the diagram again, leaving out the compasses. Draw magnetic field lines round the solenoid.

3 Figure 3.12 represents a wire placed vertically, with the current coming out of the paper towards you. Copy it and draw the magnetic field lines round the wire, showing how the field decreases with distance.

Figure 3.12

4 Make sketches to show the shape of the magnetic fields near to each of the following when they carry a current:

a) a long straight wire

b) a flat circular coil

c) a solenoid.

6.4 The motor effect

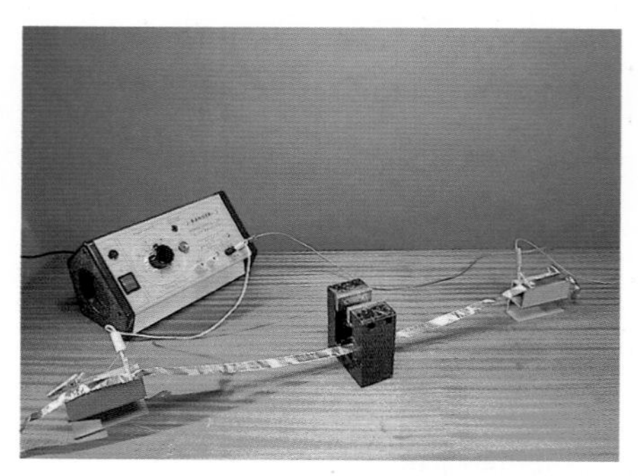

Figure 4.1 Aluminium foil carrying a current is pushed out of a magnetic field.

In Figure 4.1 you can see a piece of aluminium foil that was positioned between the poles of a strong magnet. A current through the foil has caused it to be pushed down, away from the poles of the magnet. Reversing the current makes the foil move upwards, away from the poles of the magnet. This is called the **motor effect**. It happens because of an interaction between the two magnetic fields, one from the magnet and one from the current. The magnet is very strong and could damage delicate electronics, etc., so keep watches, mobile phones and bank cards well away.

Combining two magnetic fields

In Figure 4.2 you can see the way in which the two fields combine. By itself the field between the poles of the magnet would be nearly uniform. The current through the foil produces a circular magnetic field. The magnetic field from the current squashes the field between the poles of a magnet. It is the squashing of the field that catapults the foil, upwards in this case.

The force acting on the foil is proportional to:

- the strength of the magnetic field between the poles
- the current
- the length of the foil between the poles.

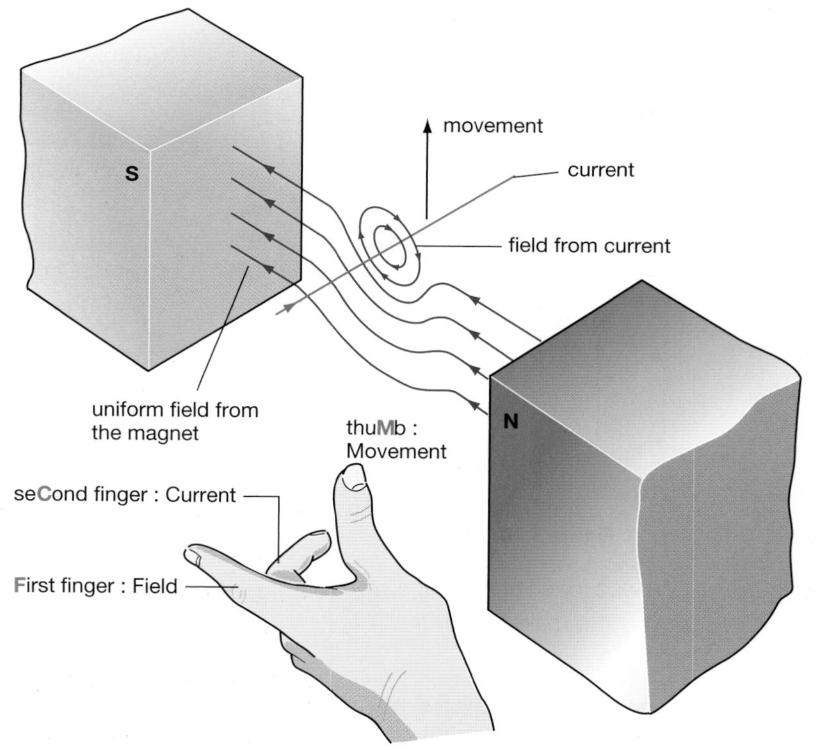

Figure 4.2 Left-hand rule: you can use the left hand to predict the direction of movement of this wire.

The deflection of a wire in a magnetic field is at right angles to both the current direction and the magnetic field direction.

The left-hand rule

To predict the direction in which a straight conductor moves in a magnetic field you can use the **left-hand rule** (Figure 4.2). Spread out the first two fingers and the thumb of your left hand so that they are at right angles to each other. Let your first finger point along the direction of the magnet's field, and your second finger point in the direction of the current. Your thumb then points in the direction in which the wire moves.

This rule works when the field and the current are at right angles to each other. When the field and the current are parallel to each other, there is no force on the wire and it stays where it is.

Ammeters

Figures 4.3 and 4.4 show the idea behind a moving-coil ammeter. A loop of wire has been pivoted on an axle between the poles of a magnet. When a current is switched on, the left-hand side of the loop moves downwards and the right-hand side moves upwards. (Use the left-hand rule to check this.) If nothing stops the loop, it turns until side DC is at the top and AB is at the bottom. However, when a spring is attached to the loop it only turns a little way. When you pass a larger current through the loop, the force is larger. This will stretch the spring more and so the loop turns further.

Figure 4.5 shows how a model ammeter can be made using a coil of wire, a spring and some magnets.

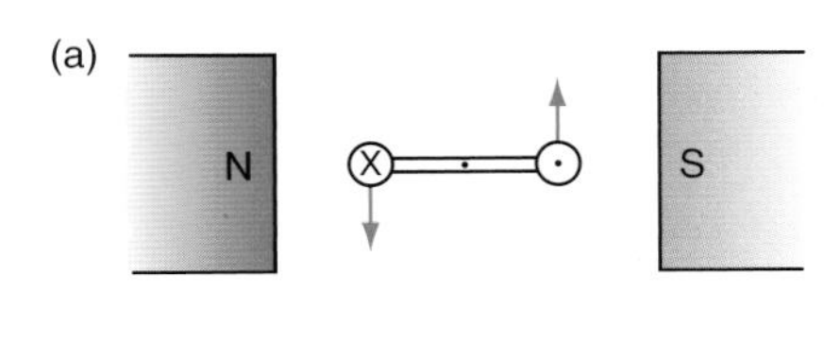

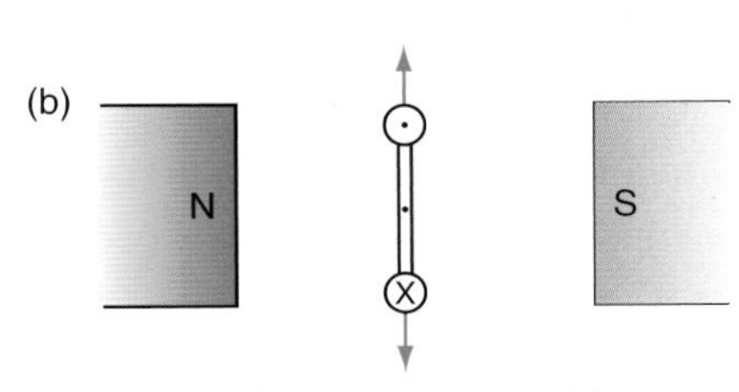

Figure 4.4 (a) The end-on view of the wire loop in Figure 4.4. The turning effect is maximum in this position. **(b)** In this position the forces acting on the coil produce no turning effect.

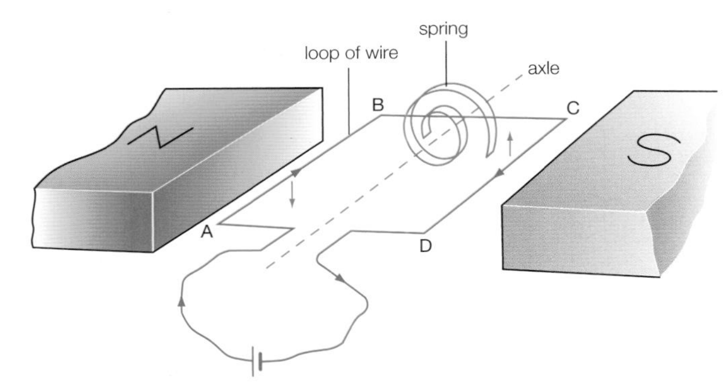

Figure 4.3 The principle of the moving-coil ammeter.

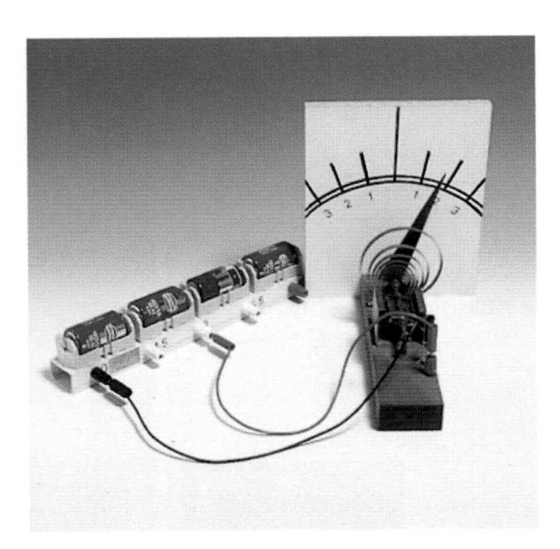

Figure 4.5 This shows a model ammeter. The coil turns in the magnetic field and is constrained by a spring.

An ammeter is sensitive if it turns a long way when there is a small current through it. The sensitivity of an ammeter will be large when:

- a large number of turns is used on the coil
- strong magnets are used
- weak springs are used.

Deflection of charged particles

An electric current is a flow of charged particles (usually carried by electrons in wires). You have just read that a wire carrying a current experiences a force in a magnetic field. In the same way, a beam of moving charged particles can be deflected by a magnetic field (Figure 4.6).

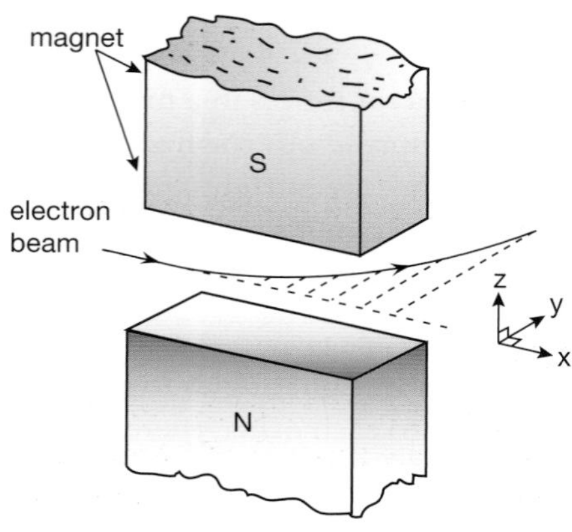

Figure 4.6 Use the left-hand rule to check that electrons are deflected by this magnetic field into the plane of the paper.

The size of the force depends on:

- the strength of the magnetic field
- the speed of the particles
- the charge on the particle.

The direction of the deflecting force on the particles can be predicted using the left-hand rule. Particles that are moving parallel to the magnetic field do not experience a deflecting force.

TIP

Remember that if the electrons are travelling to the right, the direction of the current is to the left.

STUDY QUESTIONS

1 State two ways of increasing the force on a conductor carrying a current in a magnetic field.

2 State how is it possible to position a wire carrying a current in a magnetic field so that there is no force acting on the wire.

3 Use the left-hand rule to predict the direction of the force on the wire in each of the following cases.

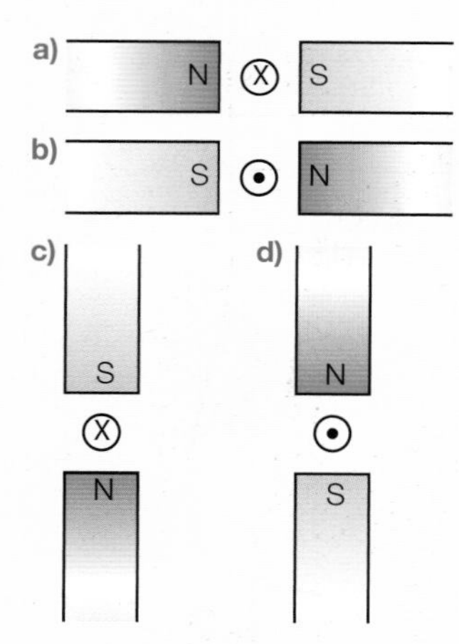

Figure 4.7

4 Look at Figure 4.3. Explain why there is no force acting on the side BC.

5 **a)** Figure 4.8 shows: (i) a pair of magnets, (ii) a wire carrying a current into the paper. Sketch separate diagrams to show the magnetic fields near each of (i) and (ii), when they are well separated.

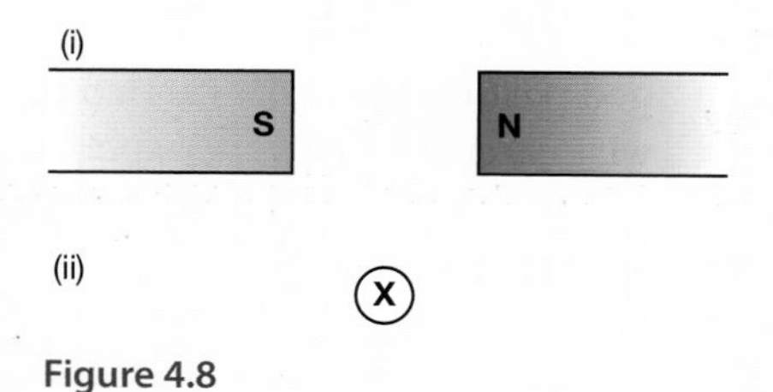

Figure 4.8

b) The wire is now placed between the poles of the magnets. Sketch the combined magnetic field. Add an arrow to show the direction in which the wire moves.

c) Which way does the wire move when:

i) the current is reversed,

ii) the north and south poles are changed round?

d) Give two ways in which the force on the wire could be increased.

6 Explain the following:

a) Beta particles are easily deflected by a magnetic field.

b) Alpha particles only show a very small deflection in a magnetic field.

c) Gamma rays cannot be deflected by a magnetic field. *[You may have to refer to chapter 7.2 to answer this question.]*

7 Figure 4.9 shows a pair of magnets. Suggest what will happen to each of the following as they pass from left to right through the magnets:

a) the positive ions

b) the negative ions

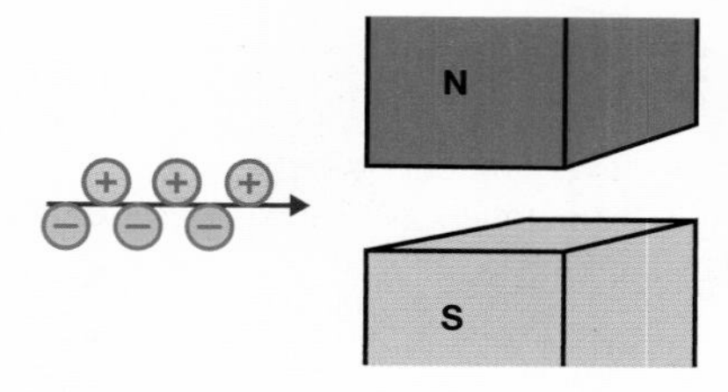

Figure 4.9

6.5 Electric motors

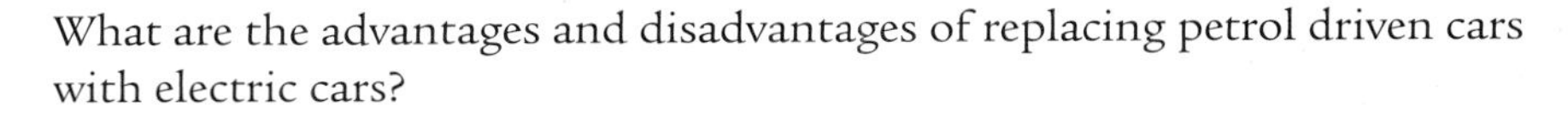

What are the advantages and disadvantages of replacing petrol driven cars with electric cars?

Figure 5.1 This electric bike carries a city worker to her office. What does it have in it to make it work?

The principle of the motor

In the last chapter you learnt that a coil carrying a current rotates when it is in a magnetic field. However, the coil can only rotate through 90° and then it gets stuck. This is no good for making a motor; we need to make a motor rotate all the time.

Figure 5.2 shows the design of a simple motor which you can make for yourself. A coil carrying a current rotates between the poles of a magnet. The coil is kept rotating continuously by the use of a **split-ring commutator**, which rotates with the coil between the carbon brush contacts. This causes the direction of the current in the coil to reverse, so that forces continue to act on the coil to keep it turning.

pivot
rotation produced
magnet
coil
A
N
S
split-ring commutator
B
carbon brushes

Figure 5.2 A design for a simple motor.

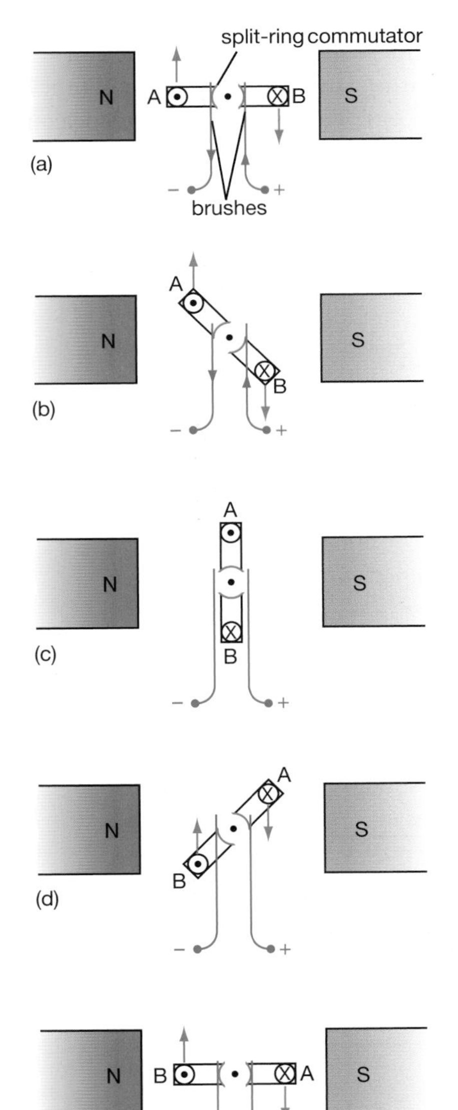

Figure 5.3

How the commutator works

Figure 5.3 explains the action of the commutator. The steps are explained on the next page.

(a) A current flows into the coil through the commutator so that there is an upwards push on side A, a downwards push on *B*.

(b) The coil rotates in a clockwise direction.

(c) When the coil reaches the vertical position there is no current through the coil. It continues to rotate past the vertical due to its own momentum.

(d) Now side A is on the right-hand side. The direction of the current has been reversed.

(e) Side A is pushed down and side B is pushed upwards. The coil continues to rotate in a clockwise direction.

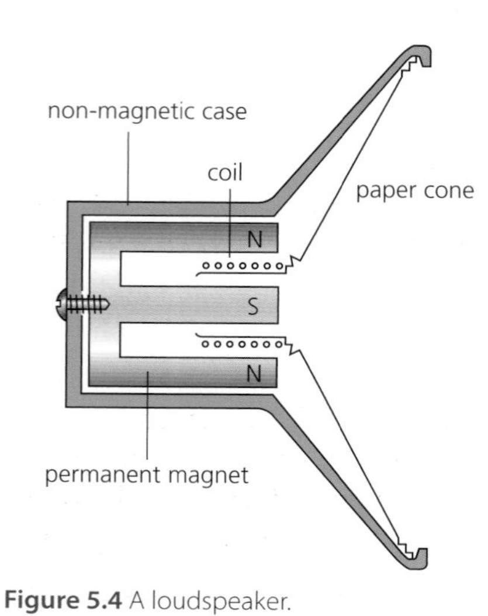

Figure 5.4 A loudspeaker.

The moving-coil loudspeaker

You can check the forces on the loudspeaker coil by looking at Figure 5.4. When the current direction in the upper side of the coil is into the paper, the current direction in the lower side of the coil is out of the paper. If you now apply the left hand rule to both sides of the coil, you will find that both sides experience a force to the left. When the current is reversed, the coil experiences a force to the right. An alternating current causes the coil (and therefore the paper cone) to vibrate in and out. This is how the loudspeaker produces sound waves.

STUDY QUESTIONS

1 Give three ways that you can increase the speed of rotation of the coil of an electric motor.

2 This question refers to the loudspeaker in Figure 5.4. The current direction is into the page as we look at the top of the coil and out of the page at the bottom.
  - **a)** State what happens to the coil when the direction of the current is reversed.
  - **b)** Explain what happens to the coil when an a.c. current is supplied to it.
  - **c)** Explain what happens to the sound produced by the loudspeaker when these changes are made to an a.c. supply to the coil:
    - **i)** the frequency is increased
    - **ii)** the size of the current is increased.

3 Figure 5.5 shows a model electric motor. In the diagram, the current goes round the coil in the direction A to B to C to D.

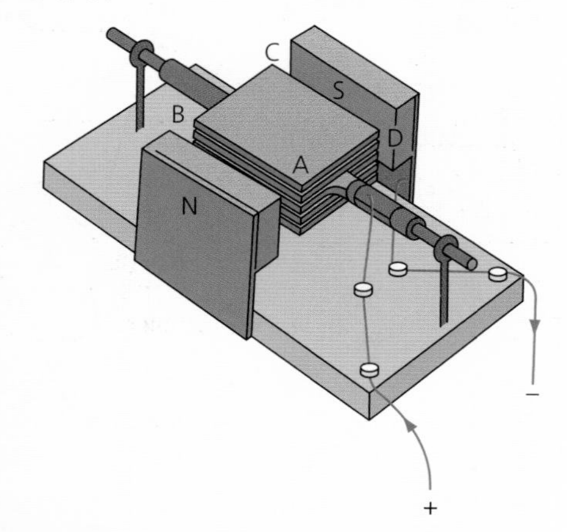

Figure 5.5

  - **a)** State the direction of the force on:
    - **i)** side AB
    - **ii)** side CD.
  - **b)** State the direction of the coil's rotation: clockwise or anticlockwise.
  - **c)** Discuss in which position is the coil most likely to stop.

4 Figure 5.6 shows an electric motor that is made using an electromagnet. The arrows show the direction of the forces on the coil.
  - **a)** The battery is now reversed. What effect does this have on:
    - **i)** the polarity of the magnet
    - **ii)** the direction of the current in the coil
    - **iii)** the forces acting on the coil?
  - **b)** Explain why this motor works using an a.c. supply as well as a d.c. supply.
  - **c)** Explain why the electromagnet and coil run in parallel from the supply rather than in series.

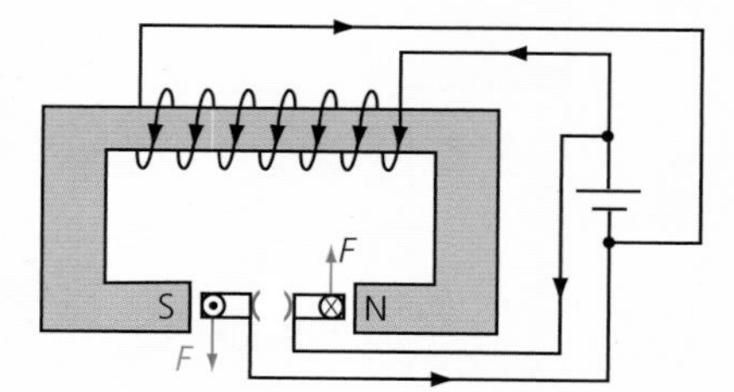

Figure 5.6

6.6 Electromagnetic induction

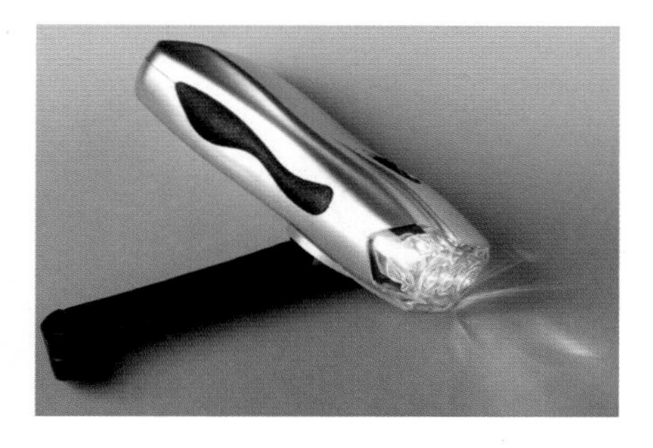

Figure 6.1

You can make the torch in Figure 6.1 work by winding the handle. Suggest how the torch works and discuss the energy transfers that allow the torch to shine line.

Induced voltage

When a conducting wire moves through a magnetic field, a voltage difference is produced across the ends of the wire. A voltage produced in this way is described as an induced voltage. A voltage will also be induced if there is a change in the magnetic field around a stationary conducting wire.

If the wire is part of a complete circuit, the voltage will cause a current in the circuit. This is an induced current.

Inducing a voltage or a current in this way is called the generator effect, because this is how we generate electricity.

Investigating induced voltages

You can investigate which direction you must move the wire in to make a current flow around the circuit in Figure 6.2. You can move the wire backwards and forwards along the directions XX', YY' or ZZ'.

You only induce a voltage and make a current when you move the wire up and down along XX'. The wire must cut across the magnetic field lines.

- Reversing the direction of motion reverses the direction of the voltage. If moving the wire up makes the meter move to the right, moving the wire down will make the meter move to the left.
- Reversing the direction of the magnetic field reverses the direction of the voltage.

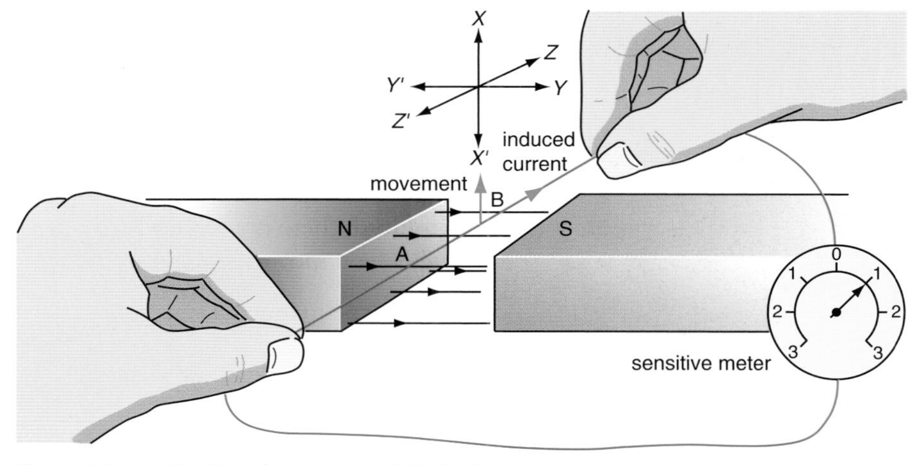

Figure 6.2 Investigating electromagnetic induction.

You can also investigate what affects the size of the induced voltage.

- Moving the wire more quickly induces a greater voltage. When the wire is stationary no voltage is induced.
- Using stronger or bigger magnets induces a greater voltage for the same speed of movement.

Coils and magnets

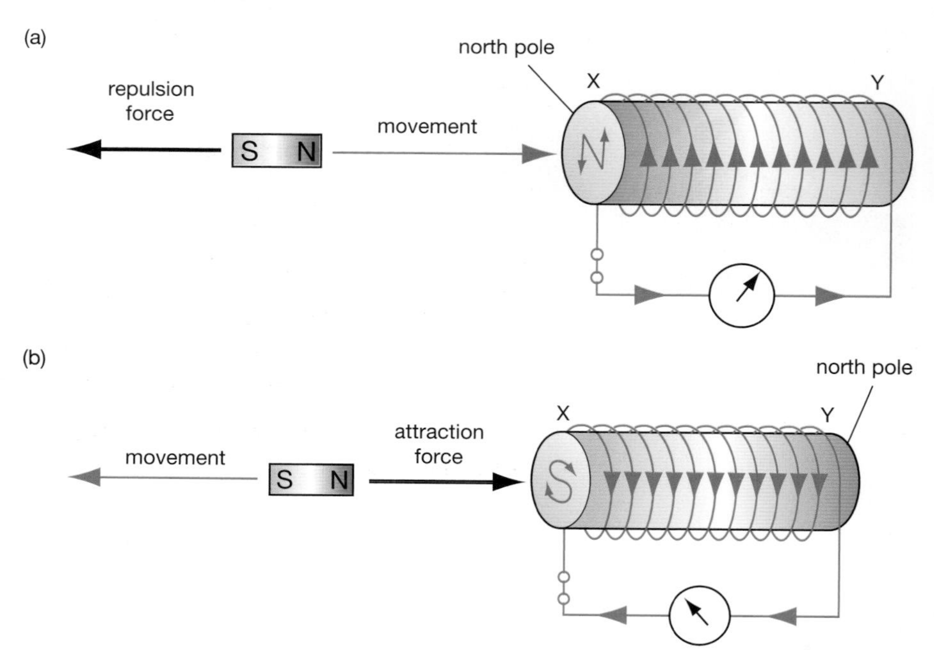

Figure 6.3(a) A north pole of a magnet is pushed into a solenoid. **(b)** The magnet is pulled out of the solenoid.

We can also induce a voltage in a coil of wire by changing the magnetic field near it. In Figure 6.3 a north pole of a magnet is pushed into a solenoid. Note that the induced current direction makes the end of the solenoid (X), next to the magnet, also behave like a north pole.

When the magnet is pulled out of the solenoid, the direction of the induced current is reversed. Now the end X behaves like a south pole.

When a current is induced in a conducting wire, the induced current itself generates a magnetic field. This magnetic field opposes whatever the original change was that caused it. This may be the movement of the conducting wire or the change in the magnetic field.

Consider Figure 6.3. In each case as the magnet is moved and a current induced, there is a force which opposes the movement of the magnet. So when you move a magnet to induce a current you do work.

When the magnet is stationary near the coil, no voltage is induced.

An electromagnetic flow meter

Figure 6.4 shows a way to measure the rate of flow of oil through an oil pipeline. A small turbine is placed in the pipe, so that the oil flow turns the blades round. Some magnets have been placed in the rim of the turbine, so that they move past a solenoid. These moving magnets induce a voltage in the solenoid which can be measured on an oscilloscope (Figure 6.4(b)). The faster the turbine rotates, the larger is the voltage induced in the solenoid. By measuring this voltage an engineer can tell at what rate the oil is flowing.

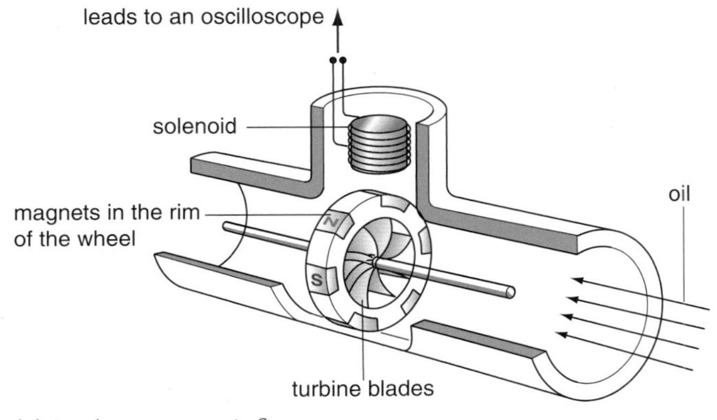

(a) An electromagnetic flow meter.

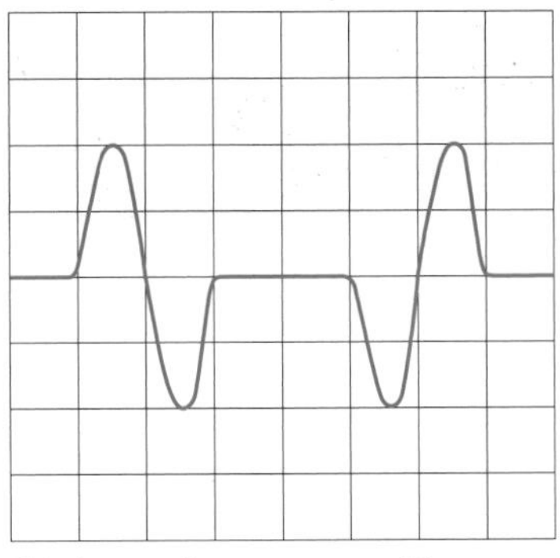

(b) The oscilloscope trace. The timebase is set so that the dot crosses the screen at a rate of one square every 0.02 seconds.

Figure 6.4

STUDY QUESTIONS

1 A wire is made to move through a magnetic field so that a voltage is induced across its ends. State two ways in which the size of the induced voltage can be increased.

2 This question refers to the magnet and solenoid shown in Figure 6.3.
  a) The magnet is stationary inside the solenoid. State the size of the induced potential difference.
  b) The magnet is now moved towards the solenoid as follows. In each case state the direction of the deflection on the meter.
    i) A south pole is moved towards X.
    ii) A south pole is moved away from X.
    iii) A north pole is moved towards Y.

3 This question refers to the flow meter shown in Figure 6.4.
  a) The poles of the magnets on the wheel are arranged alternately with a north pole then south pole facing outwards. Use this fact to explain the shape of the trace on the oscilloscope.
  b) Sketch the trace on the oscilloscope for these two separate changes:
    i) the number of turns in the solenoid is made 1.5 × as large
    ii) the oil flow rate increases, so the turbine rotates at twice the speed.

6.7 Generators

Find out how many power stations there are in your country.

Figure 7.1 The same principles of electromagnetic induction apply in power stations but the scale is much bigger. The generator shown in Figure 7.1 can deliver 200 MW of power.

TIP

Doubling the speed of rotation of a generator doubles the frequency and the maximum value of the induced voltage.

The a.c. generator (alternator)

Figure 7.2 shows the design of a very simple **alternating current (a.c.)** generator. By turning the axle you can make a coil of wire move through a magnetic field. This causes a voltage to be induced between the ends of the coil.

You can see how the voltage waveform, produced by this generator, looks on an oscilloscope screen (Figure 7.3(a)). In position (i) the coil is vertical with AB above CD (Figure 7.3(b)). In this position the sides CD and AB are moving parallel to the magnetic field. No voltage is generated since the wires are not cutting across the magnetic field lines.

When the coil has been rotated through a ¼ turn to position (ii), the coil produces its greatest voltage. Now the sides CD and AB are cutting through the magnetic field at the greatest rate.

In position (iii), the coil is again vertical and no voltage is produced. In position (iv) a maximum voltage is produced, but in the opposite direction. Side AB is moving upwards and side CD downwards.

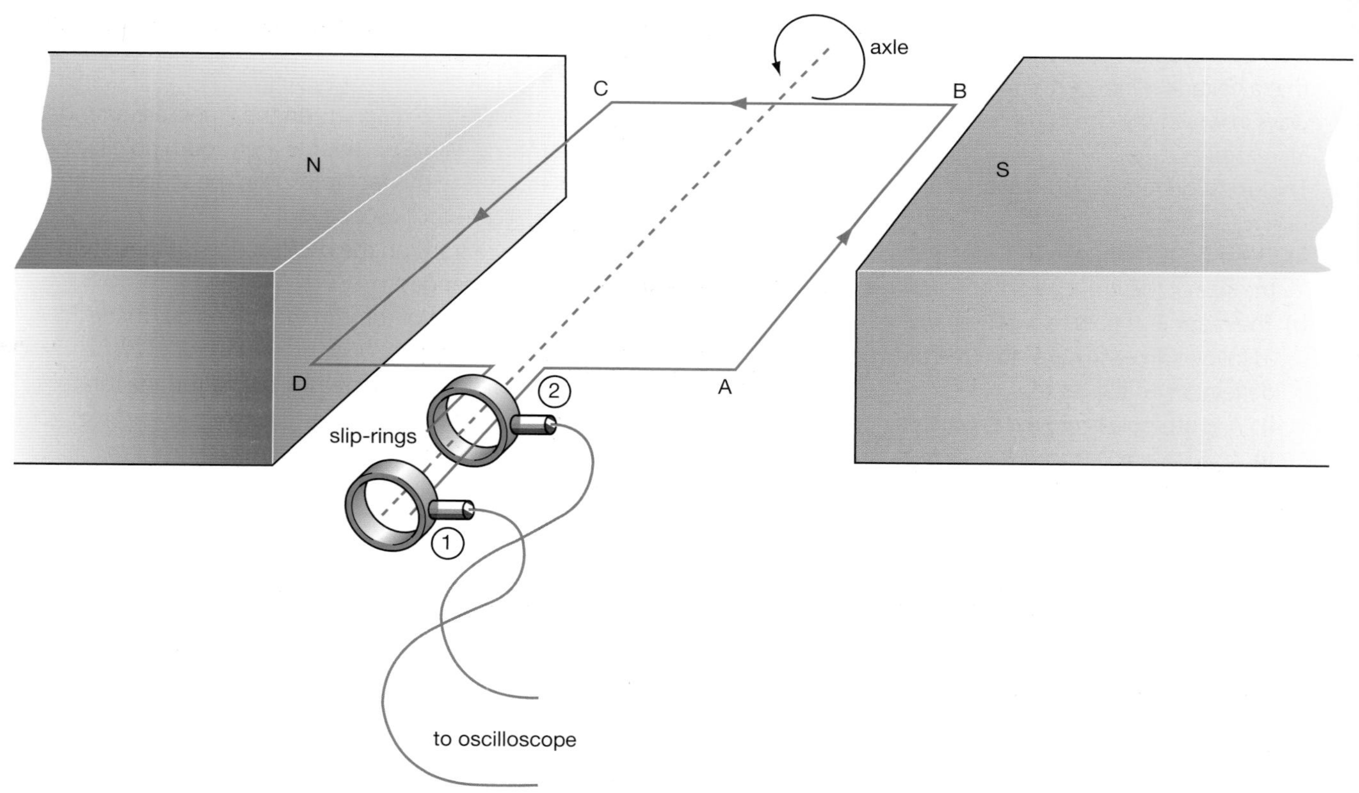

Figure 7.2 A simple a.c. generator.

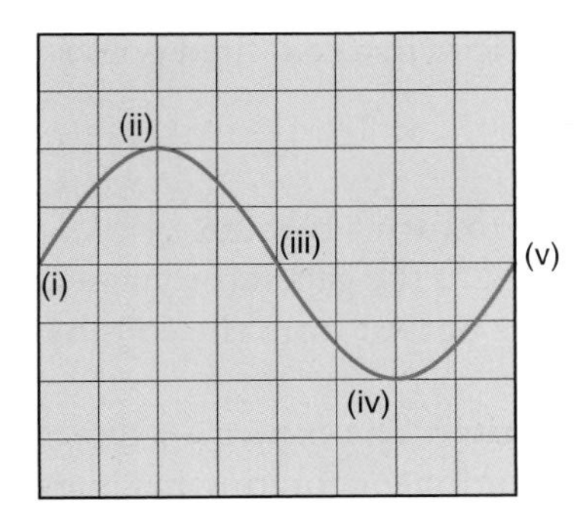

Figure 7.3(a) Voltage waveform.

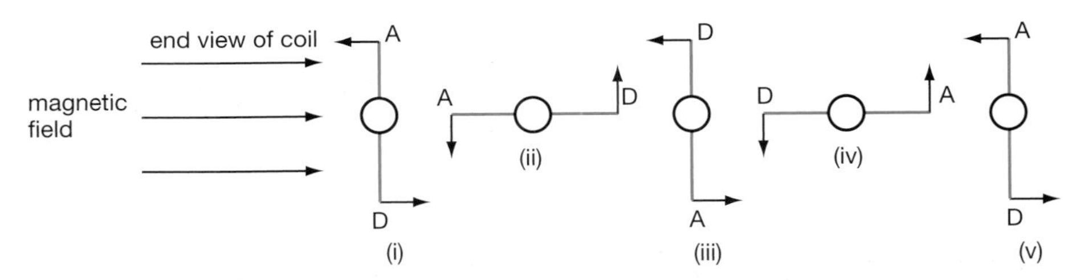

Figure 7.3(b) Position of coil.

The size of the induced voltage

The size of the induced voltage can be made larger by:

- rotating the coil faster
- using stronger magnets
- using more turns of wire
- wrapping the wire round a soft iron core.

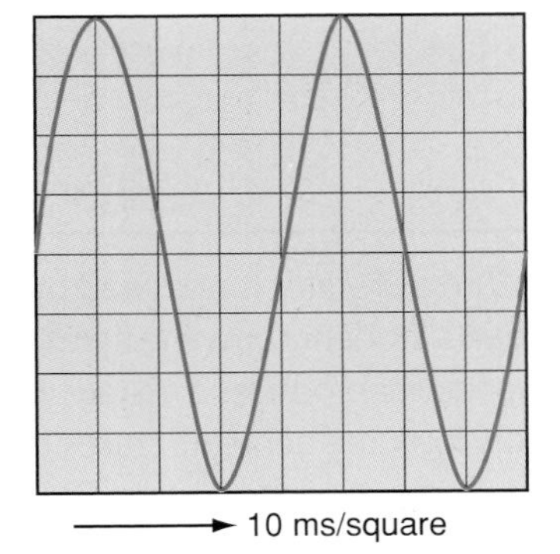

Figure 7.4 The generator is turned twice as fast.

Figure 7.4 shows the voltage waveform when the generator is rotated twice as quickly. There are two effects: the maximum voltage is twice as large; the frequency is doubled, i.e. the interval between the peaks is halved.

Producing power on a large scale

The electricity that you use in your home is produced by very large generators in power stations. These generators work in a slightly different way from the simple one you have seen so far.

Instead of having a rotating coil and a stationary magnet, large generators have rotating electromagnets and stationary coils (see Figure 7.5). The advantage of this set-up is that no moving parts are needed to collect the large electrical current that is produced.

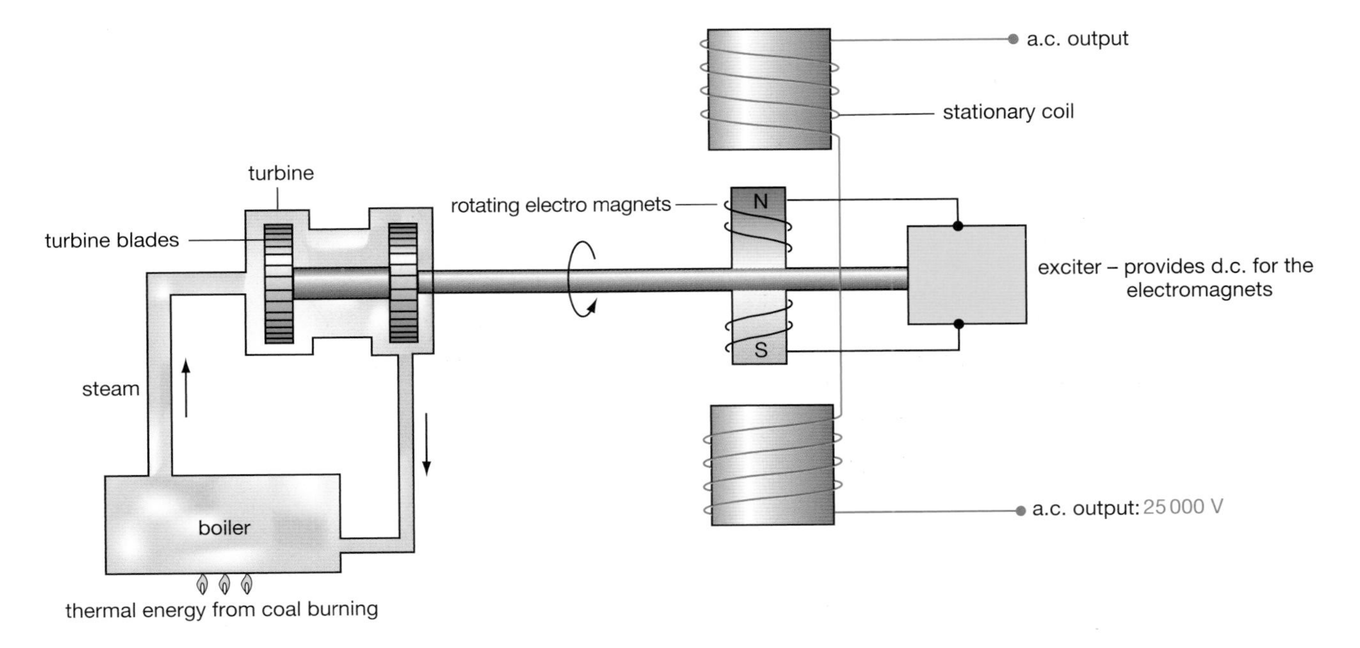

Figure 7.5 Electricity generation in a coal-fired power station.

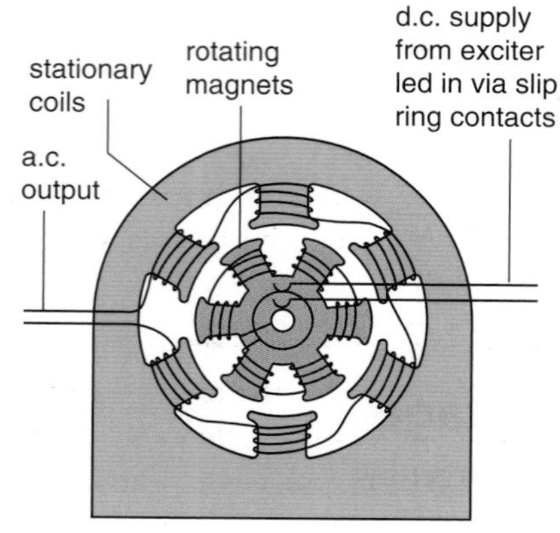

Figure 7.6 A section through a large generator.

The important steps in the generation of electricity in a coal-fired power station (Figure 7.5) are these:

1. Coal is burnt to boil water.
2. High-pressure steam from the boiler is used to turn a turbine.
3. The drive shaft from the turbine is connected to the generator magnets, which rotate near to the stationary coils. The output from the coils has a voltage of about 25 000 V.
4. The turbine's drive shaft also powers the **exciter**. The exciter is a direct current (d.c.) generator that produces current for the rotating magnets, which are in fact electromagnets.

STUDY QUESTIONS

1 The circuit diagram is for a 'shake-up' torch. A magnet is placed in a coil of wire, which is linked to a rechargeable battery and a light-emitting diode (LED). The magnet is free to slide backwards and forwards, down inside the coil.

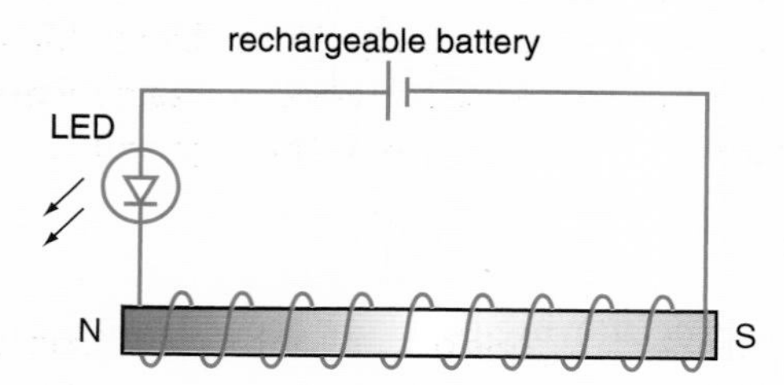

a) Explain what a light-emitting diode is.
b) Explain why a diode is used in this circuit.
c) Explain how the battery is recharged.

2 This question refers to the waveform in Figure 7.3(a). Copy the graph and add additional graphs to show what happens to the waveform as each of these changes is made. Explain your answers.
a) The coil rotation is reversed.
b) An extra turn of wire is added.
c) The coil is rotated twice as quickly.

3 In the diagram below, the generator is connected to a flywheel by a drive belt. The generator is operated by winding the handle. When a student turns the handle as fast as she can, the generator produces a voltage of 6 V. The graph shows how long the generator takes to stop after the student stops winding the handle. The time taken to stop depends on how many lamps are connected to the generator.
a) Explain the energy transfers that occur after the student stops winding the handle:
i) when the switches S_1, S_2 and S_3 are open (as shown),
ii) when the switches are closed.
b) Explain why the flywheel takes longer to stop when no lamp is being lit by the generator.
c) Make a copy of the graph. Use your graph to predict how long the flywheel turns when you make it light four lamps.
d) The experiment is repeated using lamps that use a smaller current. They use 0.15 A rather than 0.3 A. Using the same axes, sketch a graph to show how the time taken to stop depends on the number of lamps, in this second case.

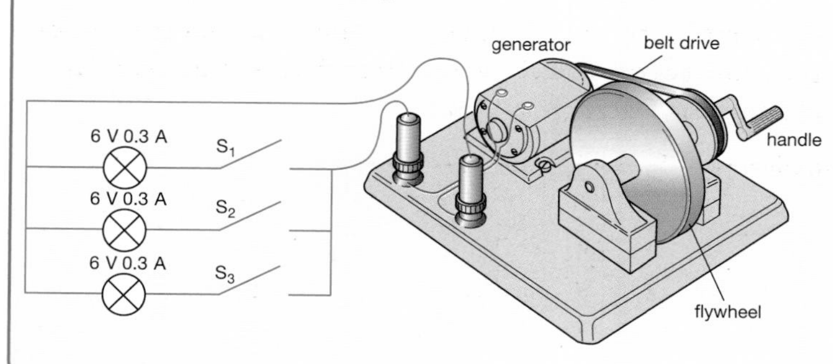

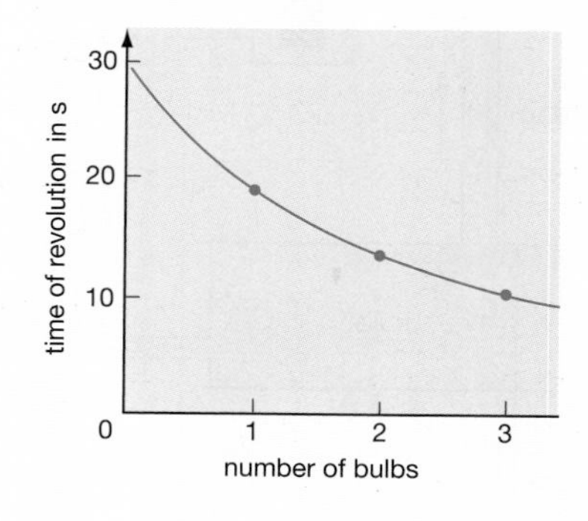

6.8 Transformers

Figure 8.1 is a familiar sight. The electricity is carried at very high voltages – as high as 400 000 V. Why is the electricity transmitted at such high voltages?

Figure 8.1 Pylons carrying electrical transmission lines.

Changing fields and changing currents

In Figure 8.2, when the switch is closed in the first circuit, the ammeter in the second circuit kicks to the right. For a moment a current flows through coil 2. When the switch is opened again the ammeter kicks to the left.

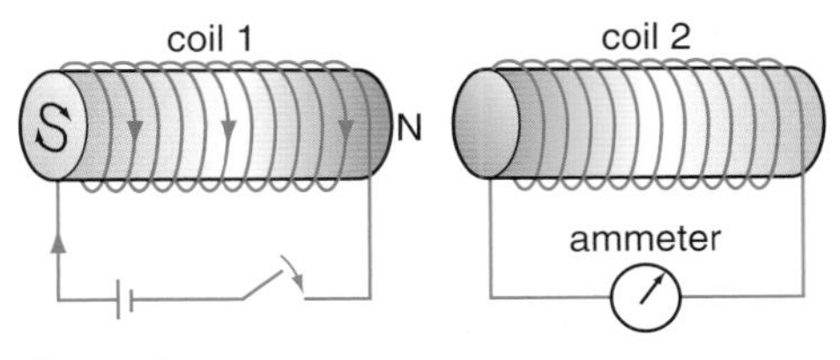

Figure 8.2

Closing the switch makes the current through coil 1 grow quickly. This makes the coil's magnetic field grow quickly. For coil 2 this is like pushing the north pole of a magnet towards it, so a current is induced in coil 2. When the switch is opened the magnetic field near coil 1 falls rapidly. This is like pulling a north pole away from coil 2. Now the induced current changes direction. The ammeter reads zero when there is a constant current through coil 1.

Transformers

A **transformer** is made by putting two coils of wire onto a soft iron core as shown in Figure 8.3. The primary coil is connected to a 2 V alternating current supply. The alternating current in the primary coil makes a magnetic field, which rises and then falls again. The soft iron core carries this changing magnetic field to the secondary coil and a changing voltage is induced in the secondary coil. In this way, energy can be transferred continuously from the primary circuit to the secondary circuit.

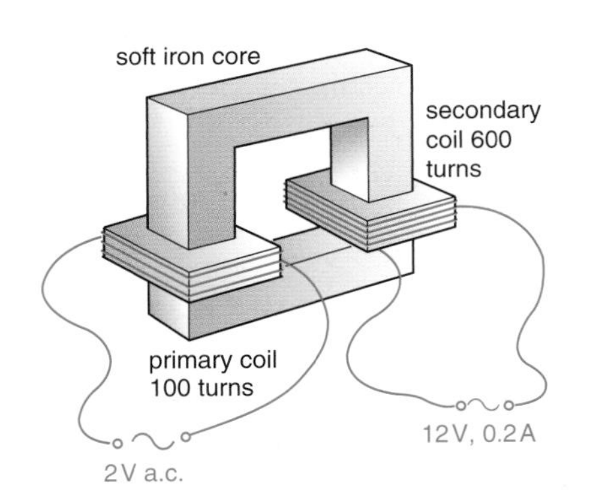

(a) A step-up transformer

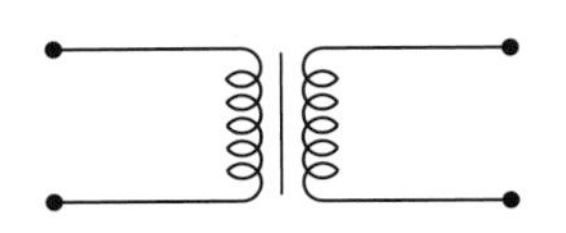

(b) The circuit symbol for a transformer

Figure 8.3

TIP

Transformers only work on a.c. supplies because an alternating current produces a changing magnetic field.

Transformers are useful because they allow you to change the voltage of a supply. For example, model railways have transformers that decrease the mains supply from 230 V to a safe 12 V; these are **step-down transformers**. The transformer in Figure 8.3 steps *up* the voltage from 2 V to 12 V; this is a **step-up transformer**.

To make a step-up transformer the secondary coil must have more turns of wire in it than the primary coil. In a step-down transformer the secondary has fewer turns of wire than the primary coil.

The rule for calculating voltages in a transformer is:

$$\frac{V_s}{V_p} = \frac{N_s}{N_p}$$

V_p = primary voltage; V_s = secondary voltage; N_p = number of turns on the primary coil; N_s = number of turns on the secondary coil.

Power in transformers

We use transformers to transfer electrical power from the primary circuit to the secondary circuit. Many transformers do this very efficiently and there is little loss of power in the transformer itself. For a transformer that is 100% efficient we can write:

power supplied by primary circuit = power used in the secondary circuit

$$V_p \times I_p = V_s \times I_s$$

The National Grid

Figure 8.4 shows how electricity, generated at power stations, is distributed around the country through the National Grid.

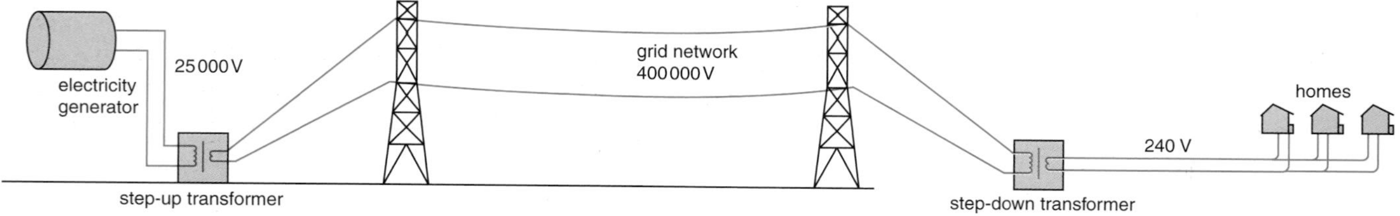

Figure 8.4 How the power gets to your home.

You may have seen a sign at the bottom of an electricity pylon saying 'Danger high voltage'. Power is transmitted around the country at voltages as high as 400 000 V. There is a very good reason for this – it saves a lot of energy. The following calculations explain why.

Figure 8.5 suggests two ways of transmitting 25 MW of power from a Yorkshire power station to the Midlands in the UK:

(a) The 25 000 V supply from the power station could be used to send 1000 A down the power cables.

(b) The voltage could be stepped up to 250 000 V and 100 A could be sent along the cables.

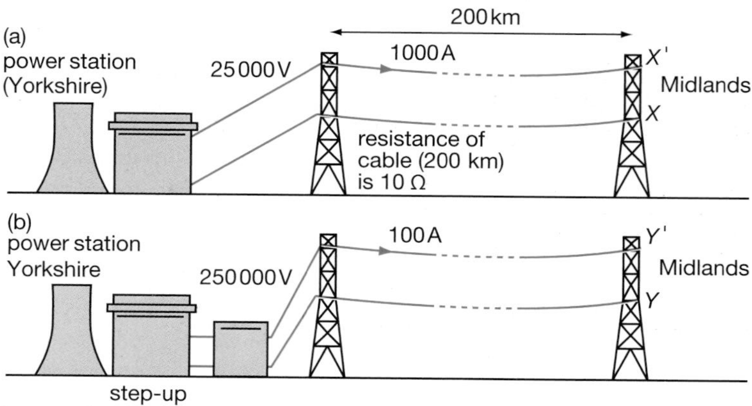

Figure 8.5

TIP

Electricity is transmitted at very high voltages and low currents to reduce energy losses.

How much power would be wasted in heating the cables in each case, given that 200 km of cable has a resistance of 10 Ω?

$$
\begin{aligned}
\text{(a) power} &= \text{voltage drop along cable} \times \text{current} \\
&= IR \times I \\
&= I^2R \\
&= (1000)^2 \times 10 \\
&= 10\,000\,000 \text{ W or } 10 \text{ MW}
\end{aligned}
$$

$$
\begin{aligned}
\text{(b) power} &= I^2R \\
&= (100)^2 \times 10 \\
&= 100\,000 \text{ W or } 0.1 \text{ MW}
\end{aligned}
$$

We waste a lot less power in the second case. The power loss is proportional to the square of the current. Transmitting power at high voltages allows smaller currents to flow along our overhead power lines.

STUDY QUESTIONS

1 a) In Figure 8.2 when the switch is opened the ammeter kicks to the left. Describe what happens to the ammeter during each of the following:
   i) The switch is closed and left closed so that there is a constant current through coil 1.
   ii) The coils are pushed towards each other.
   iii) The coils are left close together.
   iv) The coils are pulled apart.
  b) The battery is replaced by an a.c. voltage supply, which has a frequency of 2 Hz. Discuss what the ammeter shows when the switch is closed.

2 Explain why a transformer does not work when you plug in the primary coil to a battery. Why do transformers only work with an a.c. supply?

3 The question refers to Figure 8.3.
  a) Calculate the power used in the secondary circuit.
  b) Show that the current flowing in the primary circuit is 1.2 A.

4 Table 1 below gives some data about four transformers. Copy the table and fill the gaps.

Table 1

Primary turns	Secondary turns	Primary voltage in V	Secondary voltage in V	Step-up or step-down
100	20		3	
400	10000	10		
	50	240	12	
	5000	33000	11000	

5 Explain why very high voltages are used to transmit electrical power long distances across the country.

6 This is about how a transformer could be used to melt a nail (see the diagram below).

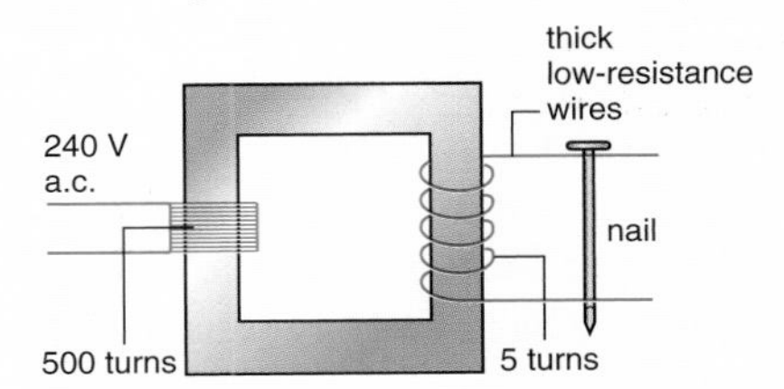

You will need to use the data provided.
- The nail has a resistance of 0.02 Ω.
- The melting point of the nail is 1540 °C.
- The nail needs 10 J to warm it through 1 °C.

  a) Calculate the voltage across the nail.
  b) Calculate the current flowing in:
   i) the secondary circuit
   ii) the primary circuit.
  c) Calculate the power in the secondary circuit that heats the nail.
  d) Estimate roughly how long it will take the nail to melt. Mention any assumptions or approximations that you make in this calculation.

Summary

I am confident that:

✓ **I can recall facts and concepts about magnetism**

- I know that magnets have a north and south pole.
- I understand that like poles repel and unlike poles attract.
- I understand the term magnetic 'field line'.
- I can draw a magnetic field pattern for a bar magnet accurately.
- I can describe experiments to plot magnetic field patterns.
- I can describe how to use two bar magnets to produce a uniform magnetic field.
- I understand that magnetism is induced in some materials when they are placed in a magnetic field.
- I can describe the properties of magnetically hard and soft materials.

✓ **I can recall facts and concepts about electromagnetism**

- I understand that an electric current in a conductor produces a magnetic field around it.
- I understand that a force is exerted on a current-carrying wire in a magnetic field.
- I understand the action of motors and loudspeakers.
- I can use the left-hand rule to predict the direction of the force when a wire carries a current in a magnetic field.
- I can describe the construction of electromagnets.
- I can draw and recognise the magnetic field patterns for a straight wire, a flat coil and a long solenoid when each carries a current.
- I understand that a moving charged particle experiences a force in a magnetic field (unless it moves parallel to the field).

✓ **I can explain electromagnetic induction:**

- I understand that a voltage is induced in a wire or coil when it moves through a magnetic field, or when the magnetic field changes its strength.
- I understand the factors that affect the size and direction of the induced voltage.
- I can describe the generation of electricity by the rotation of a magnet inside a coil of wire.
- I can describe the construction and action of a transformer.
- I can explain the action and uses of step-up and step-down transformers.
- I know and can use the relationship:
  $$\frac{V_S}{V_P} = \frac{N_S}{N_P}$$
- I know and can use the relationship:
  $V_P I_P = V_s I_s$
  for a transformer that is 100% efficient.
- I understand why very high voltages are used to transmit electricity on the national grid.

Sample answers and expert's comments

1 The diagram shows a bar magnet.

N	S

a) i) Copy the diagram. Add magnetic field lines to show the direction, shape and strength of the magnetic field near the magnet. (3)

ii) Label your diagram with the letter X to show two places where the field is strongest. (1)

b) i) Explain how the diagram shows where the field is strongest. (1)

ii) Explain how you would change your diagram to show a stronger bar magnet. (1)

(Total for question = 6 marks)

Student response Total 4/6	Expert comments and tips for success
a) i) X N S X ✔ ✔ ✔	This diagram gains 1 mark for the direction of the field and 1 mark for the shape of the field. But the diagram needs at least two extra lines to show the strength of the field properly, so the final mark is lost.
ii) See the Xs above.	The Xs are correctly placed so this gains the mark.
b) i) The field is strongest near the poles. O	This is correct, but it does not answer the question. The answer needs to link the strength of the field to the field lines being closer together.
ii) Draw more field lines. ✔	This is the correct idea.

2 The diagram shows a step-down transformer in the plug of an electric shaver, which is used in the mains socket in a bathroom.

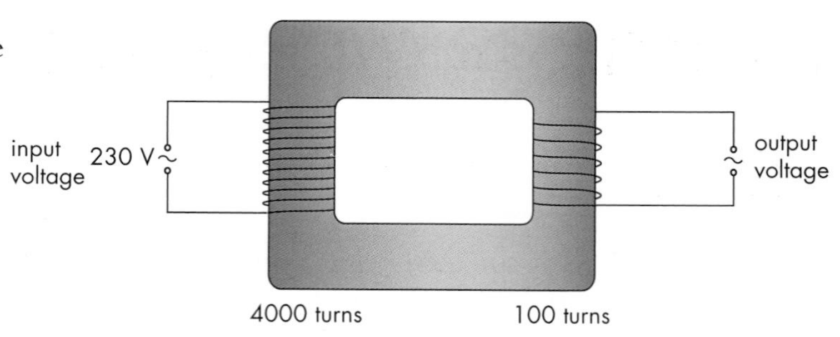

a) The manufacturer has written on the shaver that the output voltage of the transformer is 6 V.

Use the information in the diagram to show that the output voltage of the transformer is about 6 V. (3)

b) When the shaver is working normally, the current in it is 0.9 A and the voltage 6 V.

Calculate the input current from the mains, assuming that the transformer is 100% efficient in transferring electrical power from the input (primary) coil to the output (secondary) coil. (3)

(Total for question = 6 marks)

Student response Total 5/6	Expert comments and tips for success
a) $\frac{V_p}{V_s} = \frac{N_p}{N_s}$ ✓ $\frac{230}{V_s} = \frac{4000}{100}$ ✓ $\frac{230}{V_s} = 40$ $V_s = \frac{230}{40} = 5.75$ V ✓	The equation is written, the numbers substituted and the answer calculated correctly, so this gains full marks. However, since the question asks you to show the voltage is about 6 V, for completeness you should add that 5.75 V ≈ 6 V.
b) $V_pI_p = V_sI_s$ ✓ $230I_p = 6 \times 0.9$ ✓ $I_p = 0.23$ A O	Correct equation. Correct numbers (though 5.75 V could also be used). Calculator error loses a mark.

3 Plan an experiment to compare the strengths of two magnets. **(Total for question = 6 marks)**

Student response Total 6/6	Expert comments and tips for success
Put a compass near to a magnet as shown below. magnet X S N ✓ ✓ *Now place a second magnet as shown, so that the compass is equidistant* ✓ *from the two north poles.* magnet X S N magnet Y S N ✓ ✓ *This shows magnet Y is stronger than magnet X, as it repels the* ✓ *compass needle more strongly.*	The diagrams give a very clear answer, and it is far easier to explain what you are doing. When the question says 'explain', 'plan' or 'describe', you can use a diagram. If there is a diagram in a question, you can work on it or add to it.

Exam-style questions

1 Which of the following is suitable for the core of a transformer?

A brass

B steel

C iron

D copper [1]

2 Which of the following best represents the shape of the magnetic field near to a wire that is carrying a current?

A **B** **C** **D**

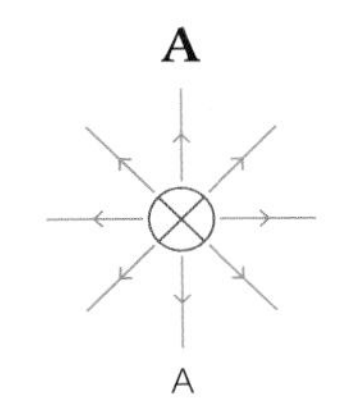

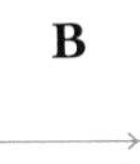

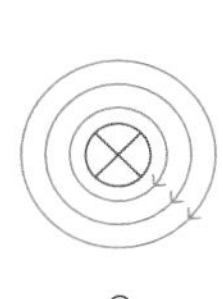

[1]

3 When S_1 and S_2 are closed, which of the following is true?

A The two coils repel each other.

B Both coils move to the right.

C The two coils attract each other.

D The attractive and repulsive forces cancel out. [1]

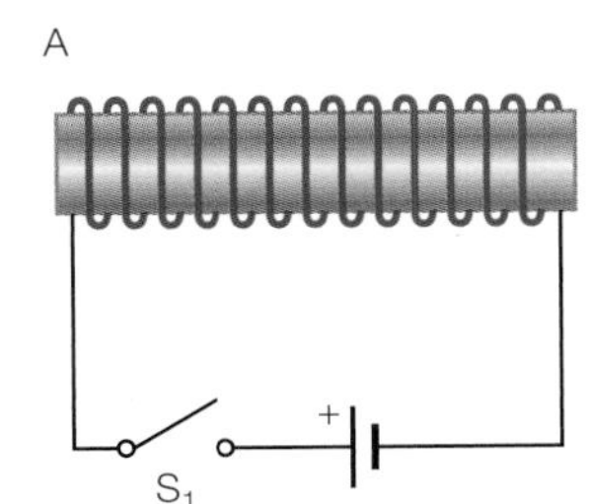

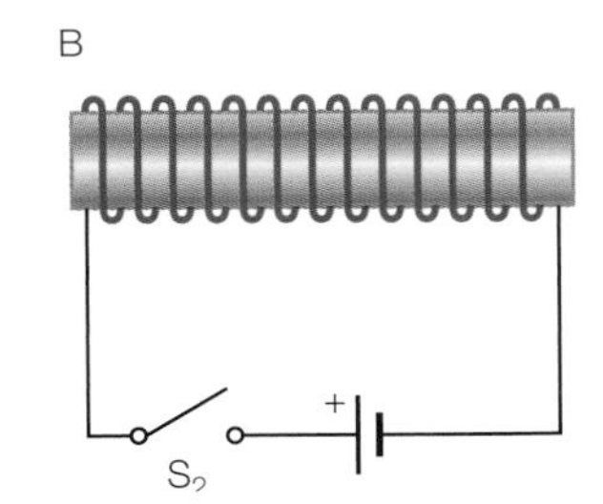

4 The switch S is closed and left closed. Which of the following could happen to the ammeter?

A The ammeter flicks to the right and then back to zero.

B The ammeter rises to a constant value.

C The ammeter flicks to the right, then to the left, and then back to zero.

D The ammeter reading rises quickly, then falls to zero after about 1 minute. [1]

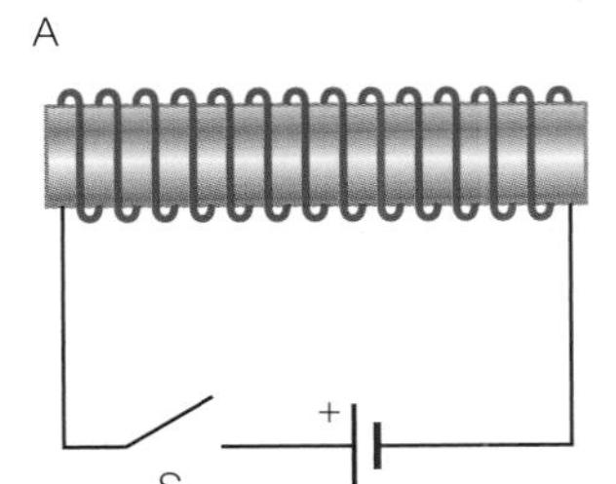

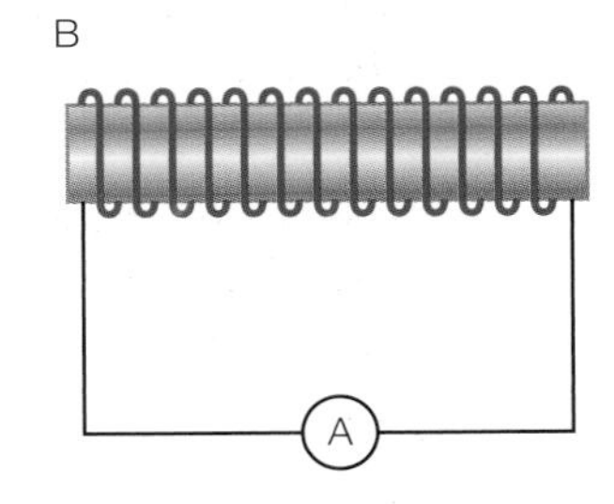

5 When the south pole of a magnet is pushed into end X of a solenoid, the ammeter kicks to the right.

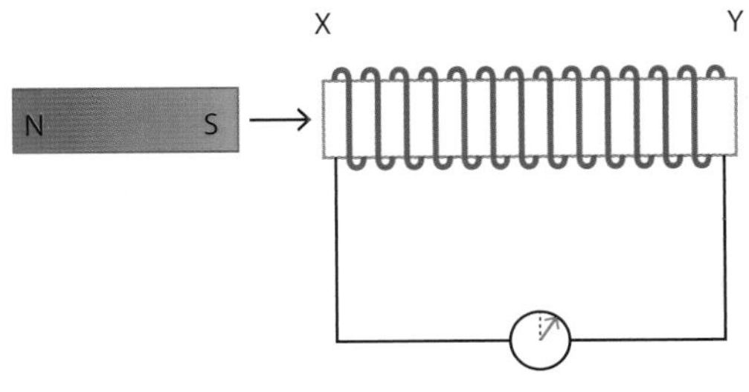

Which of the following actions would also make the ammeter kick to the right?

A moving a south pole away from X

B moving a north pole towards X

C moving a north pole towards Y

D moving a south pole towards Y [1]

6 The diagram below shows four compasses near to a magnet. Which compass is pointing in the wrong direction? [1]

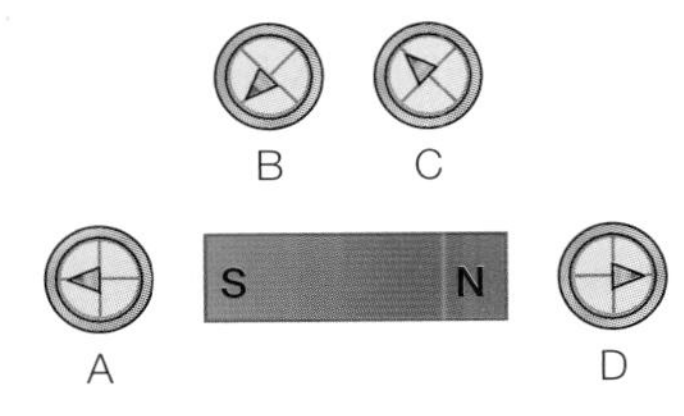

7 If the current flows in the direction of the arrow, in which direction will the conductor tend to move?

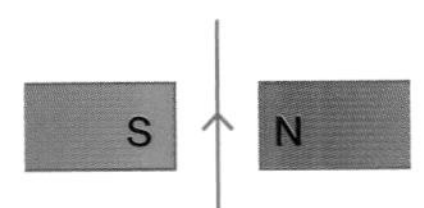

A toward the north-seeking pole of the magnet

B into the page (and perpendicular to it)

C out of the page (and perpendicular to it)

D toward the south-seeking pole of the magnet [1]

8 The diagram shows a step-up transformer.

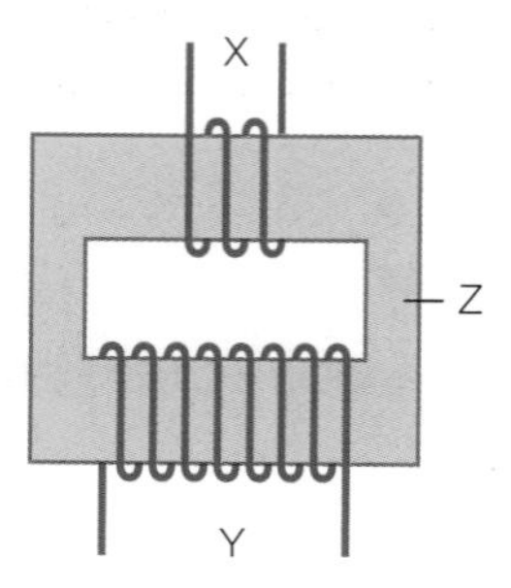

Which of the following lines is correct?

	X	Y	Z
A	secondary coil	primary coil	steel core
B	primary coil	secondary coil	steel core
C	secondary coil	primary coil	soft iron core
D	primary coil	secondary coil	soft iron core

[1]

9 The diagram shows a transformer.

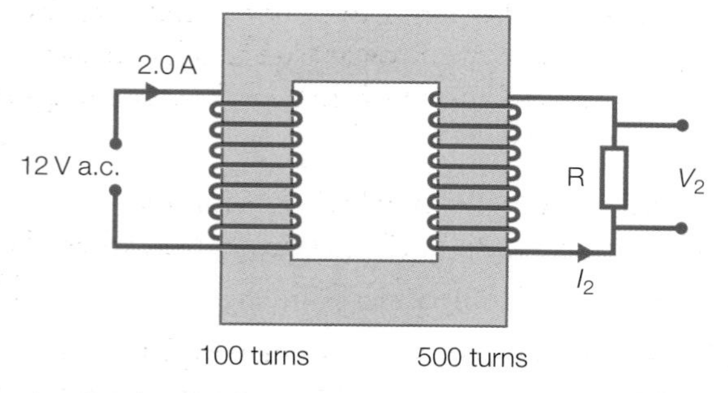

a) Which of the following is the correct value for V_2?

A 2.4 V **B** 6.0 V **C** 60 V **D** 120 V [1]

b) Which of the following is the correct value for I_2?

A 0.2 A **B** 0.4 A **C** 2.0 A **D** 10 A [1]

10 When a wire carries an electric current and it is placed in a magnetic field, a force may act on it.

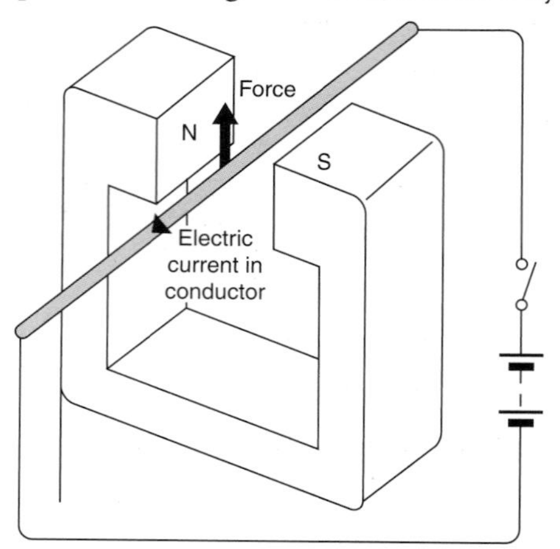

a) State **two** ways in which this force can be made larger. [2]

b) State **two** ways in which this force can be reversed. [2]

c) Describe the circumstances under which no force will act on this wire when it is carrying an electric current and it is placed in a magnetic field. [1]

11 The diagram shows part of a bicycle generator, which is in contact with the wheel.

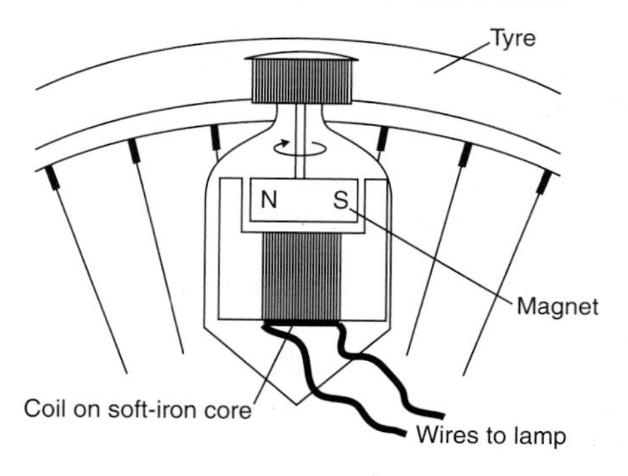

a) Explain why there is a current through the lamp when the bicycle wheel turns. [3]

b) Explain why the lamp gets brighter as the cycle moves faster. [1]

c) Explain why the lamp does not work when the bicycle is stationary. [1]

12 The diagram shows a transformer that is being used to light a 12 V bulb. The bulb only lights dimly.

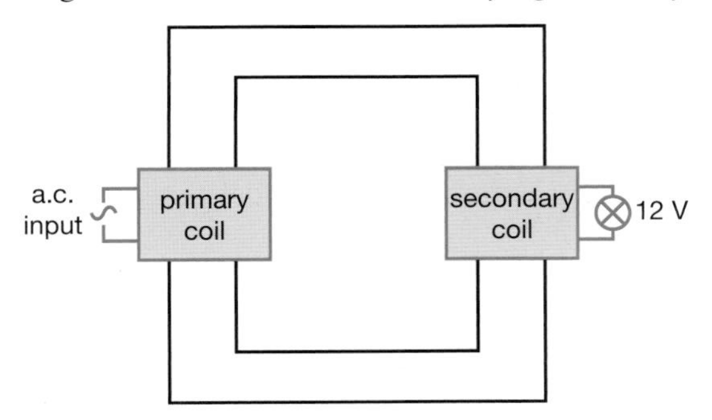

a) Describe **three** ways in which you could make the bulb light more brightly. [3]

b) Explain why a transformer only works using an a.c. supply. [3]

c) Electricity is distributed across countries using very high voltage networks.

Explain:

i) why very high voltages are used

ii) where transformers are used in the distribution of electricity. [4]

d) A power station generates 50 MW of power at 25 kV, to deliver to a transformer, which links the power station to the transmission lines. The transformer has 48 000 turns in its 400 kV secondary coil.

i) Calculate the number of turns in the transformer's primary coil.

ii) Calculate the current in the transformer's primary coil.

iii) Calculate the current in the transformer's secondary coil. [6]

13 The diagram below shows a coil connected to a sensitive meter. The meter is a 'centre zero' type: the needle is in the centre when no current flows.

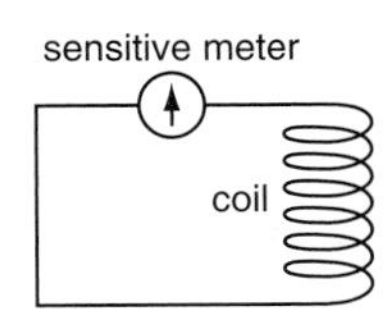

A student does two experiments.

First experiment

First the magnet is pushed into the coil. Then the magnet is removed. The metre only reads a current when the magnet is moving.

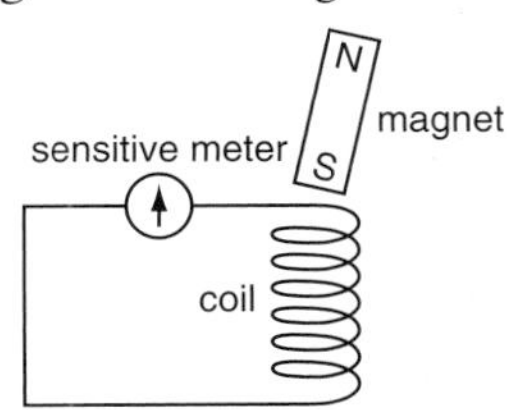

Second experiment

A second coil is brought close to the first. The student switched the current on and off in the second coil. The needle deflects for a brief time each time the current is turned on or off.

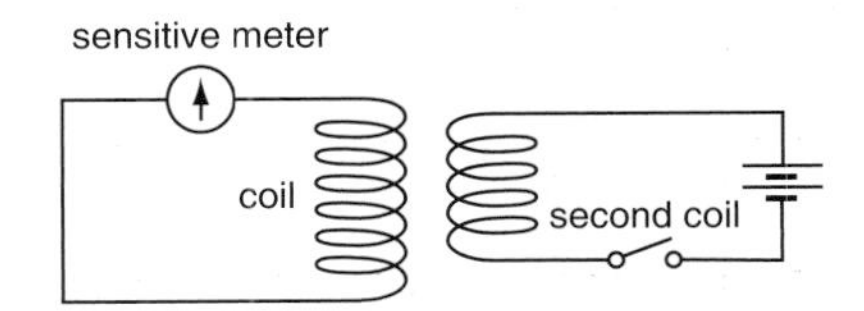

a) In the first experiment:

i) State what is happening in the coil while the magnet is moving. [1]

ii) State how the deflection of the needle, as the magnet is pushed towards the coil, compares with the deflection of the needle as the magnet is pulled away. [1]

iii) State three ways in which the student could increase the deflection on the meter, when the magnet is moved. [3]

b) Use the second experiment to help explain why transformers only work with alternating current. [3]

14 The waves from earthquakes are detected by instruments called seismometers. The diagram shows a simple seismometer.

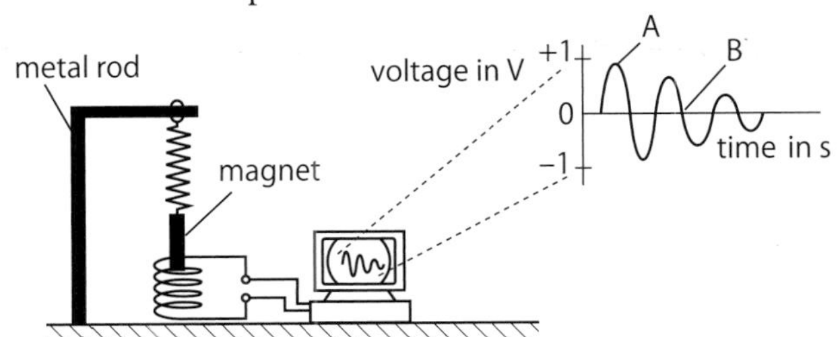

It consists of a bar magnet suspended on a spring. The spring hangs from a metal rod that transmits vibrations from the Earth. When there is an earthquake, the magnet moves in and out of the coil. A computer monitors the voltage across the coil.

a) Explain why a voltage is induced in the coil. [1]

b) Explain why the induced voltage is alternating. [1]

c) Describe the movement of the magnet when the induced voltage has its greatest value, at the point labelled A. [1]

d) Describe the movement of the magnet when the induced voltage is zero, as at point B. [1]

e) Suggest **two** ways in which the seismometer could be made more sensitive, so that it can detect smaller earthquakes. [2]

15 The diagram shows a long wire placed between the poles of a magnet. When there is a current *I* through the wire, a force acts on the wire causing it to move.

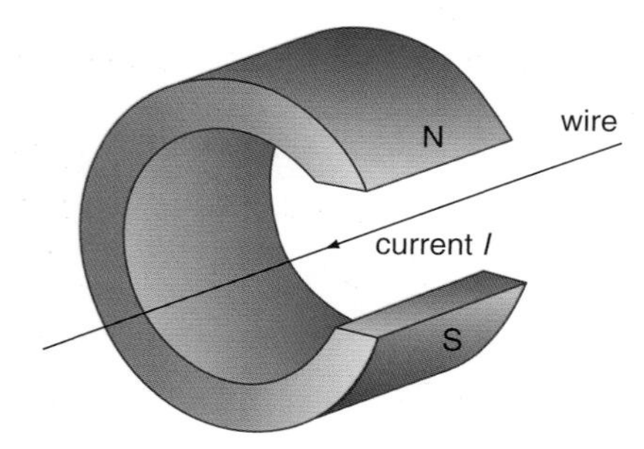

a) Use the left-hand rule to find the direction of the force on the wire. [1]

b) State what happens to the force on the wire when:

i) the size of the current through the wire is decreased [1]

ii) a stronger magnet is used [1]

iii) the direction of the current is reversed. [1]

c) Name **one** practical device that uses this effect. [1]

16 a) A student investigates how a short copper wire can be made to move in a magnetic field. The diagram shows the apparatus.

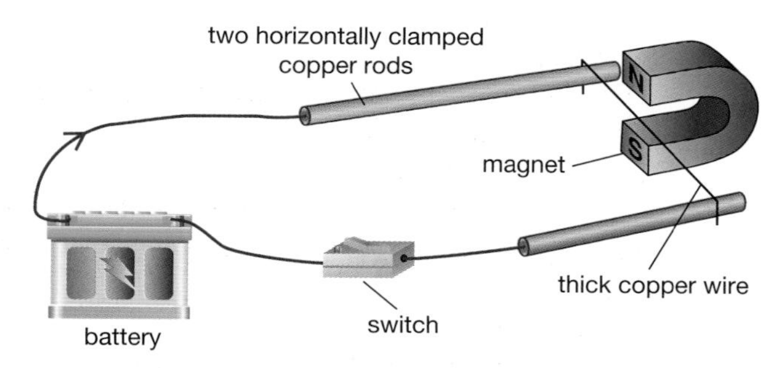

The wire is placed between the poles of the magnet.

i) Use the information in the diagram to predict the direction of motion of the wire. [1]

ii) State what happens to the motion of the wire when the magnet is turned with the north pole to the left of the wire. [1]

b) The diagram shows a model generator.

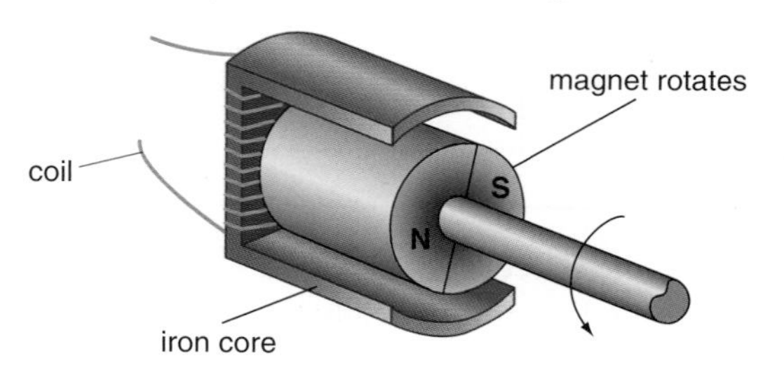

i) Explain why a voltage is induced across the ends of the coil when the magnet rotates. [2]

ii) Explain why the voltage is alternating. [1]

c) The ends of the coil are connected to a cathode ray oscilloscope (CRO). The diagram shows the trace on the screen as the magnet rotates.

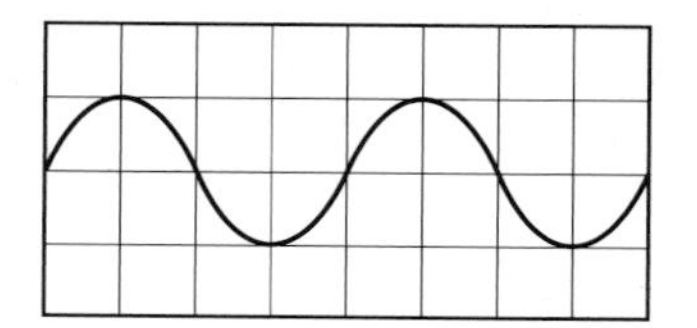

Copy the diagram and draw new traces for each of the following changes using the same scale. The settings of the oscilloscope remain the same.

i) The magnet rotates at the same speed but in the opposite direction. [1]

ii) The magnet rotates at the same speed, in the same direction as the original, but the number of turns of the coil is doubled. [2]

iii) The magnet rotates at twice the speed, in the same direction, with the original number of turns of the coil. [2]

EXTEND AND CHALLENGE

1 In the diagrams below, explain which arrangement will produce a larger deflection on the meter when the wires are moved between the poles of the magnets.

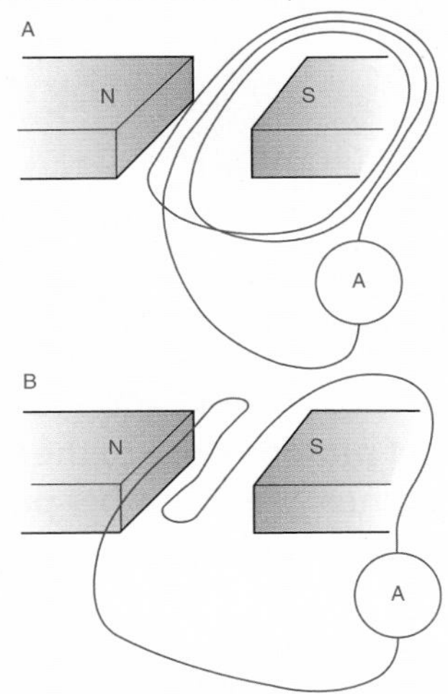

2 The diagram below shows three coils connected in series to a data logger. A magnet is dropped through the three coils. The graph shows the voltage measured by the data logger as the magnet falls. Explain the shape of the graph.

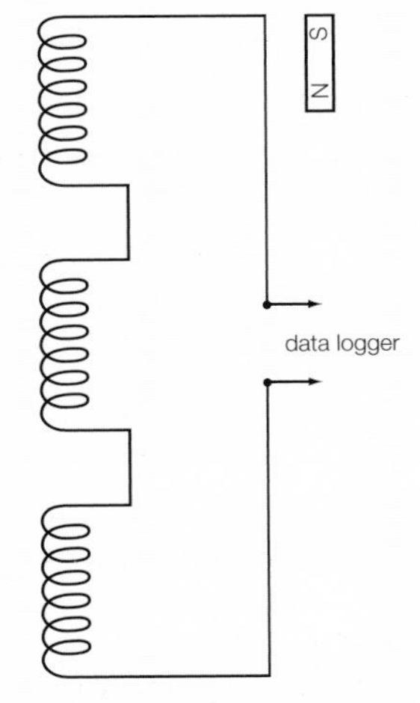

You should comment on these points

a) the height of the peaks on the graph

b) the width of the peaks

c) the gaps between the peaks

d) the direction of the peaks

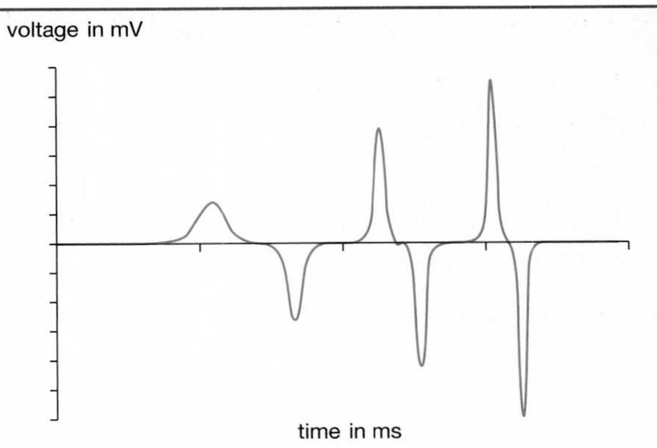

3 The diagram below shows a heavy copper pendulum suspended from a wire, which swings backwards and forwards from A to C, between the poles of a strong magnet. As the wire moves, a current is made to flow through the resistor R; a data logger is used to record how the voltage changes across R.

a) Give two reasons why you would expect the current to be greatest as the pendulum moves past B.

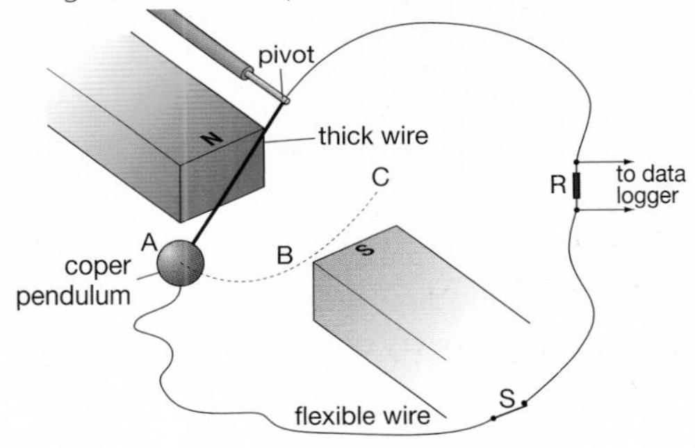

b) Sketch a graph to show what voltage the data logger records as the pendulum swings from A to C and back again. Mark your graph clearly to show the points A, B and C.

c) On the same axes sketch the voltage you would observe when:

i) the pendulum is released from a greater height

ii) the polarity of the magnets is reversed.

d) Explain the energy transfers which occur when the pendulum swings from A to B:

i) with S closed

ii) with S open.

e) It is observed that the pendulum swings take longer to die away when the switch is open, than when the switch is closed. Explain why.

f) The pendulum is replaced by a heavier one; it is released again from A. Predict what effect you will notice on:

i) the graph drawn in part b)

ii) the time taken for the swings to die away. Explain your answer.

7 Radioactivity and particles

CALCULATE • CONSIDER

1 In Section 1 you learnt that the braking distance of a car travelling at 30 m/s is about 75 m. If a car could travel at the same speed as protons in the LHC, which of the following distances would you think would be its approximate stopping distance when you apply the brakes:
   a) 7500 m
   b) 7500 km
   c) 7500 light years?
2 Explain which you think is the greatest scientific discovery in history – the discovery of the nucleus of an atom, the discovery of DNA or the working out of the periodic table? Or can you think of another more important discovery?
3 A new school costs €25 million, a new hospital €100 million. Can governments justify spending €7.5 billion on looking for a Higgs boson?

The Large Hadron Collider (LHC) at the European Centre for Nuclear Research (CERN) accelerates beams of protons close to the speed of light. When protons collide with this great energy, many other particles are produced. Nuclear physicists at CERN have discovered quarks, W and Z bosons and the Higgs boson.

By the end of this section you should:

- describe the structure of the atom in terms of protons, neutrons and electrons
- know the terms atomic, mass number and isotope
- know that alpha (α) particles, beta (β) particles and gamma (γ) rays are ionising radiations emitted from unstable nuclei in a random process
- describe the properties of α and β particles and γ rays
- describe the effect on the atomic and mass number of nuclei caused by the emission of the four main types of radiation
- understand how to balance a nuclear equation in terms of mass and charge
- know that ionising radiations may be detected by a Geiger-Müller tube or photographic film
- explain the sources of background radiation from Earth and space
- know that the activity of a source decreases with time and is measured in becquerels (Bq)
- know the definition of the term half-life and use the concept to carry out calculations on activity
- describe uses of radioactivity in industry and medicine
- describe the difference between contamination and irradiation
- describe the dangers of ionising radiation
- know that nuclear reactions, including fission, fusion and radioactive decay can be sources of energy
- understand that a nucleus of U-235 can be split by collision with a neutron (fission) and that the process releases energy
- describe how a chain reaction of fissions can be set up
- describe the role played by control rods, moderator, and shielding in a nuclear reactor
- explain the difference between nuclear fission and nuclear fusion
- describe nuclear fusion as the creation of larger nuclei resulting in a loss of mass from smaller nuclei accompanied by a release of energy
- know that fusion is the energy source for stars
- explain why fusion does not happen at low temperatures and pressures.

7.1 Atomic structure

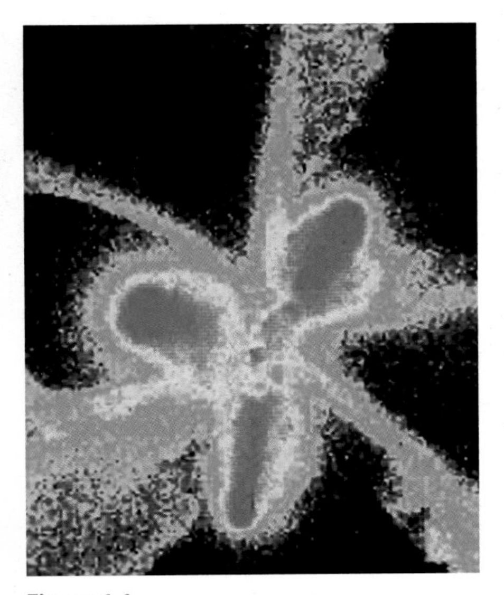

Figure 1.1

Over the last century, we have learnt how to use radioactive materials safely and to put them to good use. The image shows the concentration of radioactive sugar 2 hours after tracer molecules were fed into a plant. The red colour shows a high concentration of sugar in the young leaves of the plant, which are growing. The fast growing young leaves take the sugar from the older leaves, which appear blue. Doctors use a similar method to look at fast-growing cancer tumours with radioactive isotopes.

Neutrons, protons and electrons

Experiments done at the beginning of the twentieth century led to the nuclear model of the atom. We now understand that an atom has a very small nucleus, of diameter approximately 10^{-15} m. Inside the nucleus there are two types of particle, **protons** and **neutrons**. The protons and neutrons have approximately the same mass; a proton has a positive charge whereas the neutron is neutral. Outside the nucleus there are electrons, which orbit the nucleus at a distance of approximately 10^{-10} m. Electrons have very little mass, in comparison with neutrons or protons, and they carry a negative charge (see Table 1). Evidence for the structure of the nuclear atom is discussed in Chapter 7.5.

Table 1

Particle	Mass*	Charge*
proton	1	1
neutron	1	0
electron	1/1840	−1

* by comparison with a proton's mass and change

A hydrogen atom has one proton and one electron; it is electrically neutral because the charges of the electron and proton cancel each other. A helium atom has two protons and two neutrons in its nucleus, and two electrons outside that. The helium atom is also neutral because it has the same number of electrons as it has protons; it has four times the mass of a hydrogen atom because it has four particles in the nucleus (see Figure 1.2 and Table 2).

Table 2 Characteristics of the three smallest atoms

Element	Hydrogen, H	Helium, He	Lithium, Li
number of electrons	1	2	3
number of protons	1	2	3
number of neutrons	0	2	4
number of particles in nucleus	1	4	7
mass relative to hydrogen	1	4	7

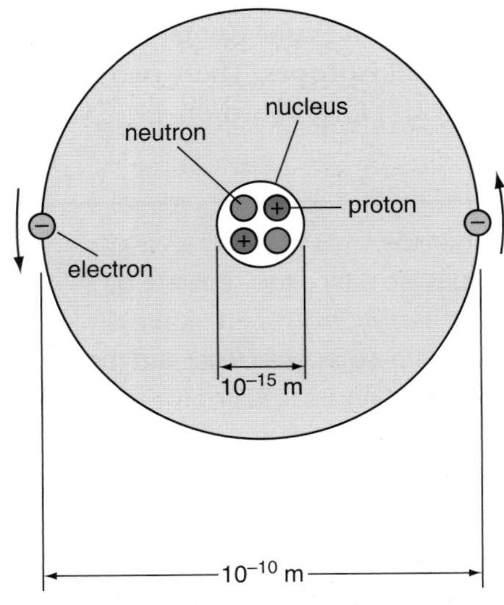

Figure 1.2 The helium atom; this is not drawn to scale – the diameter of the nucleus is about 100 000 times smaller than that of the atom itself.

Ions

Atoms are electrically neutral since the number of protons balances exactly the number of electrons. However, it is possible either to add extra electrons to an atom, or to take them away. When an electron is added to an atom a **negative ion** is formed; when an electron is removed a **positive ion** is formed. Ions are made in pairs because an electron that is removed from one atom attracts itself to another atom, so a positive and negative ion pair

TIP

Remember that nearly all of an atom's mass is in the nucleus.

is formed. Some examples are given in Table 3. The name ion is also used to describe charged molecules. The process of making ions is called ionisation.

Table 3 Examples of ions

Element	Number of protons	Number of electrons	Total charge	Ion
helium, He	2	1	+1	He^{+}
magnesium, Mg	12	10	+2	Mg^{2+}
chlorine, Cl	17	18	−1	Cl^{-}

Atomic and mass numbers

The number of protons in the nucleus of an atom determines what element it is. Hydrogen atoms have one proton, helium atoms two protons, uranium atoms 92 protons. The number of protons in the nucleus decides the number of electrons surrounding the nucleus. The number of electrons determines the chemical properties of an atom. The number of protons in the nucleus is called the **atomic** (or **proton**) **number** of the atom (symbol Z). So the proton number of hydrogen is 1, $Z = 1$.

The mass of an atom is decided by the number of neutrons and protons added together. Scientists call this number the **mass** (or **nucleon**) **number** of an atom. The name nucleon refers to either a proton or a neutron.

TIP

mass number 12 C
atomic number 6

atomic (or proton) number = number of protons
mass (or nucleon) number = number of protons and neutrons

For example, an atom of carbon has six protons and six neutrons. So its atomic number is 6, and its mass number is 12. To save time in describing carbon we can write it as $^{12}_{6}C$; the mass number appears on the left and above the symbol C, for carbon, and the atomic number on the left and below.

Isotopes

TIP

The symbols $^{12}_{6}C$ describe only the nucleus of a carbon atom.

Not all the atoms of a particular element have the same mass. For example, two carbon atoms might have mass numbers of 12 and 14. The nucleus of each atom has the same number of protons, six, but one atom has six neutrons and the other eight neutrons. Atoms of the same element (carbon in this case) that have different masses are called **isotopes**. These two isotopes of carbon can be written as carbon-12, $^{12}_{6}C$, and carbon-14, $^{14}_{6}C$.

STUDY QUESTIONS

1 a) An oxygen atom has eight protons, eight neutrons, and eight electrons. State its:
   i) atomic number
   ii) mass number
  b) Explain why the oxygen atom is electrically neutral.

2 Calculate the number of protons and neutrons in each of the following nuclei:
  a) $^{17}_{8}O$
  b) $^{238}_{92}U$
  c) $^{235}_{92}U$
  d) $^{40}_{19}K$

3 Write a paragraph to explain and describe the structure of an atom. In your answer, mention atomic number, mass number and electrons.

4 a) Explain the term 'isotope'.
  b) Explain why isotopes are difficult to separate by chemical methods.

5 Lead-209 and Lead-210 are isotopes of lead, and they both have the same atomic number, 82.
  a) State what the numbers 209, 210 and 82 represent.
  b) Explain what these two isotopes of lead have in common.
  c) Explain how the two isotopes are different.

7.2 Radioactivity

Henry Becquerel discovered radioactivity in 1896. He placed some uranium salts next to a photographic plate, which had been sealed in a thick black bag to prevent light exposing the plate. When the plate was later developed it had been affected as if it had been exposed to light. Becquerel realised that a new form of energy was being emitted from the uranium salts. In his honour, the activity of a radioactive source is measured in becquerels.

What travelled through the black bag? Where did it come from? What is radioactivity?

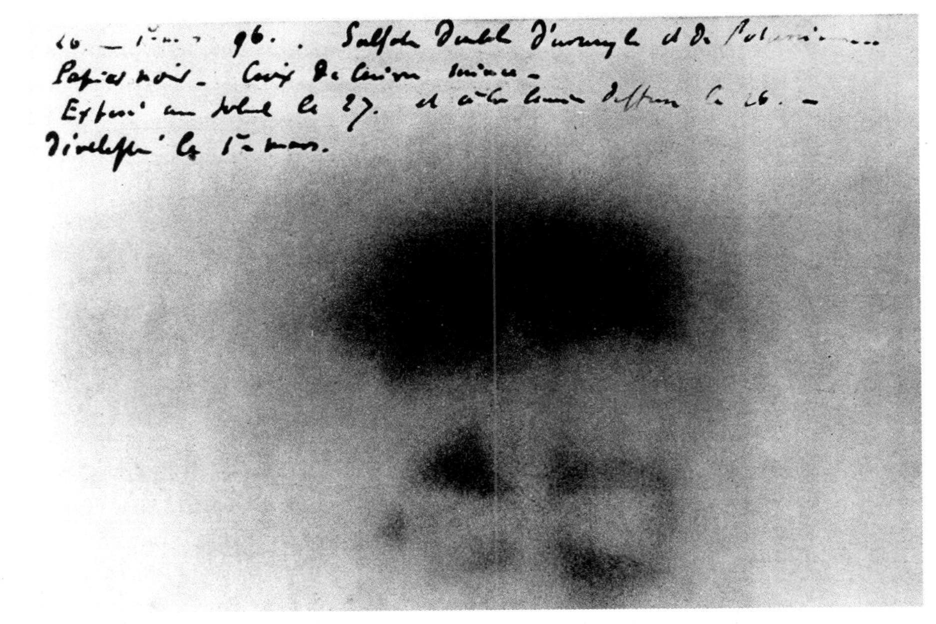

Figure 2.1 Becquerel's film.

radioactive source

black bag

photographic film

Radioactive particles pass through a lightproof bag to expose a photographic film.

Figure 2.2 How Becquerel discovered radioactivity.

Nuclear decay

The nuclei of most atoms are very stable; the atoms that we are made of have been around for thousands of millions of years. Atoms may lose or gain a few electrons during chemical reactions, but the nucleus does not change during such processes.

The activity of a source is equal to the number of particles emitted per second. This is measured in becquerel (Bq).

1 becquerel (1 Bq) = an emission of 1 particle/second

However, there are some atoms that have unstable nuclei which throw out particles to make the nucleus more stable. This is a random process that depends only on the nature of the nucleus; the rate at which particles are emitted from a nucleus is not affected by other factors such as temperature or chemical reactions. One element discovered that emits these particles is radium, and the name **radioactivity** is given to this process.

There are four types of radioactive emission:

- **Alpha particles** are the nuclei of helium atoms. The alpha particle is formed from two protons and two neutrons, so it has a mass number of 4 and an atomic number of 2. When an alpha particle is emitted from a nucleus it causes the nucleus to change into another nucleus with a mass number 4 less and an atomic number 2 less than the original one. It is usually only very heavy elements that emit alpha particles, for example:

$$^{238}_{92}\text{U} \rightarrow\ ^{234}_{90}\text{Th} +\ ^{4}_{2}\text{He}$$

uranium nucleus → thorium nucleus + alpha particle (helium nucleus)

This is called **alpha decay**.

TIP

An alpha particle is formed from two protons and two neutrons.

- **Beta particles** are electrons. In a nucleus there are only protons and neutrons, but a beta particle is made and ejected from a nucleus when a neutron turns into a proton and an electron. Since an electron has a very small mass, when it leaves a nucleus it does not alter the mass number of that nucleus. However, the electron carries away a negative charge so the removal of an electron increases the atomic number of a nucleus by 1. For example, carbon-14 decays into nitrogen by emitting a beta particle:

$$^{14}_{6}\text{C} \rightarrow\ ^{14}_{7}\text{N} +\ ^{0}_{-1}\text{e}$$

carbon nucleus → nitrogen nucleus + beta particle (electron)

This is called **beta decay**.

TIP

When a β-particle is emitted from a nucleus, a neutron turns into a proton. The mass number of the nucleus increases by 1.

TIP

Make sure you can write balanced equations for alpha and beta decay.

- When some nuclei decay by sending out an alpha or beta particle, they also give out a **gamma ray**. Gamma rays are electromagnetic waves, like radio waves or light. They carry away from the nucleus a lot of energy, so that the nucleus is left in a more stable state. Gamma rays have no mass or charge, so when one is emitted there is no change to the mass or atomic number of a nucleus (see Table 1).

- **Neutrons** are emitted from some highly unstable nuclei. The effect of this is to reduce the mass number by 1, but the atomic number does not change. Neutron emission is rare, but neutrons are a dangerous radiation. You will meet them again when you learn about fission. An example of neutron emission from helium-5 is given below:

$$^{5}_{2}\text{He} \rightarrow\ ^{4}_{2}\text{He} +\ ^{1}_{0}\text{n}$$

TIP

Make sure you know the nature of alpha, beta and gamma radiations.

Table 1 The four types of radioactive emission

	Particle lost from nucleus	Change in mass number	Change in atomic number
alpha (α) decay	helium nucleus $^{4}_{2}\text{He}$	–4	–2
beta (β) decay	electron $^{0}_{-1}\text{e}$	0	+1
gamma (γ) decay	electromagnetic waves	0	0
neutron (n) decay	neutron	–1	0

SAFETY

Students under 16 may not handle radioactive sources.

Ionisation

All three types of radiation (alpha, beta and gamma) cause **ionisation** and this is why we must be careful when we handle radioactive materials. The radiation makes ions in our bodies and these ions can then damage our body tissues (see Chapter 7.6).

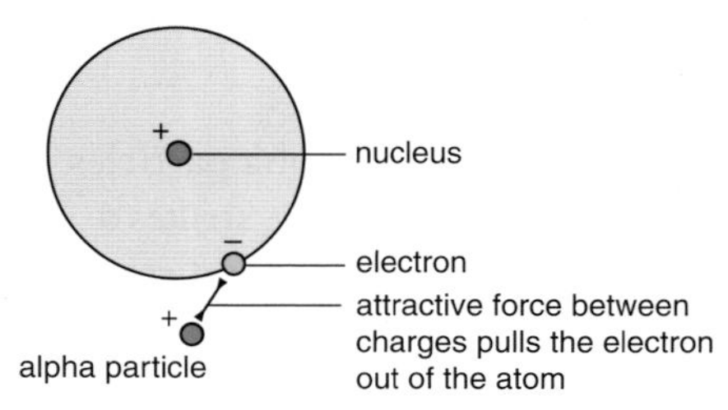

Figure 2.3 Alpha particles cause ionisation.

Your teacher can show the ionising effect of radium by holding some close to a charged gold leaf electroscope (Figure 2.4). The electroscope is initially charged positively so that the gold leaf is repelled from the metal stem. When a radium source is brought close to the electroscope, the leaf falls, showing that the electroscope has been discharged. The reason for this is that the alpha particles from the radium create ions in the air above the electroscope. This is because the charges on the alpha particles pull some electrons out of air molecules (Figure 2.3). Both negative and positive ions are made; the positive ones are repelled from the electroscope, but the negative ones are attracted so that the charge on the electroscope is neutralised. It is important that you understand that it is not the charge of the alpha particles that discharges the electroscope, but the ions that they produce.

TIP

Ionising radiations create both positive and negative ions.

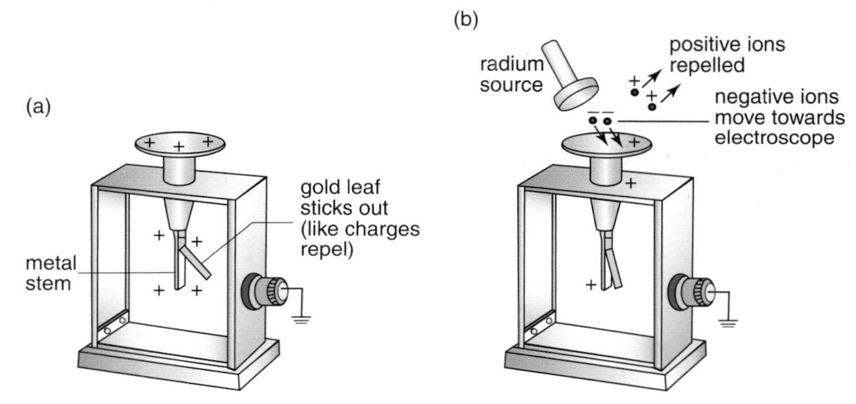

Figure 2.4 (a) Positively charged electroscope. **(b)** Negative ions neutralise the electroscope.

STUDY QUESTIONS

1. Explain what is meant by the word 'radioactivity'.
2. Justify why Becquerel concluded that he had discovered a new form of energy.
3. **a)** Explain what is meant by the word 'ionisation'.
   b) Explain why ions are always produced in pairs.
4. State the nature of the following:
   a) an alpha particle
   b) a beta particle
   c) a gamma ray
5. Complete the following radioactive decay equations.
   a) $^{3}_{?}\text{H} \rightarrow ^{?}_{2}\text{He} + ^{0}_{?}\text{e}$
   b) $^{229}_{90}\text{Th} \rightarrow ^{?}_{?}\text{Ra} + ^{4}_{2}\text{He}$
   c) $^{14}_{6}\text{C} \rightarrow ? + ^{0}_{-1}\text{e}$
   d) $^{209}_{82}\text{Pb} \rightarrow ^{?}_{83}\text{Bi} + ?$
   e) $^{225}_{89}\text{Ac} \rightarrow ^{?}_{87}\text{Fr} + ?$
   f) $^{13}_{4}\text{Be} \rightarrow ^{?}_{?}\text{Be} + ^{1}_{0}\text{n}$
6. $^{238}_{92}\text{U}$ decays by emitting an alpha particle and two beta particles; name the element produced after those three decays.
7. Explain what effect losing a gamma ray has on a nucleus.
8. Explain how a radioactive source which is emitting only alpha particles can discharge a negatively charged electroscope.
9. Beryllium-12 is an unstable isotope which decays by the emission of a beta particle and a neutron. Complete this equation to show the decay:

   $^{12}_{4}\text{Be} \rightarrow ^{?}_{?}\text{B} + ^{?}_{?}\text{e} + ^{?}_{?}\text{n}$

7.3 The nature of α, β and γ radiations

Figure 3.1 Cloud chamber tracks, shown in false colour.

In Figure 3.1 the green lines illustrate the straight paths of alpha particles in a cloud chamber. One of the alpha particles, coloured yellow, strikes a nitrogen nucleus and recoils. The nitrogen nucleus travels a short distance forward, and this is shown by the red line.

A vapour trail is left in a cloud chamber where ions have been produced by the passage of a particle. A thicker trail is left by a strongly ionising particle. A nitrogen nucleus is more ionising than an alpha particle because the nitrogen nucleus carries a greater positive charge.

Detecting particles

We make use of the ionising properties of radiations to detect them. This is done using a Geiger-Müller (GM) tube. When alpha, beta or gamma radiation enters the tube, gas atoms inside are ionised. The ions produce a pulse of current that enables a counter, which is attached to the GM tube, to count the number of ionising particles entering the tube.

REQUIRED PRACTICAL

Make sure you are familiar with this practical.

An investigation into the range and penetration of radiations

Figure 3.2 shows a practical arrangement for investigating the range and penetration of alpha, beta and gamma radiations.

Alpha particles travel about 5 cm through air but can be stopped by a sheet of paper (Figure 3.3). Alpha particles ionise the air strongly.

Beta particles can travel several metres through air. They can be stopped by a sheet of aluminium that is a few millimetres thick (Figure 3.3). Beta particles ionise the air less strongly than alpha particles.

TIP

A Geiger-Müller (GM) tube is a device which detects ionising radiation. An electronic counter can record the number of particles entering the tube.

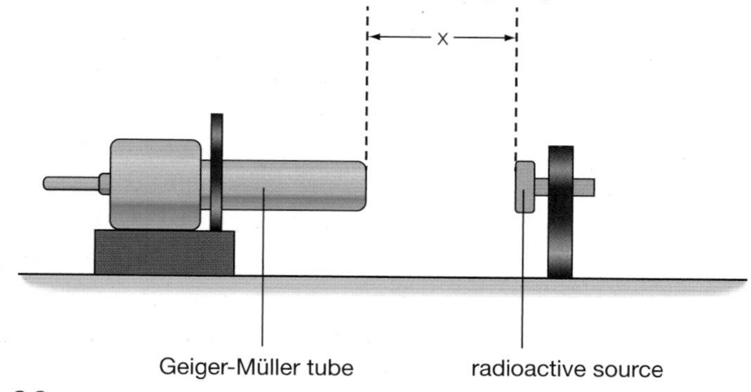

Figure 3.2

SAFETY

For safety reasons, students under the age of 16 are not allowed to handle radioactive materials, so your teacher will demonstrate this experiment, or you might use a computer simulation.

Gamma rays travel great distances through air and can only be stopped effectively by a very thick piece of lead (Figure 3.3) Gamma rays ionise air weakly.

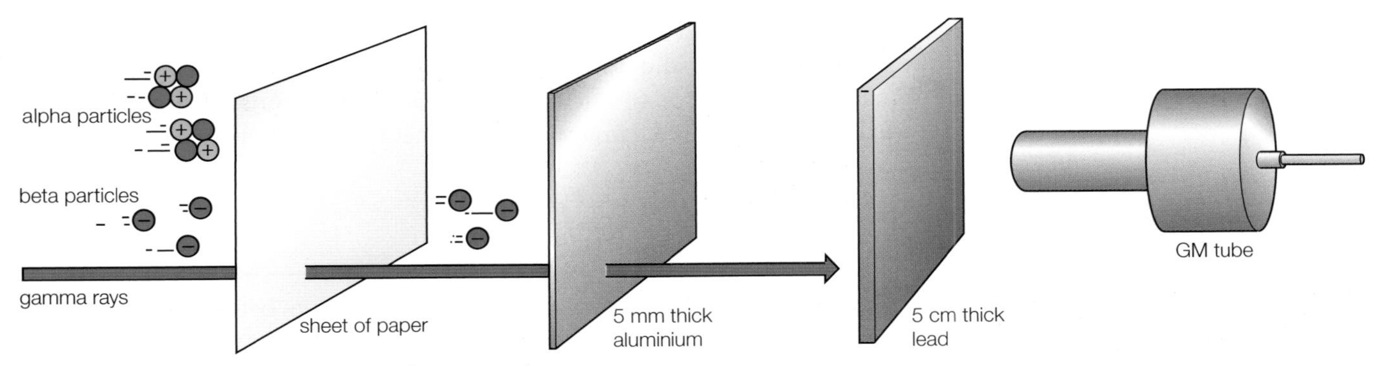

Figure 3.3 Penetration of alpha, beta and gamma radiation.

Make sure you learn which absorber stops each of alpha, beta and gamma radiation.

Table 1 A summary of radiation properties

Radiation	Nature	Ionising power	Penetrating power
alpha α	helium nucleus	very strong	stopped by paper
beta β	electron	medium	stopped by aluminium
gamma γ	electromagnetic waves	weak	stopped by thick lead

Radiation damage

If radiation gets into our body, our cells and tissues can be damaged. The ions which are produced by the radiation produce chemicals which destroy the cells they come into contact with. Alpha particles cause the most damage if they get inside the body as they are strongly ionising. This could happen if we breathed in a radioactive gas. An alpha source in school is less dangerous as the radiation only travels short distances, so does not enter our body.

Although gamma rays are less ionising than alpha particles, a gamma ray source in a laboratory is a hazard as the rays are so penetrating. A gamma ray can pass into our body from a source several metres away from us. Your teacher will always keep away from a source, by handling it with long tongs.

Background radiation

There are a lot of rocks in the Earth that contain radioactive uranium, thorium, radon and potassium, and so we are always exposed to some ionising particles. Radon is a gas that emits alpha particles. Since we can inhale this gas it is dangerous, as radiation can get inside our lungs. We are exposed to ionising radiations from the Sun and from outer space. These are called cosmic rays. The atmosphere helps to protect us from cosmic rays. These are two of the sources that make up **background radiation**. Figure 3.4 shows the contribution to the total background radiation from all sources in Britain. Fortunately the level of background radiation is quite low and in most places it does not cause a serious health risk.

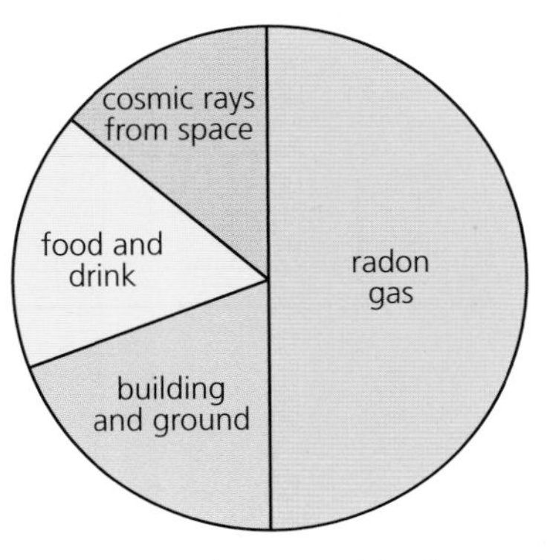

Figure 3.4 Natural sources of background radiation in Britain.

In some jobs, people are at a greater risk. X-rays used in hospitals also cause ionisation. Radiographers make sure that their exposure to X-rays is as small as possible. In nuclear power stations neutrons are produced in **nuclear reactors**. The damage caused by neutrons is a source of danger for workers in that industry.

STUDY QUESTIONS

1 Which of the following A, B or C, is a property of beta radiation?
   a) It is the most strongly ionising radiation.
   b) It travels several metres in air.
   c) It can easily be stopped by paper.

2 Who is most likely to receive the highest annual radiation dose?
   an airline pilot   a motor mechanic   an office worker

3 **a)** Explain what is meant by the term 'background radiation'.
   b) Name three sources of background radiation

Figure 3.5

   c) Explain why the astronaut in Figure 3.5 was exposed to a higher background radiation than he was on Earth.

4 A radioactive source is placed 2 cm from a GM tube and an activity of 120 Bq is measured. When a piece of paper is put between the source and the GM tube the activity reduces to 75 Bq.
   A 5 mm thick piece of aluminium between the source and tube reduces the measured activity to zero. Explain what this tells you about the source.

5 **a)** Radon gas is a product of radioactive decay in granite rocks. Radon is also radioactive and emits alpha particles. Explain why radon might be a hazard to mining engineers working underground in granite rocks.
   b) Explain why a strong gamma source left in a laboratory might be dangerous to us.

6 This question is about testing the thickness of aluminium foil. The foil moves between a β-source and Geiger counter at a speed of 0.2 m/s.
   a) Plot a graph from the data below. Explain why the count rate changes over the first 50 s. Remember that particles are emitted *randomly* from a nucleus.
   b) Explain why the count rate drops and then rises again, after 50 s. What has happened to the thickness of the foil?

Time in s	0	10	20	30	40	50	60	70	80	90	100	110	120
Count rate in Bq	75	80	77	73	76	75	63	57	50	55	67	75	77

7.4 Radioactive decay

Figure 4.1 On average, how many sixes will there be when you roll these dice? Why can you not predict what will be thrown on each occasion?

The atoms of some radioactive materials decay by emitting alpha or beta particles from their nuclei. But it is not possible to predict when the nucleus of one particular atom will decay. It could be in the next second, or sometime next week, or not for a million years. Radioactive decay is a random process.

Random process

The radioactive decay of an atom is rather like tossing a coin. You cannot say with certainty that the next time you toss a coin it will fall heads up. However, if you throw a lot of coins you can start to predict how many of them will fall heads up. You can use this idea to help you understand how radioactive decay happens. You start off with a thousand coins, if any coin falls heads up then it has 'decayed' and you must take it out of the game. Table 1 shows the likely result (on average). Every time you throw a lot of coins about half of them will turn up heads.

Table 1 Coin-tossing experiment

Throw	Number of coins left
0	1000
1	500
2	250
3	125
4	62
5	31
6	16
7	8
8	4
9	2
10	1

Decay

Radioactive materials decay in a similar way. If we start off with a million atoms then after a period of time (for example 1 hour), half of them have decayed. In the next hour we find that half of the remaining atoms have decayed, leaving us with a quarter of the original number (Table 2). The period of time taken for half the number of atoms to decay in a radioactive sample is called the **half-life**, and it is given the symbol $t_{1/2}$. It is important to understand that we have chosen a half-life here of 1 hour to explain the idea. Different radioisotopes have different half-lives.

Table 2 The number of nuclei left in a sample: half-life 1 hour

Time in hours	Number of nuclei left
0	1000 000
1	500 000
2	250 000
3	125 000
4	62 500
5	31 250
6	15 620
7	7810
8	3900
9	1950
10	980

Measurement of half-life

If you look at Table 2 you can see that the number of nuclei that decayed in the first hour was 500 000, then in the next hour 250 000 and in the third hour 125 000. So as time passes not only does the number of nuclei left get smaller but so does the rate at which the nuclei decay. So by measuring the activity of a radioactive sample we can determine its half-life.

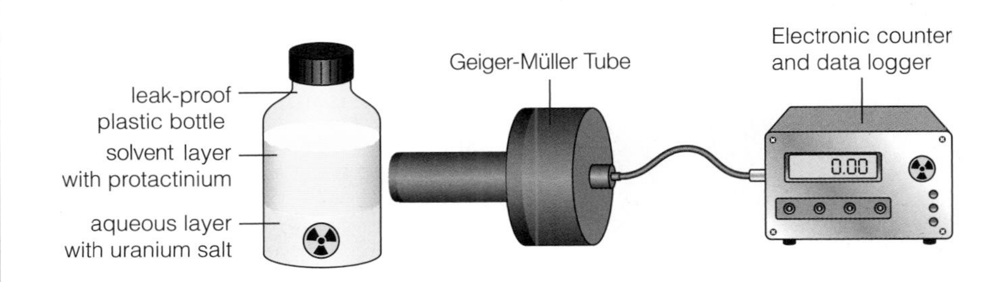

Figure 4.2 The apparatus for determining the half-life of protactinium.

TIP

After one half-life the activity of a source has halved. The activity is a quarter after two half-lives.

PRACTICAL

Correcting for the background count

A GM tube left well away from a radioactive source will still count some radioactive emissions; this is the background count. For example this might be 3 Bq. If the GM tube is placed close to a radioactive source, it will count the emissions from the source and the background count; if the count recorded is now 43 Bq, we can calculate that the count from the source is 40 Bq, since 3 Bq were due to the background count.

Figure 4.2 shows how the half-life of protactinium can be measured. Protactinium is produced in the decay of uranium, which is dissolved in an aqueous layer in a plastic bottle. When the bottle is shaken some protactinium is dissolved into a less dense upper layer. The protactinium in the upper layer then decays. A GM tube, placed near to the upper layer, is attached to a ratemeter which measures the activity every 10 seconds. The graph of the activity shows that the count rate halves every 70 seconds, so that is the half-life of protactinium.

The teacher must demonstrate this experiment, ensuring the apparatus is leak tight. Follow CLEAPSS guidance.

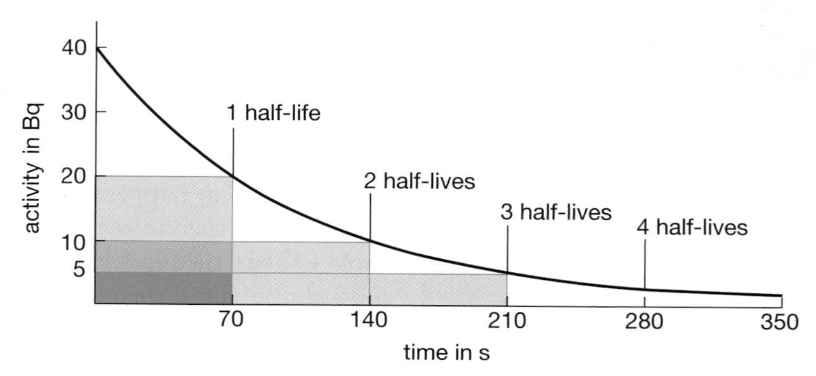

Figure 4.3 Finding the half-life from a decay curve.

Dating archaeological remains

Carbon-14, $^{14}_{6}C$, is a radioactive isotope; it decays to nitrogen with a half-life of about 5700 years. All living things (including you) have a lot of carbon in them, and a small fraction of this is carbon-14. When a tree dies, for example, the radioactive carbon decays and after 5700 years the fraction of carbon-14 in the dead tree will be half as much as you would find in a living tree. So by measuring the amount of carbon-14 in ancient relics, scientists can calculate their age (see Study Question 8).

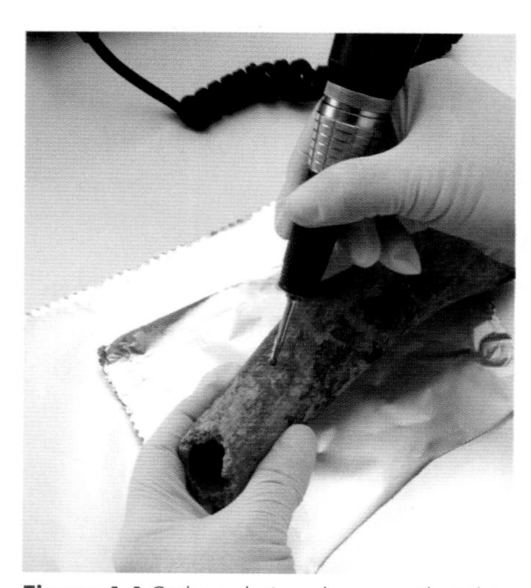

Figure 4.4 Carbon dating shows us that this piece of wood is 4000 years old.

Half-life and stability

Elements usually have several isotopes – two isotopes of an element have the same atomic (or proton) number but they have a different number of neutrons in the nucleus, so they have different mass (or neutron) numbers. Some isotopes are unstable and they undergo radioactive decay. Less stable isotopes decay more quickly, so they have shorter half-lives than the more stable isotopes. Half-lives can be very long: for example uranium-238 decays with a half-life of 4.5 billion years; by contrast other half-lives are measured in fractions of a second. Table 3 lists the half-lives of some radioactive isotopes.

Table 3

Isotope	Half-life
potassium-40	1.3 billion years
carbon-14	5700 years
caesium-137	30 years
iodine-131	8 days
lawrencium-260	3 minutes

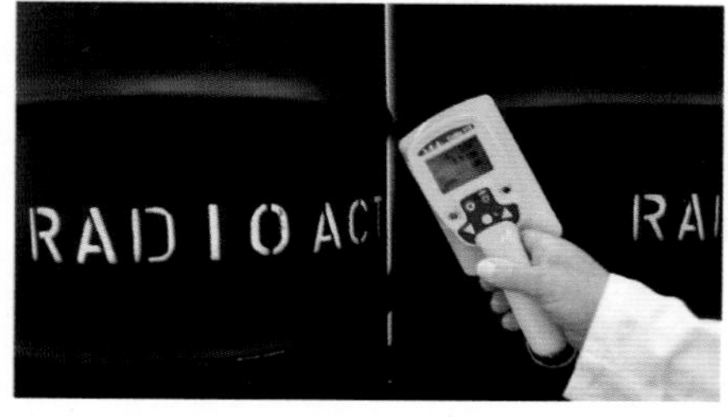

Figure 4.5 This Geiger counter is being used to determine the activity of radioactive waste which is stored underground.

STUDY QUESTIONS

1 Use a word from the box to complete the sentence.

count rate half-life reaction

The ______________________ is the number of alpha, beta or gamma emissions from a radioactive source in one second.

2 The graphs in Figure 4.6 show the decay of three different radioactive isotopes.

State which isotope has:

a) the longest half-life

b) the shortest half-life

c) the possibility of being used as a medical tracer.

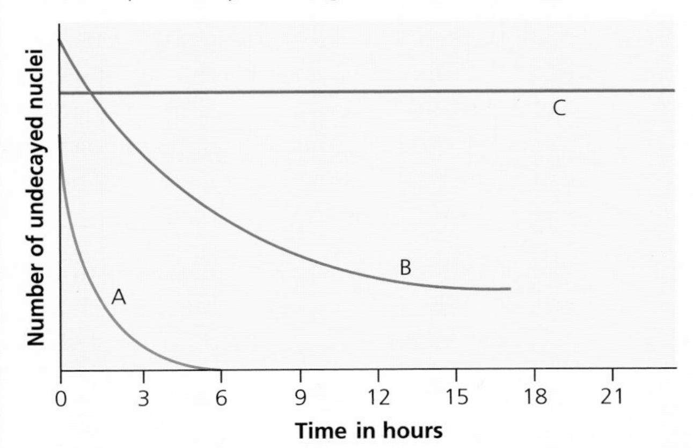

Figure 4.6

3 Explain the meaning of the word random.

4 A GM tube is placed near to a radioactive source with a long half-life. In three 10-second periods the following number of counts were recorded: 150, 157, 145. Explain why the three counts were different.

5 A radioactive material has a half-life of 2 minutes. Explain what that means. Determine how much of the material will be left after 8 minutes.

6 The following results for the count rate of a radioactive source were recorded every minute.

Count rate in Bq	Time in minutes
1000	0
590	1
340	2
200	3
120	4
70	5

a) A correction was made for background count. Explain what this means.

b) Plot a graph of the count rate (*y*-axis) against time (*x*-axis), and use the graph to determine the half-life of the source.

7 A radioisotope has a half-life of 8 hours. At 12 noon on 2 March a GM tube measures an activity of 2400 Bq.

a) Calculate the activity at 4.00 am on 3 March.

b) Determine the time at which an activity of approximately 75 Bq will be measured.

8 a) Archaeologists are analysing ancient bones from a human settlement. They discover that a sample of bone has one-sixteenth of the carbon-14 of modern human bones. Determine the age of the settlement. (The half-life of carbon-14 is 5700 years.)

b) The limit of carbon dating is about 50 000 years. Explain why.

9 The age of rocks can be estimated by measuring the ratio of the isotopes potassium-40 and argon-40. We assume that when the rock was formed it was molten, and that any argon would have escaped. So, at the rock's formation there was no argon. The half-life of potassium-40 is 1.3×10^9 years, and it decays to argon-40. Analysis of two rocks gives these potassium (K) to argon (A) ratios:

$$\text{Rock A } \frac{K}{A} = \frac{1}{1} \quad \text{Rock B } \frac{K}{A} = \frac{1}{7}$$

Calculate the ages of the two rocks.

7.5 Uses of radioactive materials

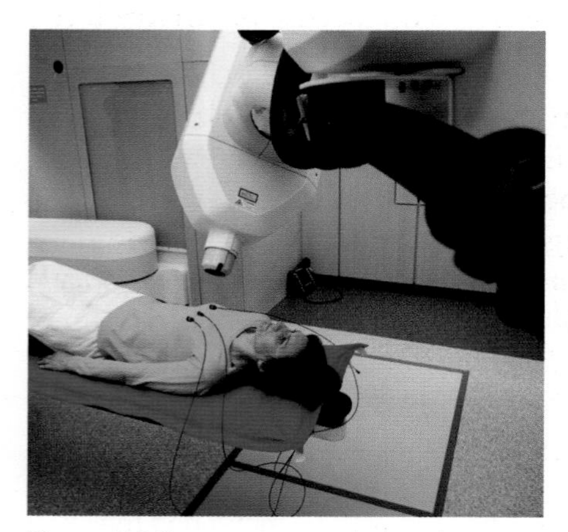

Figure 5.1 Gamma rays can be used to destroy cancer cells. This photograph shows a patient with Hodgkins disease (cancer of the lymph nodes) being treated by radiotherapy. What precautions must the radiographer take when working with radiation?

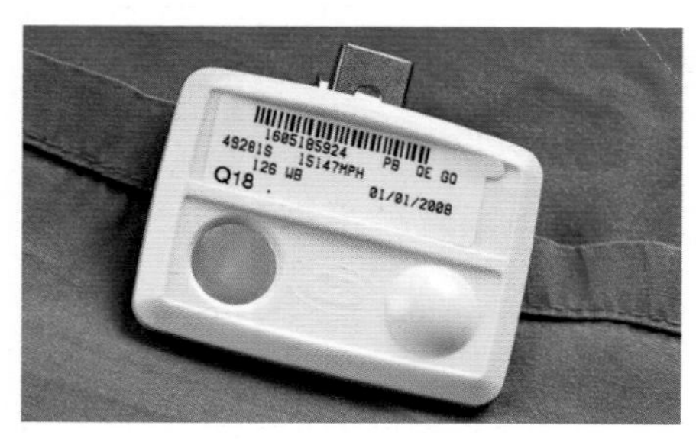

Figure 5.2 To check the amount of radiation that workers in a nuclear power station are exposed to, they wear special radiation-sensitive badges, like the ones in this photograph. At the end of each month the sensitive film in the badges is developed and examined.

Radioactive materials have a great number of uses in medicine and industry. People who work with radioactive materials must wear radiation badges that record the amount of radiation to which they are exposed.

Medicine

Radioactive **tracers** help doctors to examine the insides of our bodies. For example, iodine-123 is used to see if our thyroid glands are working properly. The thyroid is an important gland in the throat that controls the rate at which our bodies function. The thyroid gland absorbs iodine, so a dose of radioactive iodine (the tracer) is given to a patient. Doctors can then detect the radioactivity of the patient's throat, to see how well the patient's thyroid is working. The tracer needs to be an isotope with a short half-life.

Cobalt-60 emits very energetic gamma rays. These rays can damage our body cells, but they can also kill bacteria. Nearly all medical equipment such as syringes, dressings and surgeons' instruments is first packed into sealed plastic bags, and then exposed to intense gamma radiation. In this way all the bacteria are killed and so the equipment is sterilised.

The same material, cobalt-60, is used in the treatment of cancers. Radiographers direct a strong beam of radiation on to the cancerous tissue to kill the cancer cells. This radiotherapy treatment is very unpleasant and can cause serious side-effects, but it is often successful in slowing down the growth or completely curing the cancer.

Industry

Radioactive tracers may be used to detect leaks in underground pipes (Figure 5.3). The idea is very simple; the radioactive tracer is fed into the pipe and then a GM tube can be used above ground to detect an increase in radiation levels and hence the leak. This saves time and money because the whole length of the pipe does not have to be dug up to find the leak.

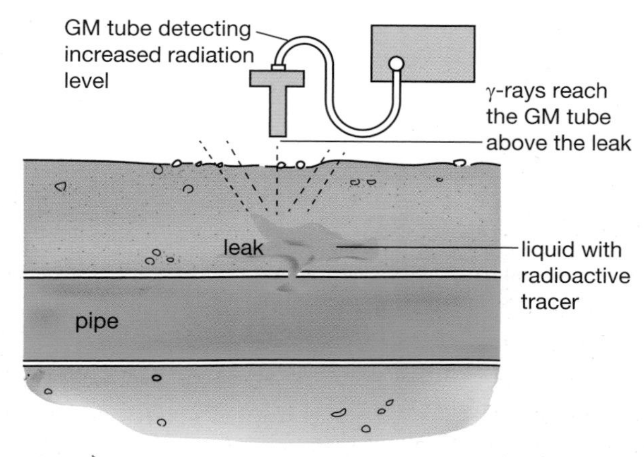

Figure 5.3 How to find leaks in pipelines without digging.

Irradiation and contamination

When a teacher brings a radioactive source into a laboratory to demonstrate to her pupils, she irradiates the surroundings (with a very small safe dose) but does not cause any contamination. When the source is near a GM tube, the tube is irradiated. This means that radiation from the source is entering the tube. However, as soon as the source is removed and put away, the GM tube will not be radioactive because there are no radioactive nuclei inside it to emit radiation.

Radioactive isotopes cause contamination when they get into places where we do not want them. For example, iodine-131 and caesium-137 were emitted in the Fukushima and Chernobyl disasters. Iodine-131 has a half-life of 8 days, which means that it is highly active for a short time. Iodine disperses in the atmosphere and gets into the food chain. This poses a very high risk, but for a short period. Caesium-137, however, has a half-life of 30 years. Consequently, the ground and water in the region close to the two nuclear reactors will be contaminated for many years to come.

TIP

Make sure you can name and explain at least three uses of radioactive materials.

Both irradiation and contamination are potentially hazardous to us. However, when a hospital patient is exposed to a dose of radiation, the dose will be carefully calculated and controlled. The problem with contamination is that a person is exposed to radiation in a way which is unknown, and it is possible that a large and very dangerous dose of radiation is consumed by mistake.

STUDY QUESTIONS

1 Explain the meanings of the words 'irradiation' and 'contamination'.

2 a) Explain what a radioactive tracer is.
   b) Do radioactive tracers show any chemical differences from other isotopes of the same element? Explain your answer.

3 Figure 5.4 shows a gamma ray unit in a hospital. Box A is a container of the radioactive material cobalt-60 ($^{60}_{27}Co$).This isotope has a half-life of 5 years.

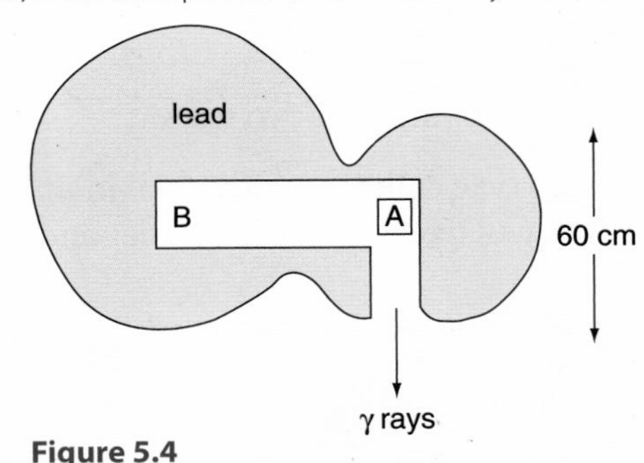

Figure 5.4

   a) Explain why the container is made mainly from lead.
   b) The unit can be made 'safe' by moving the container to position B. Explain why.
   c) State how many protons, neutrons and electrons are there in one atom of cobalt-60.
   d) The activity of the source was 2 million Bq in January 2013. Determine the activity in January 2023.
   e) Suggest **three** safety precautions that someone should take when removing a container of cobalt-60 from the unit.

4 The table shows information about some radioactive isotopes. You are a medical physicist working in a hospital. Advise the doctors which isotopes are suitable for the following tasks:
   a) Checking for a blockage in a patient's lungs.
   b) Directing a strong dose of radiation deep into a patient to treat cancer.

   Explain in each case what apparatus you would use, explain why you have chosen a particular isotope and what safety precautions you would take to protect the doctors and the patient.

Isotope	Solid, liquid or gas at 20°C	Type of radiation	Half-life
hydrogen-3	gas	beta	12 years
cobalt-60	solid	gamma	5 years
strontium-90	solid	beta	28 years
xenon-133	gas	gamma	5 days
terbium-160	solid	beta	72 days
actinium-227	solid	alpha	22 years
americium-241	solid	alpha	430 years

7.6 The hazards of radiation

The Fukushima Daiichi power plant in Japan was devastated by the tsunami produced by the Tohoku earthquake of 11 March 2011. This led to the second biggest nuclear disaster in history, producing at its peak a local activity of 400 million million Bq. While nuclear power has its risks, supporters of the nuclear industry point out that the environmental risk posed by coal-fired power stations is greater. There is the immediate risk posed by pollution through burning coal, and the long-term risk of global warming due to the emission of greenhouse gases.

In this section, you will learn about the hazards of using radioactive materials.

Figure 6.1 The Fukushima nuclear power plant disaster.

TIP

In the early days of research into radioactivity, scientists did not understand the hazards of radiation. For example, Marie Curie (an early pioneer) died from an illness most likely caused by long term exposure to radiation. We now understand the impact of radiation because scientists have published their findings which have been checked and validated by others. As a result, we now have clear procedures to protect ourselves from the effect of radiation.

Radiation damage

Radiation entering our bodies causes damage in two ways.

- Direct damage is caused by a particle colliding with a cell in our body. An alpha particle, for example, behaves like a miniature bullet and it destroys body tissue.
- Indirect damage is caused by ionisation. Radiation produces ions which can make strong acids in our bodies. These acids can then destroy our cells or cause mutations in our genes.

Types of radiation

Inside the body, the most damaging radiation is the alpha particle. This is because the alpha particle does not travel far through body tissues and it transfers its energy in a very small space, so damage to tissue is localised.

Beta and gamma radiations spread their energy over greater distances in the body, so they are not as damaging as alpha radiation.

In nuclear reactors, neutrons are emitted. These particles are also dangerous to us as they penetrate the body and collide with atoms, thus damaging cells.

Radiation dose

When we assess the dangers of radiation, we need to know how much we have been exposed to. We measure the radiation dose in **sieverts**, Sv. The greater the radiation dose, the more likely we are to suffer ill effects.

Doses are often quoted in thousandths of sieverts – millisieverts mSv.

TIP

The damage caused by an alpha particle inside our bodies is about 20 times the damage caused by beta or gamma radiation of the same energy.

How dangerous is radiation?

In the UK, there is about a 40% chance that we will suffer from cancer at some stage of our life. It is thought that about 1% of cancers are caused by background radiation which sends thousands of particles through our bodies each second. Table 1 below gives some examples of radiation doses and their risk to us.

Low doses

Low doses of radiation, below 10 mSv, are unlikely to cause us harm. However, it is possible that any exposure to radiation will increase our chances of cancer. We are exposed to background radiation all the time, in the air, from rocks and from the food we eat.

Table 1

Dose in mSv	Source of dose	Effect of dose
0.0001	Airport security scan; Eating a banana	Low risk
0.005	Dental X-ray	Low risk
0.04	Transatlantic flight	Low risk
0.1	Chest X-ray	Low risk
0.2	Release limit from a nuclear plant per person per year	Low risk
0.4	Yearly dose from food	Low risk
2.4	Average from background radiation in UK	Low risk
3	Mammogram	Low risk
10	Average computer tomography (CT) scan	Low risk
50	Maximum yearly dose permitted for radiation workers	Medium risk
100	Lowest annual dose where increased risk of cancer is evident	Medium risk
1000	Highly targeted dose used in radiotherapy (single dose)	High risk, but balanced by it likely destroying cancer cells
5000	Extremely severe dose, received in a nuclear accident	Death probable within 6 weeks
10 000	Maximum radiation dose per day found at the Fukushima plant in 2011	Fatal dose; death within 2 weeks
50 000	10 minutes exposure to the Chernobyl reactor meltdown in 1986	Death within hours

You will not be expected to remember the size of doses from various sources, or that the doses are measured in sieverts.

Moderate doses

Moderate doses of radiation below 1000 mSv (1 Sv) are unlikely to kill someone, but the person will be very unwell. Damage will be done to cells in the body, but not enough to be fatal. The body will be able to replace dead cells and the person is likely to recover completely. However, studies of the survivors from the Hiroshima and Nagasaki bombs, and of survivors from the Chernobyl reactor disaster of 1986, show that there is an increased chance of dying from cancer some years after the dose of radiation.

High doses

High doses of about 4000 mSv (4 Sv) are likely to be fatal, and a dose of 10 Sv will definitely be fatal. A high dose damages the gut and bone marrow so much that the body cannot work normally. About 30 people died of acute radiation syndrome in the Chernobyl disaster.

Disposal of radioactive waste

Nuclear power stations produce radioactive waste materials, some of which have half-lives of hundreds of years. These waste products are packaged up in concrete and steel containers and are buried deep underground or are dropped to the bottom of the sea. This is a controversial issue; some scientists tell us that radioactive wastes produce only a very low level of radiation, and that the storage containers will remain intact for a very long time. Others worry that these products will contaminate our environment and believe it is wrong to dump radioactive materials that could harm future generations.

STUDY QUESTIONS

1 Name four types of nuclear radiation that are dangerous to us.

2 Explain why radiation workers are only allowed to receive a maximum annual dose of about 50 mSv – refer to Table 1.

3 A patient receives a dose of 10 mSv from a CT scan. Use Table 1 on page 233 to work out how many years of background radiation are equivalent to the dose from a CT scan.

4 Table 2 shows the results of research into the number of deaths caused by various types of radiation.

  a) Discuss whether the table supports the suggestion that alpha particles are more dangerous than gamma rays.

  b) What conclusion can you draw about the relative dangers of neutrons?

  c) A student comments that the table does not provide a fair test for comparison of the radiations. Evaluate this comment.

Table 2

Source of radiation	Type of radiation	Number of people studied	Extra number of cancer deaths caused by radiation
Uranium miners	Alpha	3400	60
Radium luminisers	Alpha	800	50
Medical treatment	Alpha	4500	60
Hiroshima bomb	Gamma rays and neutrons	15 000	100
Nagasaki bomb	Gamma rays	7000	20

5 Discuss the problems arising from the disposal of radioactive waste and how the associated risks can be reduced.

7.7 Nuclear fission

Radioactive zirconium-coated fuel rods produce a blue glow in the water surrounding them. The radiation is caused by high-speed charged particles passing through the water. Why are there traces of zirconium in uranium fuel rods?

Figure 7.1 Nuclear reactor radiation.

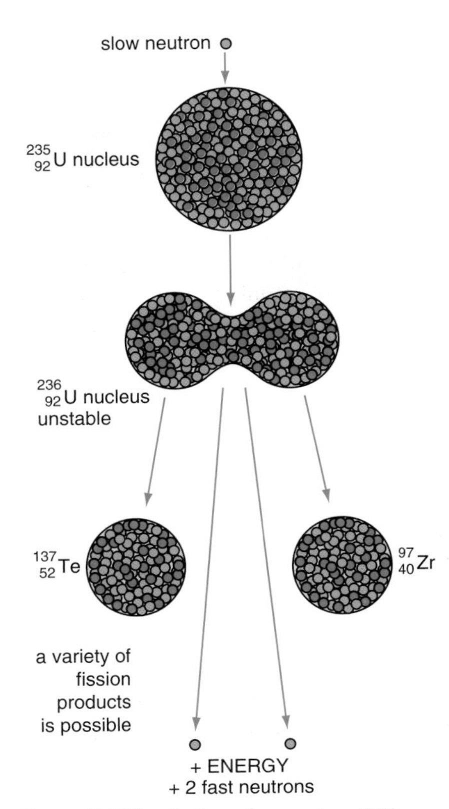

Figure 7.2 The fission of a uranium-235 nucleus.

Fission

You read in Chapter 7.2 that the nuclei of some large atoms are unstable and that to become more stable they lose an alpha or a beta particle. Some heavy nuclei, uranium-235 for example, may also increase their stability by **fission**. Figure 7.2 shows how this works. Unlike alpha or beta decay, which happens at random, the fission of a nucleus is usually caused by a neutron hitting it. The ^{235}U nucleus absorbs this neutron and turns into a ^{236}U nucleus, which is so very unstable that it splits into two smaller nuclei. The nuclei that are left are very rarely identical; two or three energetic neutrons are also emitted. The remaining nuclei are usually radioactive and will decay by the emission of beta particles to form more stable nuclei.

Nuclear energy

In Figure 7.2, just after fission has been completed, the tellurium 137 ($^{137}_{52}Te$) and the zirconium-97 nucleus ($^{97}_{40}Zr$) are pushed apart by the strong electrostatic repulsion of their nuclear charges. In this way the nuclear energy is transferred to the kinetic energy of the fission fragments. When these fast-moving fragments hit other atoms, this kinetic energy is transferred to the atoms' thermal store of energy.

The fission process releases a tremendous amount of energy. The fission of a nucleus provides about 40 times more energy than the release of an alpha particle from a nucleus. Fission is important because we can control the rate at which it happens, so that we can use the energy released to create electrical energy.

Chain reaction

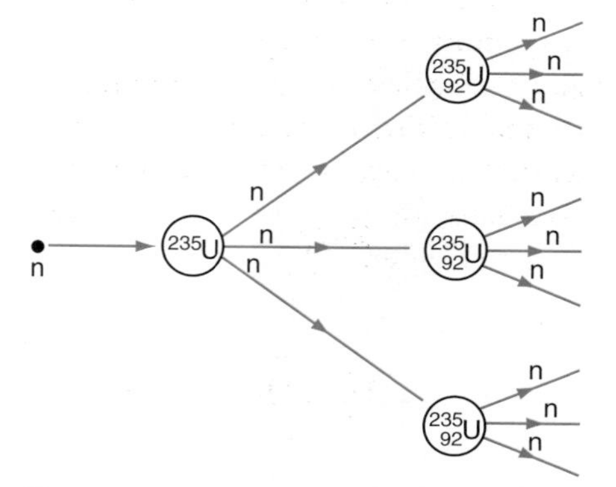

Figure 7.3 A chain reaction in uranium-235 (the fission fragments have been omitted for clarity).

Once a nucleus has divided by fission, the neutrons that are emitted can strike other neighbouring nuclei and cause them to split as well. This **chain reaction** is shown in Figure 7.3. Depending on how we control this process we have two completely different uses for it. In a controlled chain reaction, on average only one neutron from each fission will strike another nucleus and cause it to divide. This is what we want to happen in a power station.

In an uncontrolled chain reaction most of the two or three neutrons from each fission strike other nuclei. This is how nuclear ('atomic') bombs are made. It is a frightening thought that a piece of pure uranium-235 the size of a tennis ball has enough stored energy to flatten a town.

Nuclear power stations

Figure 7.4 shows a gas-cooled **nuclear reactor**. The energy released by the fission processes in the uranium fuel rods produces a lot of thermal energy. This thermal energy is carried away by carbon dioxide gas, which is pumped around the reactor. The hot gas then boils water to produce steam, which can be used to work the electrical generators.

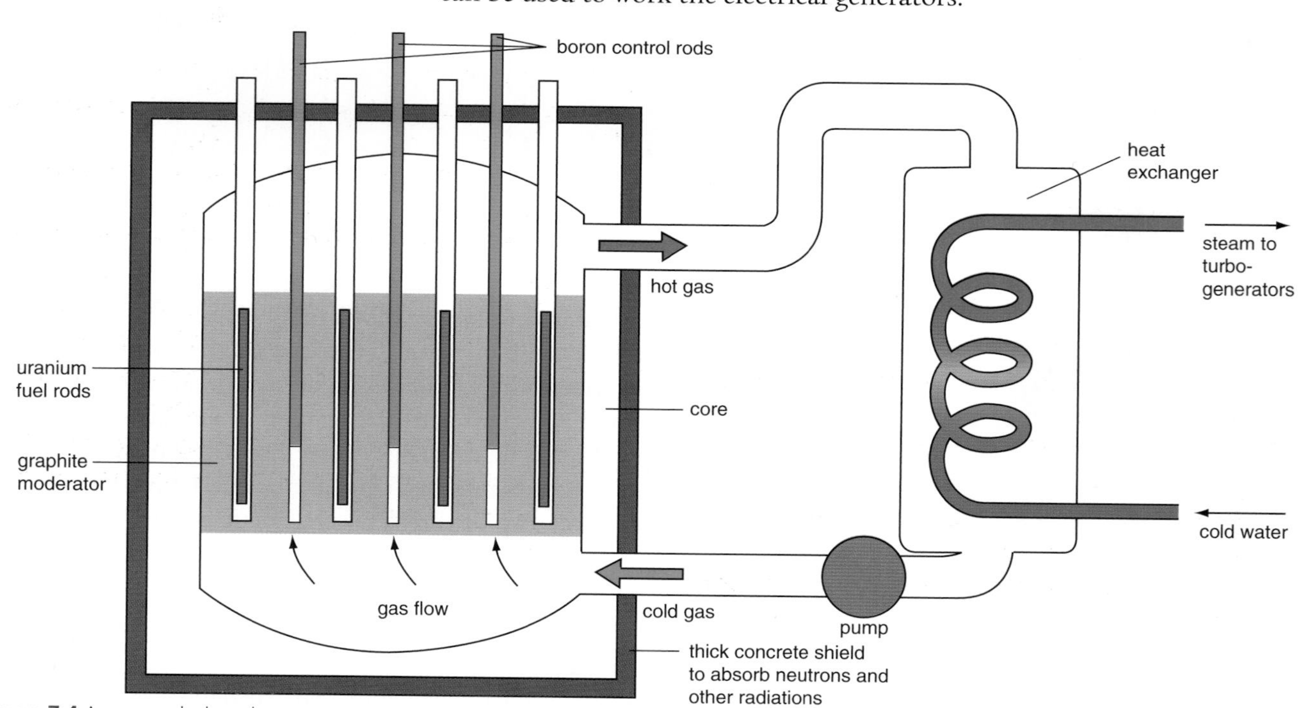

Figure 7.4 A gas-cooled nuclear reactor.

- The **fuel rods** are made of uranium-238, 'enriched' with about 3% uranium-235. Uranium-238 is the most common isotope of uranium, but it is only uranium-235 that will produce energy by fission. The fuel rods are embedded in graphite, which is called a **moderator**. The purpose of a moderator is to slow down neutrons that are produced in fission. A nucleus is split more easily by a slow-moving neutron. The fuel rods are long and thin so that neutrons can escape. Neutrons leave one rod and cause another nucleus to split in a neighbouring rod.

- The rate of production of energy in the reactor is carefully regulated by the **boron control rods**. Boron absorbs neutrons very well, so by lowering the control rods into the reactor the reaction can be slowed down. In the event of an emergency the rods are pushed right into the core of the reactor and the chain reaction stops completely.
- The concrete shielding around the reactor provides vital protection for workers. Neutrons are a dangerous and penetrating radiation. They are stopped most effectively by light nuclei such as carbon or hydrogen in our body. Neutrons do a lot of damage to our body tissues. Concrete provides an effective shield because it contains a lot of water.

Nuclear equations

Figure 7.2 showed a possible fission of a uranium nucleus. This fission reaction can be described by an equation:

$$^{1}_{0}\text{n} + ^{235}_{92}\text{U} \rightarrow ^{137}_{52}\text{Te} + ^{97}_{40}\text{Zr} + 2^{1}_{0}\text{n}$$

The mass and atomic numbers on each side balance. However, very accurate measurement shows that the mass on the left-hand side of the equation is slightly more than the mass on the right-hand side. This small reduction in mass is the source of the released nuclear energy.

STUDY QUESTIONS

1 Name the particle that the nucleus of a uranium-235 atom must absorb before fission can happen.
electron neutron proton alpha particle

2 a) Explain what is meant by nuclear fission.
b) In what way is fission
i) similar to radioactive decay?
ii) different from radioactive decay?

3 State what is meant by a chain reaction. Explain how the chain reaction works in a nuclear bomb and in a nuclear power station.

4 The following questions are about the nuclear reactor shown in Figure 7.4.
a) Explain the purpose of the concrete shield surrounding the reactor.
b) Explain why carbon dioxide gas is pumped through the reactor.
c) State which isotope of uranium undergoes fission in the fuel rods.
d) Will the fuel rods last for ever? Explain your answer.
e) Explain the purpose of the graphite moderator.
f) State what you would do if the reactor core suddenly got too hot.

5 After absorbing a neutron, uranium-235 can also split into the nuclei barium-141 ($^{141}_{56}\text{Ba}$) and krypton-92 ($^{92}_{36}\text{Kr}$). Write a balanced symbol equation to show how many neutrons are emitted in this reaction.

6 Boron has two stable isotopes: 80% $^{11}_{5}\text{B}$ and 20% $^{10}_{5}\text{B}$. Write a nuclear equation to show why boron is used in nuclear reactor control rods.

7.8 Nuclear fusion

Here you can see the inside of the Joint European Torus. This is an experiment to try to use nuclear fusion to generate electrical power. Nuclei of tritium and deuterium are confined inside the torus by a strong magnetic field and heated to very high temperatures.

Figure 8.1 Inside the JET tokamak device.

Nuclear fission is a process whereby energy is released by the splitting of a larger nucleus. Energy can also be released by **nuclear fusion**; in this process two small nuclei combine or fuse to form a single larger nucleus.

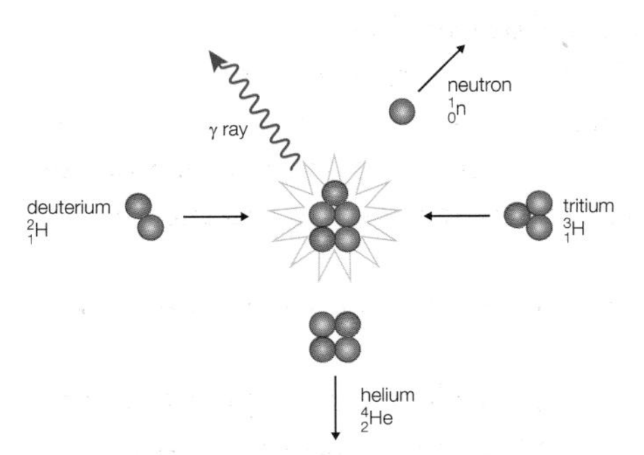

Figure 8.2 The fusion of deuterium and tritium.

Figure 8.2 shows the fusion of two isotopes of hydrogen, deuterium ^2_1H, and tritium ^3_1H. The fusion process can be described by this nuclear equation:

$$^3_1H + ^2_1H \rightarrow ^4_2He + ^1_0n + \gamma \text{ ray}$$

When the two small nuclei join together, they form the large ^4_2He nucleus, and a neutron is ejected.

In addition, a gamma ray is produced which carries away energy from the process. The combined mass of the ^3_1H and ^2_1H nuclei is slightly greater than the combined mass of the ^4_2He nucleus and the neutron. This is similar to nuclear fission in that a small amount of mass has been converted to energy, which, in this case, is carried away by the gamma ray.

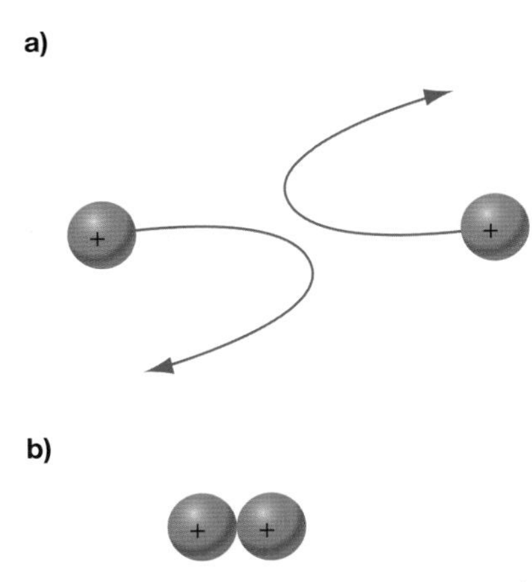

Figure 8.3 a) At low temperatures (less than 15 million kelvin), two protons repel each other b) At high temperatures, two protons have enough kintetic energy to overcome the electrostatic repulsion of their charges, and fusion takes place.

Fusion at high temperatures

Nuclear fusion is sometimes referred to as thermonuclear fusion, as the process can only occur at very high temperatures. Figure 8.3 explains why. The average kinetic energy of a group of particles depends on their temperature – the higher the temperature, the greater the kinetic energy.

In Figure 8.3(a), two protons approach each other at a low temperature, but their charges repel and they do not collide. In Figure 8.3(b), the protons have a high enough kinetic energy to overcome the electrostatic repulsion of their charges, and fusion takes place.

Fusion occurs because there is a very short-range attractive force between the protons called the strong nuclear force. This nuclear force does not extend outside the nucleus. At a short range, the strong nuclear force is much stronger than the electrostatic repulsion of the protons, but at a long range, the electrostatic repulsion is stronger than the strong nuclear force.

Nuclear fusion is the energy source for the Sun and all other stars. Although our Sun has a surface temperature of about 5800 K, the temperature of the core is about 15 000 000 K. At such high temperatures, fusion can take place, and hydrogen nuclei are fused to make helium nuclei. The energy released by fusion in the core is then dissipated at the Sun's surface. The pressure in the core of the Sun is very high – about 500 million times larger than the Earth's atmospheric pressure. At such high pressures, collisions between the nuclei are very frequent, and fusion is more likely to occur.

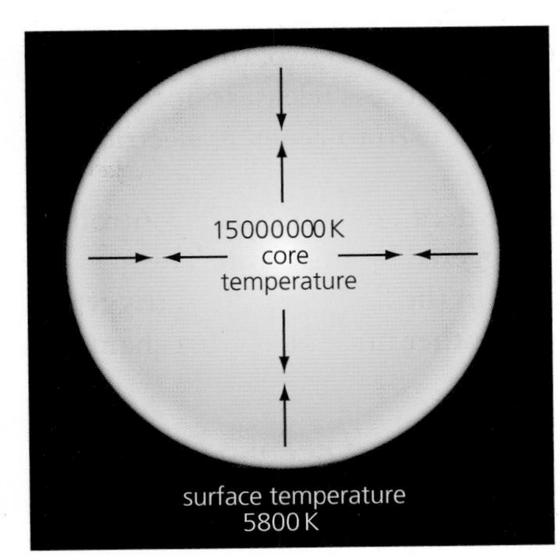

Figure 8.4 A star is a battle ground. The forces of gravity try to collapse a star, but these inward forces are balanced by the enormous outward pressure exerted by the hot core. The pressure at the centre of the star is about 500 million times bigger than the Earth's atmospheric pressure.

Nuclear fission and nuclear fusion

Nuclear fission and nuclear fusion are similar because:

- they are both sources of energy
- both release energy due to a small loss of mass.

Nuclear fission and nuclear fusion are different because:

- fusion is the creation of a larger nucleus from smaller nuclei
- fission is the creation of smaller nuclei from a larger nucleus
- fusion occurs only at very high temperatures and pressure
- fission is triggered by the collision between a neutron and a large nucleus
- fission releases neutrons which can cause further fissions and then a chain reaction
- the products of fission have kinetic energy which can transfer energy to other stores
- fission can be controlled and used to generate electricity in power stations
- fusion cannot be controlled by us, and only occurs naturally in stars.

STUDY QUESTIONS

1. a) Explain what is meant by nuclear fusion.
   b) Discuss what nuclear fusion has in common with nuclear fission, and how the processes differ.
2. Explain why nuclear fusion does not occur at low temperatures and pressures.
3. Much research has been done to try to make nuclear fusion a source of energy to try to generate electricity. Discuss the advantages and disadvantages of this idea.

Summary

I am confident that:

✓ I can describe the structure of the atom

- The atom is electrically neutral: each proton in the nucleus carries a positive charge, which is balanced by the negative charge on an equal number of electrons.
- The mass of the atom is concentrated in the nucleus: a proton and a neutron have approximately the same mass – an electron has a very small mass by comparison.

✓ I can recall definitions and use symbols correctly

- Atomic (proton) number: the number of protons in the nucleus.
- Mass (nucleon) number: the number of protons and neutrons in the nucleus.
- An ion is formed by either adding or removing electrons from a neutral atom.
- An isotope of an atom is another atom that has the same number of protons, but a different number of neutrons.

✓ I can explain radioactivity

- An alpha particle is a helium nucleus ($^{4}_{2}He$)
- A beta particle is a fast-moving electron ($^{0}_{-1}e$).
- A gamma ray is a short-wavelength electromagnetic wave.
- An alpha particle travels 5 cm through air and is stopped by paper.
- Beta particles travel many metres through air and can be stopped by aluminium sheet a few millimetres thick.
- Gamma rays travel great distances through air and can only be absorbed effectively by very thick lead sheets – several centimetres thick.

✓ I understand nuclear transformation

- On ejecting an α-particle, a β-particle, or a neutron, a nucleus is transformed into another nucleus (a γ-ray makes no change to the mass or charge of its nucleus). Three examples are:

α-particle decay	$^{238}_{92}U \rightarrow ^{234}_{90}Th + ^{4}_{2}He$
β-particle decay	$^{234}_{90}Th \rightarrow ^{234}_{91}Pa + ^{0}_{-1}e$
neutron decay	$^{5}_{2}He \rightarrow ^{4}_{2}He + ^{1}_{0}n$

- β-decay causes the atomic number to increase by 1.

✓ I can calculate radioactive decay

- Radioactive decay is a random process – like throwing a lot of dice or tossing a lot of coins.
- In one half-life, half of the radioactive nuclei will decay. In a further half-life, a further half of what is left will decay.
  - After 1 half-life, ½ the sample is left
  - after 2 half-lifes, ¼ of the sample is left
  - after 3 half-lifes, ⅛ of the sample is left, and so on.

✓ I can discuss hazards of radiation

- Alpha particles, beta particles and gamma rays are ionising radiations. These radiations knock electrons out of atoms thereby making ions.

✓ I understand nuclear fission

- A large nucleus can split into two; this is called nuclear fission. This process releases a lot of energy, which can be used to generate electricity in power stations.
- Unlike radioactive decay, fission can be controlled.
- Fission can be triggered by a neutron. The neutrons released in the fission reaction trigger further fissions in other nuclei. This is a chain reaction.

$$^{1}_{0}n + ^{235}_{92}U \rightarrow ^{137}_{52}Te + ^{97}_{40}Zr + 2^{1}_{0}n$$

- The fission reaction can be controlled in a nuclear reactor by boron rods.

✓ I understand nuclear fusion.

- Two small nuclei can fuse together to form a larger nucleus. This is called nuclear fusion and the process releases a lot of energy.
- The energy is released as a result of a loss of mass from the smaller nuclei.
- Fusion is the source of energy in stars.
- Fusion only occurs at very high temperatures and pressures due to electrostatic repulsion between nuclei.
- I can explain the differences between nuclear fission and nuclear fusion.

Sample answers and expert's comments

1 a) When an atom of fermium-257 decays an alpha particle is emitted from the nucleus. An atom of californium is left behind. The alpha particle is a helium nucleus.

i) Copy and complete the equation below which describes this decay. (1)

$$^{257}_{100}Fm \rightarrow Cf + {}^{4}_{2}He$$

ii) Calculate how many neutrons there are in a nucleus of fermium-257. (1)

b) When an atom of caesium-137 decays a beta particle is emitted from the nucleus. An atom of barium is left behind. The beta particle is a fast-moving electron.

i) Copy and complete the equation below which describes this decay. (1)

$$^{137}_{55}Cs \rightarrow Ba + {}^{0}_{-1}e$$

ii) How many protons are there in a caesium nucleus? (1)

(Total for question = 4 marks)

	Student response Total 3/4	Expert comments and tips for success
a) i)	$^{257}_{100}Fm \rightarrow {}^{253}_{98}Cf + {}^{4}_{2}He$ ✔	This is balanced correctly. The top line of numbers is the mass or nucleon numbers. A nucleon is a neutron or a proton. The bottom line is the atomic or proton number of each particle. The numbers in these two lines must balance.
ii)	257 O	Incorrect. You need to be very clear about what the numbers mean in the symbols. Here 257 is the sum of protons and neutrons. Fermium-257 has 257 nucleons and 100 protons. So there are 257 – 100 = 157 neutrons.
b) i)	$^{137}_{55}Cs \rightarrow {}^{137}_{56}Ba + {}^{0}_{-1}e$ ✔	Correct. With beta decay a neutron turns into a proton. So barium's atomic or proton number is one higher than caesium's. Note the numbers always balance: 55 = 56 – 1
ii)	55 ✔	55 is the atomic or proton number, which is the number of protons in the nucleus.

2 Phosphorus-32 is a radioactive isotope, which decays by emitting beta particles.

a) State the nature of a beta particle. (1)

b) The graph shows how the count rate for a sample of phosphorus-32 changes with time.

i) Use the graph to determine the half-life of phosphorus-32. To get full marks you must show your working. (2)

ii) Use your answer to i) above to calculate how long it takes the count rate to reach 80 counts per minute. (2)

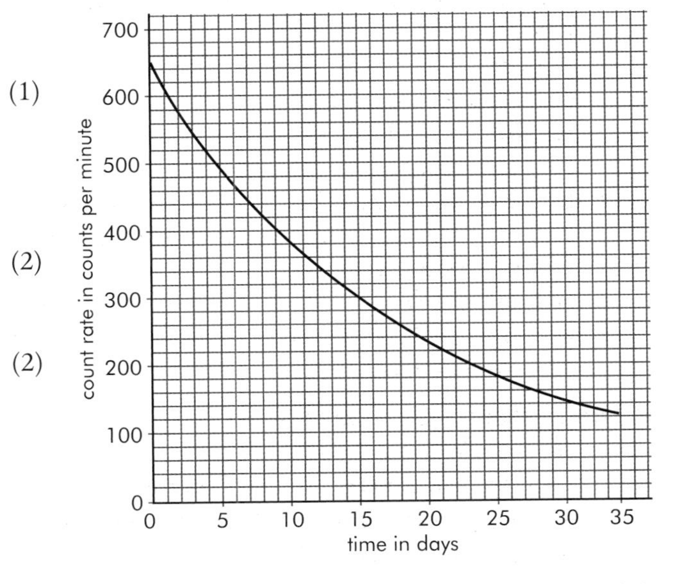

(Total for question = 5 marks)

Student response Total 4/5	Expert comments and tips for success
a) *A fast electron.* ✔	Correct.
b) i) *14 days* ✔	The correct answer gains 1 mark, but does not the get the mark available for showing working on the graph. The initial count rate is about 640 counts/min. After one half-life the count will be 320 counts/min, which corresponds to a time of 14 days on the *x*-axis.
ii) *320, 160, 80* ✔ *3 × 14 = 42 days* ✔	This answer lists the count rates after one, two and three half-lives, but does not explain clearly what is being done. However, there is enough for the examiner to see that the student understands how to work out the answer correctly. You should try to set your work out as clearly as possible, for example: After 1 half-life the count rate is 320 After 2 half-lives the count rate is 160 After 3 half-lives the count rate is 80 This is 3 × 14 = 42 days.

3 The diagram below shows the process of nuclear fission.

a) Name the particles which are labelled X, Y and Z. (3)

b) Once the process of fission has been started it continues with a chain reaction.

i) Explain what is meant by a chain reaction. (2)

ii) Copy the diagram and add the next stage of the chain reaction. (2)

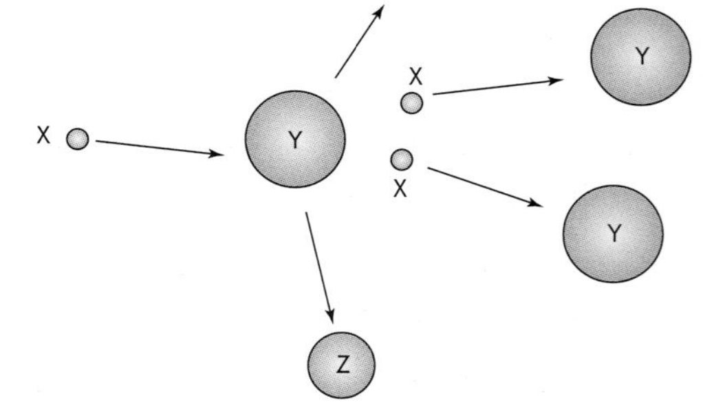

(Total for question = 7 marks)

Student response Total 6/7	Expert comments and tips for success
a) *X neutron* ✔ *Y large nucleus which splits* ✔ *Z small nucleus* ✔	These answers show the student understands the idea, and gains the marks. A better answer would use the correct technical terms: X neutron Y fissionable nucleus Z fission fragments or smaller nucleus or split nucleus
b) i) *The reaction goes on and never stops.* ✔ *The neutrons make it happen.*	This answer gets a mark for explaining that the reaction continues. To gain the second mark it needs to explain that the neutrons hit another nucleus, which releases more neutrons.
ii)	The diagram is excellent. It clearly shows the next nucleus splitting into two and the neutrons being released.

Exam-style questions

1 The diagram represents an atom of beryllium-9. Which of the following is the atomic number of beryllium?

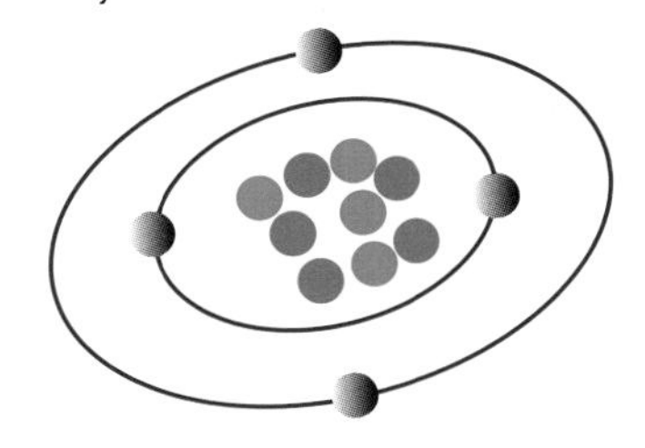

A 4 B 5 C 9 D 13 [1]

2 Beryllium-10 is radioactive isotope of beryllium. Which of the following statements is true?

A Beryllium-10 has one more proton and one more electron than beryllium-9.

B Beryllium-10 has one more neutron and one more electron than beryllium-9.

C Beryllium-10 has one more neutron than beryllium-9.

D Beryllium-10 has one more proton than beryllium-9. [1]

3 Beryllium-10 decays by emitting beta particles. Which of the following equations correctly describes this radioactive decay?

A $^{10}_{4}\text{Be} \rightarrow {}^{10}_{5}\text{B} + {}^{0}_{1}\text{e}$

B $^{10}_{4}\text{Be} \rightarrow {}^{10}_{4}\text{Be} + {}^{0}_{0}\text{e}$

C $^{10}_{4}\text{Be} \rightarrow {}^{10}_{5}\text{Be} + {}^{0}_{-1}\text{e}$

D $^{10}_{4}\text{Be} \rightarrow {}^{10}_{5}\text{B} + {}^{0}_{-1}\text{e}$ [1]

4 Gadolinium-159 has a half-life of 18 hours. A sample of mass 64 g is prepared in a laboratory experiment. How much gadolinium-159 remains after 3 days?

A 48 g B 24 g C 8 g D 4 g [1]

5 Fresh strawberries are sometimes irradiated before being exported to another country. Which of the following statements is true?

A The strawberries are contaminated by the radiation.

B The strawberries cannot be eaten for a few days after irradiation.

C Irradiated strawberries do not become radioactive.

D Radioactive particles settle on the strawberries. [1]

Several people who work in industry and hospitals are exposed to different types of radiation. Four ways to check or reduce their exposure to radiation are:

A wearing a badge containing photographic film.

B wearing a lead apron.

C working behind a thick glass screen with remote handling equipment.

D having a medical check-up every month.]

6 Which of the above should be used by a hospital radiographer? [1]

7 Which of the above should be used by a nuclear power worker? [1]

8 Alpha, beta, gamma and neutron emission are types of radiation. Which has the shortest range in air?

A Beta

B Alpha

C Neutron

D Gamma [1]

9 What type of particle must a uranium-235 nucleus absorb before nuclear fission can happen?

A Neutron

B Alpha particle

C Proton

D Electron [1]

10 Which of the following statements about nuclear fusion is not true?

A Fusion occurs in stars.

B Two small nuclei join together to make a larger nucleus.

C Two uranium nuclei can fuse to form a larger nucleus.

D Fusion only occurs at very high temperatures. [1]

11 a) Copy and complete the table about atomic particles.

Atomic particle	Relative mass	Relative charge
proton		+1
neutron	1	0
electron	negligible	

[2]

b) Read the following passage about sodium.

Sodium is an element with an atomic number of 11.

Its most common isotope is sodium-23, ${}^{23}_{11}Na$.

Another isotope, sodium-24, is radioactive.

i) State the number of protons, neutrons and electrons in sodium-23. [2]

ii) Explain why sodium-24 has a different mass number from sodium-23. [1]

iii) What is meant by the word 'radioactive'? [1]

iv) Atoms of sodium-24 change into atoms of magnesium by beta decay. Write an equation to show the atomic and mass numbers of this isotope of magnesium. [2]

v) Give the name, or symbol, of the element formed when an atom of sodium-24 (proton number = 11) emits gamma radiation. [1]

c) i) Name a suitable detector that could be used to show that sodium-24 gives out radiation. [1]

ii) Name a disease that can be caused by too much exposure to a radioactive substance such as sodium-24. [1]

12 The table shows the half-life of some radioactive isotopes.

Radioactive isotope	Half-life
aluminium-29	7 minutes
technetium-99	6 hours
rubidium-83	86 days
cobalt-60	5 years

a) Which **one** of the isotopes shown in the table is the most suitable for use as a medical tracer? Explain your choice. [2]

b) Sketch a graph to show how the number of radioactive atoms present in the isotope cobalt-60 will change with time. [3]

13 a) Copy and complete the following table for an atom of actinium-225 (${}^{225}_{89}Ac$).

mass number	225
number of protons	89
number of neutrons	

[1]

b) Explain what is meant by the terms 'atomic number' and 'mass number'. [2]

c) An atom of actinium-225 (${}^{225}_{89}Ac$) decays to form an atom of francium-221 (${}^{221}_{87}Fr$).

i) What type of radiation – alpha, beta or gamma – is emitted by actinium-225? [1]

ii) Why does an atom that decays by emitting alpha or beta radiation become an atom of a different element? [1]

14 The radioactive isotope, carbon-14, decays by beta (β) particle emission.

a) Plants absorb carbon-14 from the atmosphere. The graph shows the decay curve for 1 g of modern linen made from a flax plant. The radioactivity is caused by a small fraction of carbon-14 in the linen sample.

Use the graph to find the half-life of carbon-14. [2]

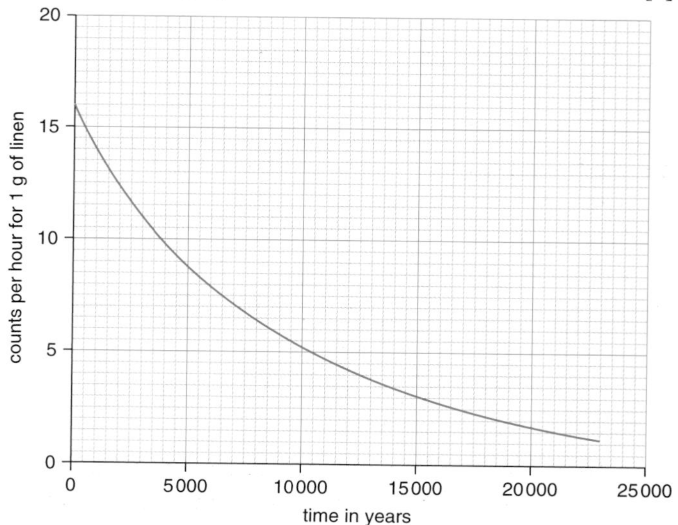

b) Linen has been used to make clothes for thousands of years. It contains radioactive carbon-14, which decays. An old sample of linen has a smaller fraction of carbon-14 than a new sample of linen. A museum has a shirt that may be 2000 years old. The carbon-14 in a 1 g sample of the shirt produces 300 counts in a day.

Is the shirt really 2000 years old? Justify your answer. [3]

c) Suggest why carbon dating is unsuitable for dating samples thought to be:

i) 100 years old

ii) 50 000 years old. [2]

15 The table gives information about some of the radioactive substances released into the air by the explosion at the Fukoshima nuclear plant in 2011.

Radioactive substance	Half-life	Type of radiation emitted
iodine-131	8 days	beta and gamma
caesium-134	2 years	beta
caesium-137	30 years	beta

a) State how the structure of a caesium-134 atom is different from the structure of a caesium-137 atom. [1]

b) Explain what beta particles and gamma rays are. [2]

c) A sample of soil is contaminated with some iodine-131. Its activity is 40 000 Bq. Calculate how long it will take for the activity to drop to 1250 Bq. [2]

d) Which of the three isotopes will be the most dangerous 50 years after the accident? Justify your answer. [2]

16 Radiation workers wear a special badge to monitor the radiation they receive.

The badge is a lightproof plastic case containing a piece of photographic film. Each worker is given a new badge to wear each month. Part of the film is covered with a thin sheet of aluminium foil.

a) Explain how the badge can show:

i) how much radiation a worker receives each month [2]

ii) what type of radiation a worker has received. [2]

b) Explain why workers in the nuclear power industry monitor their levels of radiation exposure. [2]

17 A radiation detector and counter are used to measure the background radiation. The background activity is 10 counts per minute. The same detector and counter are then used to measure the radiation from some radioactive gas. The graph shows how the number of counts per minute changes with time.

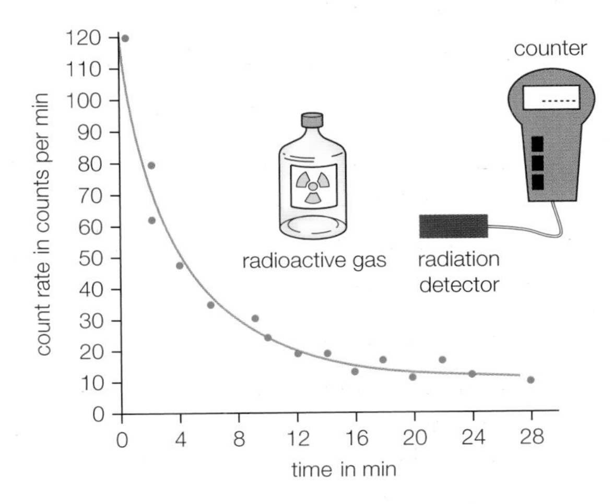

a) Although the readings on the counter are accurately recorded, the points do not exactly fit a smooth curve. Explain why. [2]

b) Explain why the count rate is almost constant after 20 minutes. [2]

c) Use the graph to estimate the half-life of the radioactive gas. [2]

18 a) The diagram shows what can happen when the nucleus of a uranium atom absorbs a neutron.

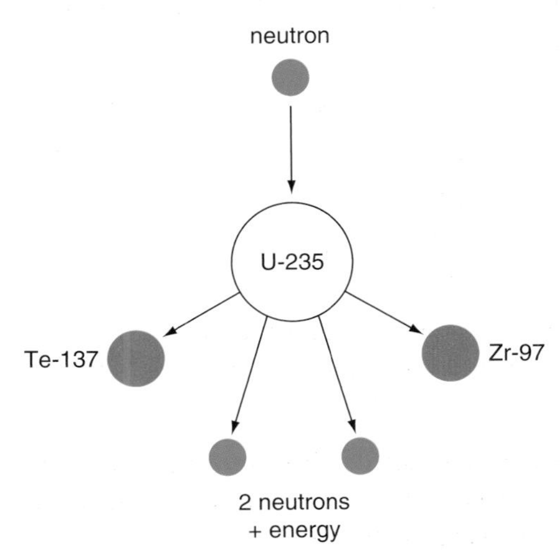

i) What name is given to the process shown in the diagram? [1]

ii) Explain how this process could lead to a chain reaction. You may copy and add to the diagram to help your answer. [3]

iii) State how the mass number of an atom changes when its nucleus absorbs a neutron. [1]

b) Uranium-235 is used as a fuel in some nuclear reactors. The reactor core contains control rods used to absorb neutrons. Explain what happens when the control rods are lowered into the reactor. [3]

19 A radioactive source is placed close to a Geiger-Müller tube, and the count rate is recorded as 520 Bq. Various absorbers are placed between the source and the Geiger-Müller tube, and the count rate is recorded in the table.

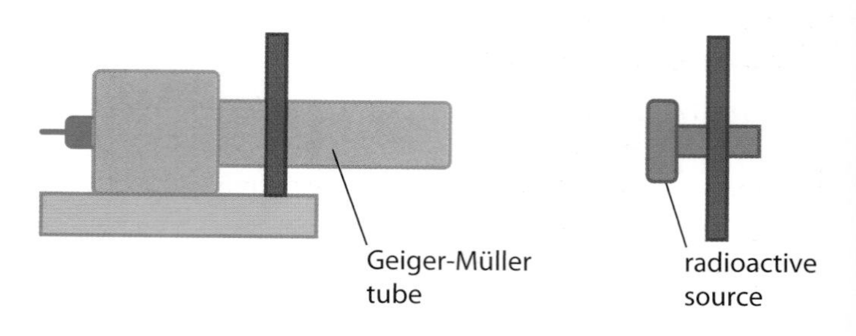

Use the data to identify the radiations emitted. Justify your conclusions. [5]

Absorber	Thickness of absorber in mm	Count rate in Bq
none		520
paper	0.2	328
aluminium	0.5	194
aluminium	2.0	8
lead	5.0	7
lead	10.0	8

EXTEND AND CHALLENGE

1 The diagram shows the piston in the cylinder of a car engine. When the car engine is running, the piston moves up and down inside the cylinder many times each second, which causes the cylinder wall to become worn.

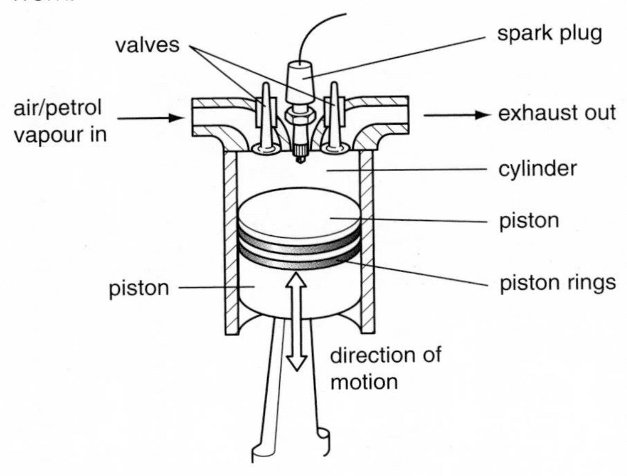

A car manufacturer wants to run tests to measure the wear of the cylinder wall due to the piston movement. A radioactive isotope of chromium with a short half-life is used.

a) Explain what is meant by an isotope.

b) A very thin layer of the radioactive isotope is placed on the inside wall of the cylinder and the car engine run continuously. A detector is placed outside the cylinder to measure the count rate.

i) Which type of emission from the radioactive isotope would be needed to reach the detector through the metal wall of the cylinder?

ii) State the name of a detector that could be used to measure the count rate.

c) The half-life of the emitter is 23 hours and the count rate taken at the start of the test was 600 counts/hour.

i) Predict what you would expect the count rate to be after 46 hours.

ii) The count rate measured after 46 hours was actually 120 counts/hour. This was explained by assuming that part of the layer of radioactive isotope on the cylinder wall had been worn away. Calculate the fraction that had been worn away. State one assumption you have made in your calculation.

iii) Explain why the half-life of the radioactive isotope has to be short but not too short.

2 All elements have some isotopes which are radioactive.

a) State what is meant by the terms

i) isotopes

ii) radioactive.

b) The graph shows how the count rate for a sample of the radioactive isotope curium-243 changes with time.

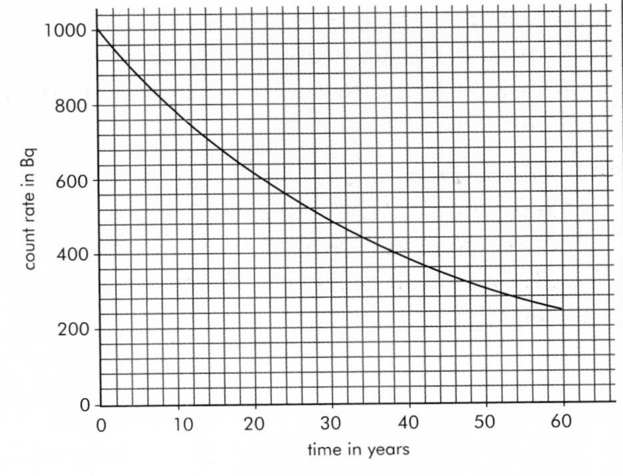

Use the graph to determine the half-life of curium-243. Show clearly on the graph how you obtain your answer.

c) Space scientists plan to send a spacecraft to the planet Mars to investigate its two moons. The voyage will last 10 years.

The electricity to power the instruments on the spacecraft will be generated using thermal energy transferred from the decay of a quantity of curium-243. The total power generated will be 500 W.

i) Curium-243 decays by emitting alpha particles. Describe what an alpha particle is

ii) During the voyage the output from the generators will decrease from the initial value of 500 W. Explain why the power will decrease.

iii) Show from the graph that the power generated will fall to about 310 W after 20 years.

d) One scientist said it would be a good idea to use the isotope americium-241 to power the generators. Americium-241 emits alpha particles and has a half-life of 432 years.

Explain why americium-241 is not as good a choice as curium-243 for this voyage.

8 Astrophysics

CALCULATE • CONSIDER

1 Explain why the moon changes its shape throughout a month.
2 Name two constellations of stars you have seen.
3 The mean mass of a star is about 10^{30} kg. Calculate the mass of the Andromeda Galaxy.

The Andromeda Galaxy and its two dwarf companion galaxies are 2.5 million light years away from our galaxy, the Milky Way. Andromeda contains approximately 1 trillion (10^{12}) stars, and is bright enough to be seen by the naked eye. Andromeda and the Milky Way are amongst billions of galaxies in the Universe.

By the end of this section you should:

- know about galaxies and their distribution in the Universe
- know that our Sun is part of the Milky Way galaxy
- understand why gravitational field strength, g, varies, and that g is different on other planets and moons
- be able to explain that the gravitational force causes planets and other bodies to orbit
- be able to describe the differences in the orbits of comets, moons and planets
- be able to use the relationship between orbital speed, orbital radius and time period
- understand how stars can be classified according to colour
- know that a star's colour is related to its surface temperature.
- be able to describe the evolution of stars
- understand the idea of a star's absolute magnitude
- be able to draw the main components of the Hertzsprung-Russell diagram
- be able to describe the evidence for the Big Bang theory
- describe that if a wave source is moving relative to an observer there is a change in observed frequency and wavelength
- know the equation relating the change in wavelength to the velocity of a galaxy
- be able to explain why red shift provides evidence for the expansion of the Universe.

8.1 Earth's place in the Universe

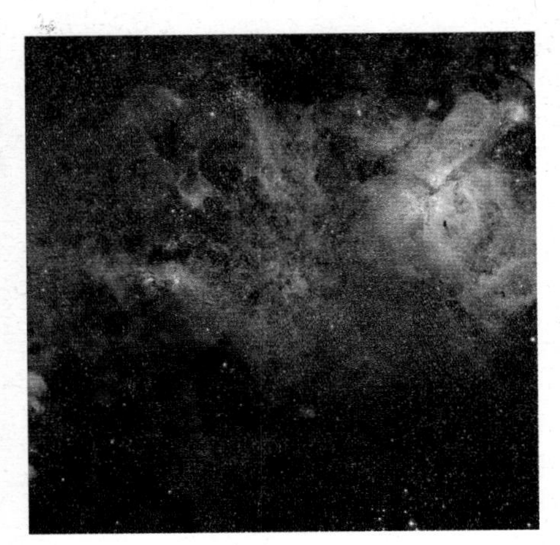

Figure 1.1 This bright cluster of stars is known as the Running Chicken Nebula.

We live in an enormous universe that contains billions of moons, planets, stars and galaxies. In the daytime, the sky is dominated by our Sun, but at night you can see lots of stars and our moon, and with a telescope you can identify planets and galaxies.

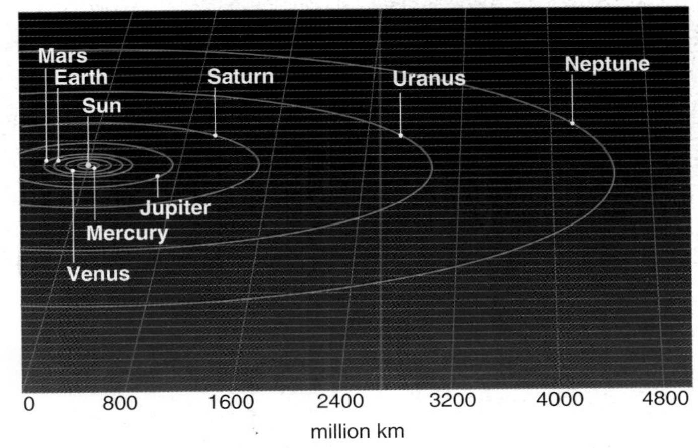

Figure 1.2 The orbits of the planets around the Sun. The four inner planets are very close to the Sun; the gaps between the outer planets are very large.

The Solar System

Figure 1.2 shows the familiar orbits of the eight planets that travel around our Sun. There are also five dwarf planets (Ceres, Pluto, Haumea, Makemake and Eris), many comets, and other fragments of rock of various sizes, all of which are in orbit around the Sun. A dwarf planet is a very small planet – still spherical in shape. The pull of gravity also keeps moons in orbit around the six outer planets.

Figure 1.3 Jupiter with its four largest moons – Io, Europa, Ganymede and Callisto. These can be seen with a small telescope.

Stars and galaxies

Our Sun is one of billions of stars. The force of gravity acts over great distances to bring stars together into a large group called a galaxy. The Sun is part of the Milky Way galaxy, which contains about 200 billion stars. On a dark night, you can see the Milky Way stretching across the sky. (Figure 1.4).

The Milky Way is a spiral galaxy. Figure 1.5 illustrates the position of our Sun in the galaxy and the distribution of stars.

Figure 1.4 When we look into the plane of our own galaxy, we see thousands of stars. We call this the Milky Way.

(a) Side view of our galaxy

(b) Top view of our galaxy

Figure 1.5 The Sun (red circle) moves around the centre of the galaxy in an orbit taking 220 million years.

Groups of galaxies

Our galaxy is part of a group of galaxies called the local group. The Andromeda Galaxy (see Chapter introduction page) is one of our near neighbours in a group of about 50 galaxies. Our nearest cluster of galaxies is the Virgo Cluster, which is about 50 million light years away from us. This sounds a long way, but the most distant galaxies are over 13 000 million light years from us. The Virgo Cluster is much larger than our local group; it contains some 1500 galaxies (Figure 1.6).

TIP

Make sure you know these facts:

- A galaxy is a large group of stars.
- A star is a body which emits energy powered by nuclear fusion.
- A planet orbits a star.
- A moon orbits a planet.

Note: you will not be examined on light years, for example question 3.

Figure 1.6 The Virgo Cluster of galaxies lies 50 million light years away from us.

STUDY QUESTIONS

1 Explain what is meant by each of the following terms:
   a) moon
   b) planet
   c) dwarf planet
   d) star
   e) galaxy
   f) group of galaxies

2 a) Name our galaxy.
   b) Which of the following best describes the number of stars in a galaxy:
   - lots
   - thousands
   - millions
   - hundreds of billions.

3 a) A light year is the distance light travels in one year. Calculate a light year in metres, using the speed of light 3×10^8 m/s.
   b) Proxima Centauri is our nearest star at a distance of 4.25 light years. Calculate this distance in metres.
   c) Proxima b is a planet in orbit around Proxima Centauri, which is thought to have a similar temperature to the Earth. Modern spacecraft fly at 30 km/s. Calculate how long it would take to travel to Proxima b.
   d) Discuss the viability of space travel. Include both time and energy in your considerations.

8.2 Orbits

Figure 2.1 What force keeps these people in their circular path at the funfair?

How is it possible for something to travel at a constant speed and to be accelerating all the time? This happens when something follows a circular path. Although the speed is constant, there is a change of direction. So the velocity changes and therefore there is an acceleration.

Gravitational field strength

Every planet, moon or star has a gravitational field at its surface. On the surface of the Earth, the **gravitational field strength**, g, is about 10 N/kg. This means that a mass of 1 kg is pulled downwards with a force of 10 N. This force is the object's weight.

The strength of the gravitational field for a planet depends on its radius and density. A larger field is produced by a planet with a larger density and larger radius.

Table 1 shows some examples of gravitational field strengths for various objects in our Solar System. The gravitational field strength is different for each planet and moon.

Table 1

Object	Gravitational field strength in N/kg
Earth	10
Mars	3.7
Jupiter	24.8
Sun	270
The Moon	1.6
Phobos (a moon of Mars)	0.001

Throwing the hammer

Figure 2.2 shows an athlete preparing to throw a weight – this event is called throwing the hammer. He spins the weight up to a large speed before releasing it. While the weight is moving in its circular path, the athlete exerts a force, F, towards the centre of the circle. When he lets go of the weight it flies off, landing a large distance away.

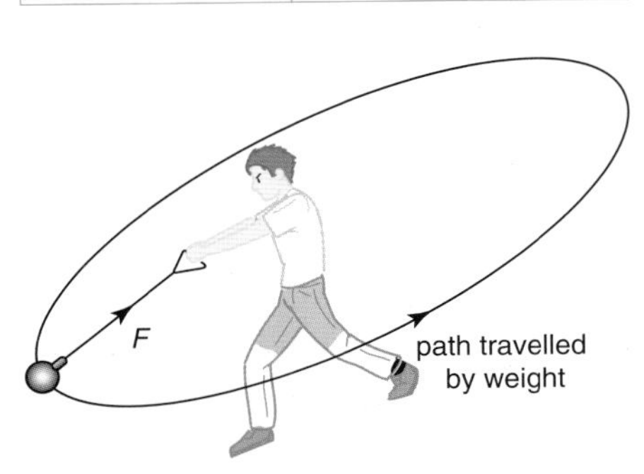

Figure 2.2 A hammer thrower spins the weight around in a circle before releasing it. To keep the weight in its circular path he must exert a force, F, towards the centre of the circle.

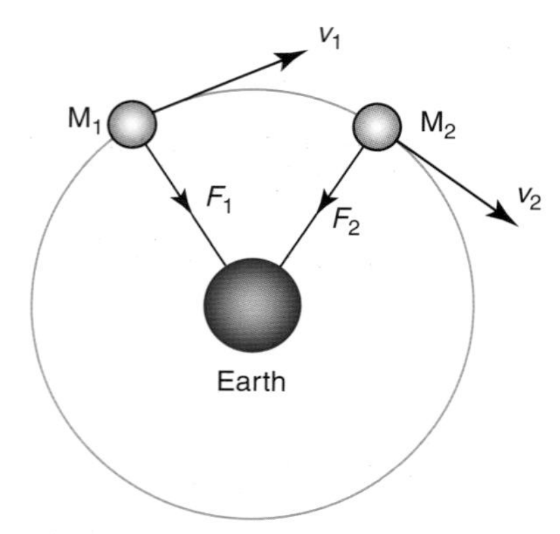

Figure 2.3 The force of gravity directed towards the Earth keeps the Moon in orbit.

The pull of gravity and orbits

You understand that gravity acts close to the Earth to pull things to the ground. However, gravity is also a force that stretches an enormous distance out into space. Gravity acts on all stars, planets and moons.

- Gravity causes planets and comets to orbit the Sun.
- Gravity causes the Moon and artificial satellites to orbit the Earth.
- Gravity causes stars to orbit the centre of their galaxies.

Figure 2.3 shows the pull of gravity from the Earth, which keeps the Moon in orbit. In position M_1, the Moon has a velocity v_1. The pull of gravity deflects its motion. Later, it has moved to its new position M_2, with a velocity v_2. The force of gravity does not make the Moon travel any faster, but the force changes the direction of motion. The Moon will stay in its orbit for billions of years.

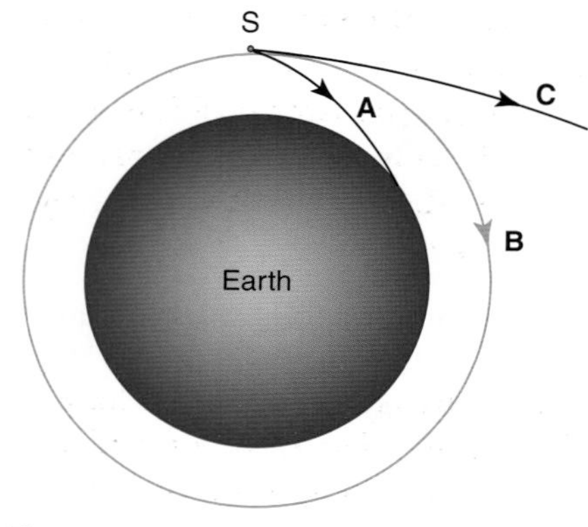

Figure 2.4 Speed C > speed B > speed A.

Speed of orbit

Figure 2.4 shows the importance of the speed of orbit. S is a satellite travelling around the Earth.

- If its speed is too great, it follows path C and disappears into outer space.
- If its speed is too slow, it follows path A and falls to Earth.
- When the satellite moves at the right speed, it follows path B and stays in a steady circular orbit.

Planets and moons move in elliptical orbits. However, many of the orbits for planets and moons in our Solar System are nearly circular. The speed of orbit of a moon, planet or satellite can then be related to the radius and time of orbit using the following equation.

$$\text{orbital speed} = \frac{2 \times \pi \times \text{orbital radius}}{\text{time period}}$$

$$v = \frac{2 \times \pi \times r}{T}$$

The equation is a special example of the more familiar equation:

$$\text{average speed} = \frac{\text{distance}}{\text{time}}$$

The distance travelled in one orbit is the circumference of a circle, which is $2 \times \pi \times r$.

Orbits near and far

If a planet is close to the Sun, the pull of gravity from the Sun is strong. This causes the speed of the orbit to be high. For a planet further away, the Sun's gravity pulls less strongly and the planet moves more slowly.

We can use data to calculate the orbital speeds for Earth and Mercury: the Earth orbits the Sun with an orbital radius of 150 million kilometres, and

TIP

Note that Mercury takes 88 days to orbit the Sun.

Mercury orbits the sun with an orbital radius of 58 million kilometres.

$$\text{orbital speed of Mercury} = \frac{2 \times \pi \times r}{T}$$

$$= \frac{2 \times \pi \times 58 \text{ million km}}{88 \text{ days}}$$

$$= 4.1 \text{ million km/day}$$

$$\text{orbital speed of Earth} = \frac{2 \times \pi \times r}{T}$$

$$= \frac{2 \times \pi \times 150 \text{ million km}}{365 \text{ days}}$$

$$= 2.6 \text{ million km/day}$$

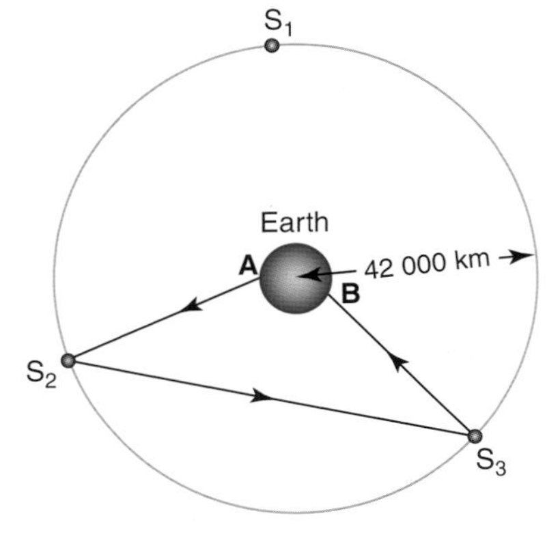

Figure 2.5 Satellites can be put into orbit of radius 42 000 km, around the Equator. These satellites have a period of rotation of 24 hours.

Geostationary orbits

Figure 2.5 shows three satellites in **geostationary orbits** above the Equator. When the radius of the orbit is about 42 000 km, the time of orbit is 24 hours. This means that a satellite always appears in the same position above the Earth's surface. This is important for satellite communication. Positions A and B on the Earth's surface, on the opposite side of the Earth, can always keep in contact by sending signals via satellites S_2 and S_3.

Comets

Comets, like asteroids, are rocks that orbit the Sun. However, comets have very elongated elliptical orbits (Figure 2.7). We think that at the edge of the Solar System there is a cloud of such comets. Occasionally, one is disturbed in its orbit so that it falls inwards towards the Sun. We notice a comet as it gets close to the Earth and Sun. Energy from the Sun melts ice contained in the rocks. The Sun emits a stream of charged particles called the Solar wind. This wind blows the melted ice away from the Sun, producing the comet's tail; the tail always points away from the Sun.

Figure 2.6 Heat from the Sun can melt the ice in a comet. Sometimes this causes the comet to develop a spectacular tail.

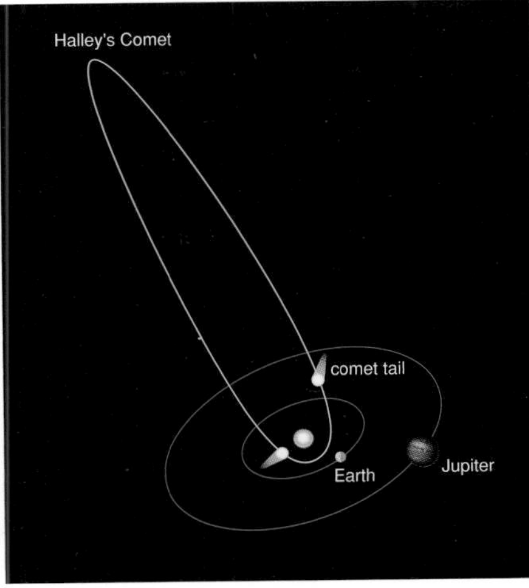

Figure 2.7 Comets come from the far edges of the Solar System.

TIP

You should be able to explain how the speed of a comet changes in its orbit.

Elliptical orbits

If a planet orbits the Sun in a circular orbit, it moves at a constant speed. In Figure 2.8(a) you can see that the Sun's gravitational pull is always at right angles to the planet's path.

However, a comet or asteroid in an elliptical orbit does not move at a constant speed. In Figure 2.8(b), the gravitational pull of the Sun on a comet at A is not at right angles to its path. The force does two things: it deflects the comet towards the Sun and speeds it up. At C, the pull of the Sun slows the comet down again. At D, the comet has low kinetic energy but high gravitational potential energy. As the comet falls towards the Sun it increases its kinetic energy to a maximum at B, and decreases its gravitational potential energy.

(a)

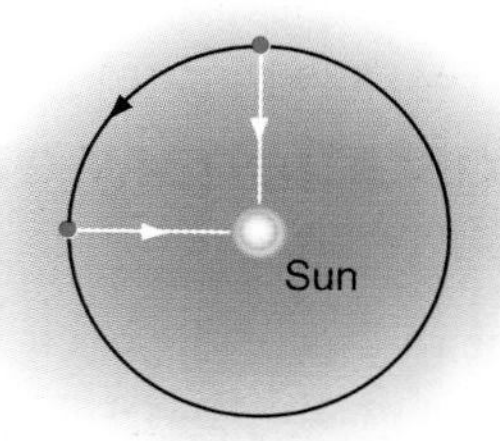

(b)

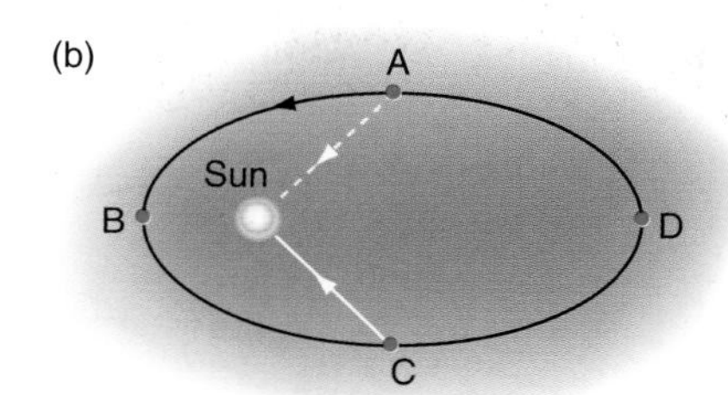

Figure 2.8 Circular and elliptical orbits.

STUDY QUESTIONS

1 a) Name the force which keeps the Moon in orbit around the Earth.
 b) Explain how this force keeps the Moon moving in a circular orbit.

2 Describe how a comet's orbit differs from that of a planet.

3 a) Make a sketch to show the orbit of a comet around the Sun.
 b) Mark the points on the orbit where the comet has:
  i) the most kinetic energy
  ii) the most gravitational potential energy.
 c) Show on your diagram where the pull of the Sun is strongest on the comet.
 d) Show where the comet travels most slowly.

4 In Figure 2.5 the satellites have an orbital speed of 3054 m/s. Their orbital radius is 42 000 km.
 a) Calculate the time it takes the satellites to orbit the Earth once.
 b) Explain why this time period is useful for communication.
 c) Some satellites go in low orbits over the poles, some 100 km above the Earth's surface.
  i) Suggest how the time period of these satellites compares with the ones in Figure 2.5.
  ii) Explain a use for such satellites.

5 A minor planet is an irregularly shaped rock in orbit around the Sun. There are about 300 000 minor planets. Some of these are only about 1 km in diameter. Others are larger, with diameters of about 1000 km.
The table shows the distances from the Sun and the time periods for the orbits of the planets Uranus and Neptune, and some minor and dwarf planets in orbit close to them. The orbital radii are quoted in Astronomical Units (AU). 1 AU is equal to the Sun–Earth distance, which is 150 million kilometres.

Planet or minor planet	Radius of orbit in AU	Time of orbit in years
Chiron	14	51
Bienor	17	68
Uranus	19	84
Neptune	32	165
Pluto	39	248
Haumea	43	283
Makemake	46	310
Eris	68	557

 a) Plot a graph of time period of the orbits against the radius of the orbits.
 b) An astronomer has made a mistake in recording radius of orbit for one of the planets. Identify which one. Suggest what the radius of the orbit should be.
 c) Astronomers think they have discovered a new minor planet with a radius of orbit 55 AU. Use the graph to predict its time period.
 d) Sedna is a minor planet thought to orbit the Sun at a distance of 519 AU, with a period of about 11 400 years. Calculate Sedna's orbital speed in:
  i) millions of km per year
  ii) m/s.

8.3 Stellar evolution

Figure 3.1 This bright cluster of stars is called the Jewel Box. Most of the stars here are blue giants.

The stars shown in the photo are millions of times more luminous than our Sun, but they live for a much shorter time.

Star temperature and colour

Our Sun is classified as a yellow dwarf star. This makes the Sun sound rather small, and whilst it is in comparison with some stars, the Sun is actually brighter than about 95% of all stars. The amount of energy emitted per second by a star is called its **luminosity**. The Sun has luminosity of about 4×10^{26} W. The brightest stars have a luminosity of more than a million times that of the sun, and the least luminous stars are nearly 10 000 times duller than the Sun.

Stars can be classified according to their colour. The colour of a star is linked to its temperature; the hottest stars are blue and the coolest stars are red. Table 1 shows how the temperature of a star is linked to its colour.

Table 1

Star colour	Surface temperature in K
blue	>11 000
blue-white	7500–11 000
white	6000–7500
yellow-white	5000–6000
orange	3500–5000
red	< 3500

Figure 3.2 Stars come in all sizes.

The diameters of stars also vary considerably. The largest stars have diameters 1000 times that of the Sun, whereas the smallest have diameters about 100 times smaller than the Sun's. See Figure 3.2.

Table 2 gives some examples of stars with surface temperatures and the luminosities, masses and diameters relative to the Sun.

Table 2

Star	Surface temperature in K	Diameter relative to the Sun	Mass relative to the Sun	Luminosity relative to the Sun
Antares A	3400	900	12	57 000
Rigel A	11 000	80	23	120 000
Beta Centauri A	25 000	6	11	42 000
Vega	9600	2.3	2.1	40
Sun	5800	1	1	1
61 Cygni	4500	0.7	0.7	0.15
Sirius B	25 000	0.03	1.0	0.06
Barnard's Star	3100	0.2	0.14	0.004

Life cycle of a star

Our Sun is a star which was formed about 4.6 billion years ago. The Earth and planets were formed at about the same time. Figure 3.3 illustrates how we think stars are formed.

Birth of a star

Figure 3.3a) shows the Eagle Nebula. This is a cloud of cold hydrogen gas and dust which is collapsing due to the pull of gravity. As the cloud collapses, the atoms and molecules move very fast. As molecules collide with each other, their store of kinetic energy is transferred to the internal energy store of the gas, and the temperature rises to several million degrees Celsius. The contracting and heating ball of gas is called a **protostar**.

> TIP
>
> A **protostar** is a large ball of gas that contracts to form a star.

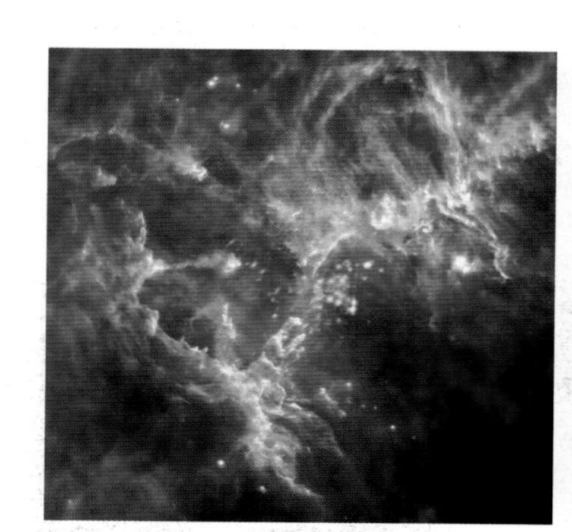

Figure 3.3 a) Stars begin to form giant clouds of dust and hydrogen. **b)** A young star has been born. **c)** A solar system condenses. This is the H L Tauri, 450 light years away from our Solar System.

15000000 K
core
temperature

surface temperature
5800 K

Figure 3.4 Inside of a star.

The temperature in the gas becomes so high that hydrogen nuclei (protons) begin to collide, and fusion begins. Now a star has been born which releases energy for millions of years from nuclear fusion. Figure 3.3b) shows a star which has just begun to shine out from its cloud of dust and hydrogen.

The stable period of a star

The Sun is about half way through its life of about 10 billion years. For most of its life, the Sun will release a constant output of energy. The Sun is described as a **main sequence** star, which means that it releases energy from the fusion of hydrogen to form helium. The inside of a star is a battleground (Figure 3.4). The inward forces of gravity, which tend to collapse the star, are balanced by the outward forces created by the pressure inside the hot core. The pressure at the centre of the Sun is about 100 billion times greater than atmospheric pressure.

Main sequence stars vary considerably in their masses and their brightness. The brightest main sequence stars are a million times brighter than the Sun, and the dullest about 1000 times duller.

> TIP
>
> A **main sequence star** releases energy by fusing hydrogen to form helium.
>
> A red giant is a very large star which fuses helium into heavier elements.
>
> At the end of its life, after fusion stops, a main sequence star collapses into a **white dwarf**.

Stars about the size of the Sun

Towards the end of a star's life as a main sequence star, the supply of hydrogen begins to run out. The star becomes unstable. Without the fusion of hydrogen, the pressure inside the star drops, the outward forces decrease, and the star begins to collapse. As the star collapses, the temperature of the core increases even further, reaching as high as 100 million kelvin. At such high temperatures, helium begins to fuse to make heavier elements such as carbon and oxygen. The hot core causes the star to swell up into a **red giant**, which has about 100 times the radius of our Sun and about 1000 times the brightness. It is possible that the Sun might grow so large that it swallows up the Earth.

Eventually, the star is no longer able to fuse helium and its core cools down. At this stage, the star collapses into a **white dwarf** star, not much larger than the Earth. The surface of a white dwarf is hot, with a temperature of 50 000 K or more. Fusion stops and the star's life is over. At this stage, the white dwarf star cools down and it will eventually become a dark cold star known as a **black dwarf**.

TIP

A **supernova** is caused by runaway fusion reactions in a very large star.

A **neutron star** is a very dense small star made out of neutrons.

A **black hole** is the most concentrated state of matter from which even light cannot escape.

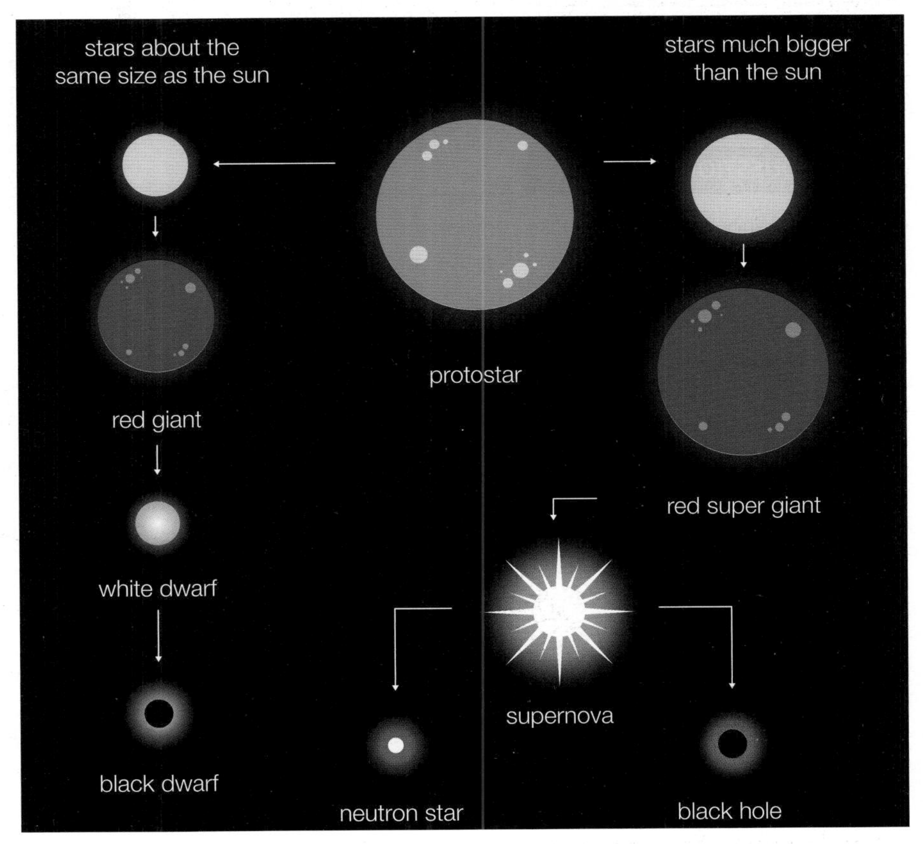

Figure 3.5

Stars much larger than the Sun

Large stars with a mass of over 10 times that of the Sun have much more dramatic ends to their lives. When such a star reaches the end of its main sequence stage, it too begins to collapse.

But now the star grows into a **red super giant**, which could have a radius 1000 times that of the Sun and be 100 000 times brighter than the Sun.

The fusion energy in red super giants is sufficient for heavier elements to be made in the fusion process. Iron is the heaviest element made inside stars due to fusion.

After millions of years as a red super giant, the star runs out of its nuclear fuel. The core cools down and the inwards forces of gravity overcome the outwards pressure from the centre of the star. There is a very rapid collapse of the star and two things happen:

- The rapid collapse heats the inside of the star to such high temperatures that there is a runaway nuclear reaction. The star explodes like a cosmic nuclear bomb. Such is the energy that elements heavier than iron are made. The remnants of a **supernova** are spread out into space. Eventually these remnants form part of another cloud of gas, which collapses to form new stars. Our planet has elements heavier than iron, so it must have once been part of a large star.
- At the same time, the great gravitational forces cause the centre to collapse into a highly condensed form of matter. The core might be left as a **neutron star** a few kilometres across, which is made only of neutrons. The collapse can be so complete that the star disappears into a microscopic point and it has become a **black hole**. A black hole is so dense that nothing, including light, can escape from it.

Figure 3.6 The Crab Nebula is the remnant of a supernova which was recorded by Chinese astronomers in 1054.

STUDY QUESTIONS

1 a) The various stages of the life cycle of a star like the Sun are listed below. Put them in the correct order.
   A main sequence star
   B white dwarf
   C red giant
   D protostar
   b) i) Name the current stage of the Sun's life cycle.
   ii) Name the final stage of the Sun's life cycle.

2 Name the process that provides the energy for a star to emit its own light.

3 a) Explain what is meant by the term main sequence star.
   b) Name the two forces which act on a main sequence star to keep it stable.

4 Explain why a white dwarf eventually becomes a black dwarf.

5 a) Explain why a red super giant collapses to produce a supernova.
   b) Discuss what may happen to the core of a star after a supernova.

6 Justify the suggestion that the atoms in the Earth were once part of a very large star.

7 Refer to Table 2 on page 255. From this table, identify one of the following types of star, justifying your choice.
   A red dwarf
   B white dwarf
   C red giant

8 Describe the life cycle of a star much larger than the Sun.

8.4 Brightness of stars and absolute magnitude

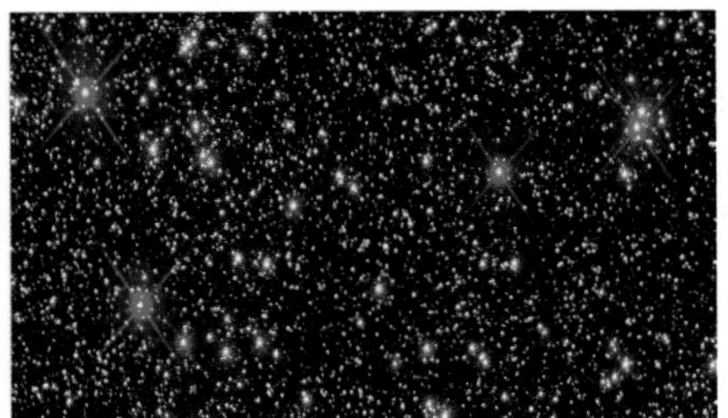

Figure 4.1 This photograph shows a panoramic view of a colourful assortment of 100 000 stars in the crowded core of the star cluster Omega Centauri. Red, orange and blue giant stars shine out above the other duller stars.

When you go outside on a dark night, well away from street lights, you can see thousands of stars. If you look at the night sky with binoculars, you can see thousands more stars. The stars you see have a wide range of brightness.

Apparent magnitude

Hipparchus was a Greek astronomer who lived about 2200 years ago; he was the first person to classify stars by brightness. Hipparchus called the brightest stars, first magnitude stars. Then he listed stars as second magnitude and so on down to sixth magnitude stars, which were the dullest that could be seen (by eye – of course he did not have binoculars). Modern measurements show that a first magnitude star turns out to be about 2½ times brighter than a second magnitude star, and a second magnitude star is 2½ times brighter than a third magnitude star. This works out that a first magnitude star is 100 times brighter than a sixth magnitude star.

This scale has now been adopted in modern times. However, the scale extends beyond 6 for dull stars which can be seen with telescopes, and below 1 for very bright stars or planets.

For example, Venus might have an apparent magnitude of –4 at a particular time; this means Venus is 5 magnitudes brighter (100 times) than a magnitude 1 star.

Table 1 lists the apparent magnitudes of some bright stars.

Table 1

Star	Apparent magnitude
Sirius	−1.5
Canopus	−0.7
Vega	0.0
Belelgeuse	0.4
Spica	1.0
Polaris	2.0

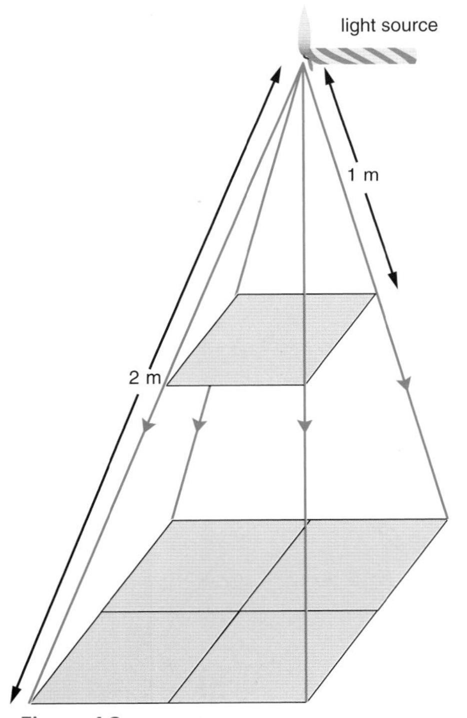

Figure 4.2

A parsec is an astronomical distance equivalent to 3.26 light years.

The absolute magnitude is the apparent magnitude that a star would be if placed a distance of 10 parsecs (32.6 light years) away from Earth.

Absolute magnitude

How bright a star appears to us depends on two things; first how much energy the star emits per second, and secondly how far the star is away from us. Figure 4.2 illustrates how distance affects the apparent brightness of a source. In this figure light is spreading out from a candle. The light spreads out on to a larger area at a distance of 2 m than it does at a distance of 1 m; so the candle appears brighter at a closer distance.

To help us classify stars more consistently, astronomers use the idea of **absolute magnitude** which measures how bright the star would appear at an agreed set distance away from us. The agreed distance is 10 parsecs, which is equivalent to 32.6 light years.

The idea of absolute magnitude allows us to draw comparisons between stars. Our Sun is very bright because it is so close to us, but it has an absolute magnitude of 4.6. Table 2 shows some stars with their absolute magnitudes, together with their brightness relative to the Sun (if placed the same distance away from us).

TIP

You will not be expected to recall the standard distance of 32.6 parsecs, nor the absolute or apparent magnitudes of any stars.

Table 2

Star	Absolute magnitude	Brightness relative to the Sun
Sun	4.6	1
Barnard's star	13.2	0.0004
61 Cygni A	6.8	0.15
Procyon A	2.7	6
Arcturus	0.2	58
Canopus	−2.5	700
Rigel	−8.1	120 000

Hertzsprung-Russell diagram

The variation in star brightness can be shown using the diagram shown in Figure 4.3. This diagram is called the Hertzsprung-Russell (HR) diagram.

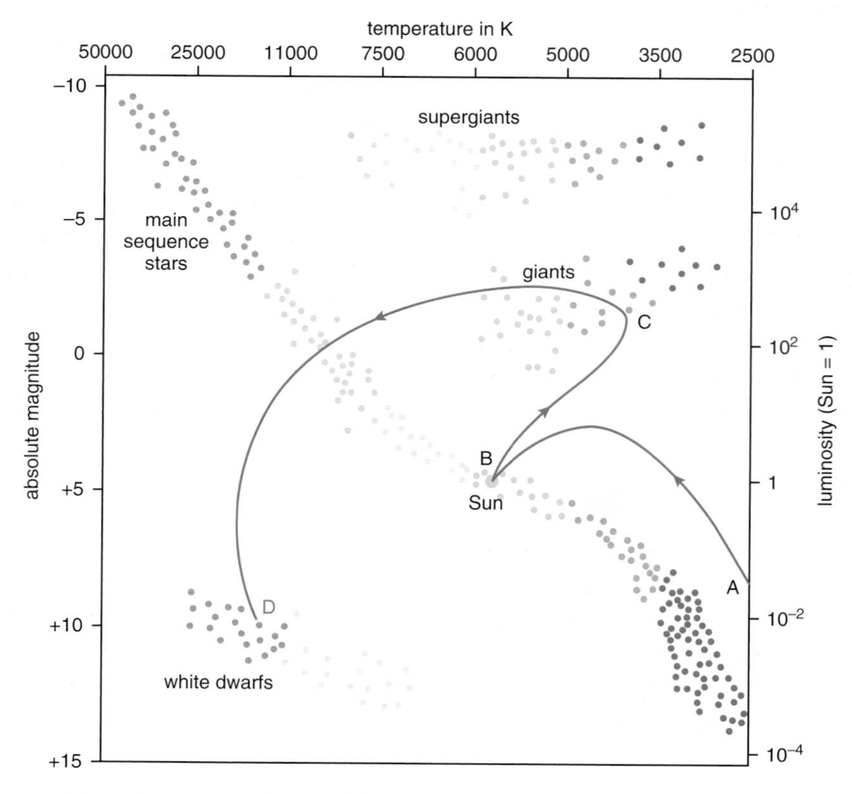

Figure 4.3 The Hertzsprung-Russell (HR) diagram.

The y-axis of the HR diagram shows the absolute magnitude of a star and its luminosity. The x-axis of the HR diagram shows the star's temperature and colour.

There are three important branches of the diagram.

- Main sequence stars. Any star which is generating its energy by fusing hydrogen to helium is a main sequence star, including the Sun. These stars have a great range of brightness or luminosity. The brightest stars have surface temperatures of 50 000 K, and the dullest surface temperatures of 2500 K. Main sequence stars can be a million times brighter than the Sun or 10 000 times duller. The main sequence runs from the top left hand side of the diagram to the bottom right.
- Giant and supergiant stars. These stars are found at the top right hand corner of the HR diagram. These stars can be relatively cool, with temperatures of 3000 K, but they are very bright because they are so large. In these stars, nuclear fusion of helium and larger nuclei occurs.
- White dwarfs. These stars are found at the bottom left hand corner of the HR diagram. Although these stars have very hot surfaces, they are relatively dull because they are so small. These stars are at the end of their lives: nuclear fusion has stopped, the stars have collapsed and are now cooling down. (Note that some white dwarfs are actually blue in colour.)

TIP

Luminosity is the total power of a star, and this measures the energy released in all wavelengths. The Sun has a luminosity of 4×10^{26} W.

We use the word **brightness** to describe what we can see – this is a measure of energy emitted in visible wavelengths.

Some hot stars emit energy in ultra-violet and X-ray wavelengths, which we cannot see. And cold stars emit infrared radiation, which is also invisible to the eye.

The Sun's evolutionary path

In Section 8.3, you read about the evolutionary path of stars. The evolutionary path of the Sun is marked on to the HR diagram.

- At A, the Sun begins to condense from a cold cloud of gas.
- At B, the cloud of gas has collapsed to be a main sequence star. Its life in this phase is about 10 billion years.
- At C, after the main sequence phase, the Sun will expand into a giant phase.
- At D, the Sun collapses to be a white dwarf after fusion reactions cease in its core. The Sun will cool down over billions of years, eventually becoming a black dwarf.

STUDY QUESTIONS

1 a) Explain what is meant by the terms:
   i) apparent magnitude of a star
   ii) absolute magnitude of a star.
   b) State the factors that affect:
   i) apparent magnitude
   ii) absolute magnitude

2 a) Make a sketch of the Hertzsprung-Russell diagram and label the main branches of the different categories of star.
   b) The brightness of a star is affected by its radius and temperature. Explain how this can be used to explain the positions of the giant and dwarf branches in the Hertzsprung-Russell diagram.

8.5 The evolution of the Universe

Figure 5.1 The Abell 1689 superclusters of galaxies.

Light from the galaxies in the Abell 1689 supercluster of galaxies has taken two billion years to reach us. The Universe has been evolving for about 13.8 billion years.

The Big Bang Theory

The most widely accepted theory for the origin of our Universe is that it started with a 'Big Bang' about 13.8 billion years ago. This theory suggests that all of the matter in the Universe originated from a point, and that ever since that primordial explosion, matter has been moving outwards in an ever-expanding Universe.

The origin of our existence has fascinated mankind since the dawn of civilisation, and our interest is as great as ever today. Cosmologists have theories that allow them to understand the Universe as close as 10^{-36} s after the Big Bang. However, what happened right at the start of the Universe remains an unsolved mystery.

Figure 5.2(a) 13.8 billion years ago, the Universe was born with a Big Bang.

- Up to 0.1 second: the temperature of the very early Universe was in excess of 10^{12} K. At these temperatures the building blocks of today's Universe, protons and neutrons, did not exist. The early Universe was populated by highly energetic particles and photons (particles of electromagnetic energy such as gamma rays). Research being done in the large Hadron collider at CERN attempts to replicate conditions in the early Universe: by accelerating and colliding protons at very high energies, physicists are able to discover other particles that existed just after the Big Bang.
- After a few seconds, the Universe had cooled sufficiently for the familiar protons, neutrons and electrons to exist.
- After a time of 3 minutes the Universe had expanded sufficiently to cool to a temperature of 10^9 K, and over the next 30 minutes it cooled further to 10^8 K. While these temperatures are still very high, such temperatures exist today in the cores of very large stars. During this brief period of time nuclear fusion occurred, so that protons fused to form deuterium and helium-4 nuclei. At the end of the period of fusion, the Universe's nuclear material comprised 75 % protons and about 25 % helium nuclei.

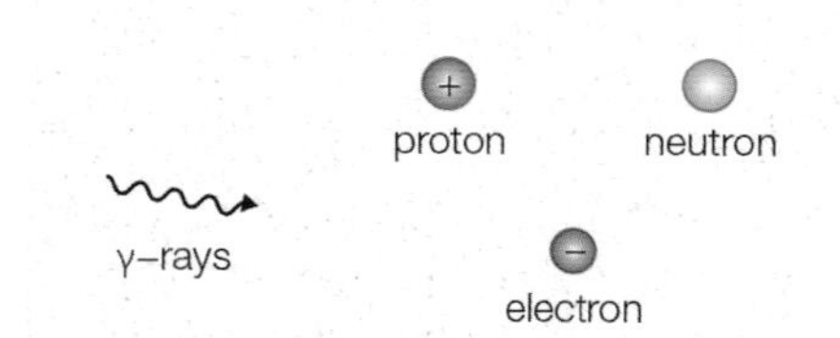

Figure 5.2(b) After a few seconds the Universe saw its first protons, neutrons and electrons, along with other particles and lots of electromagnetic energy.

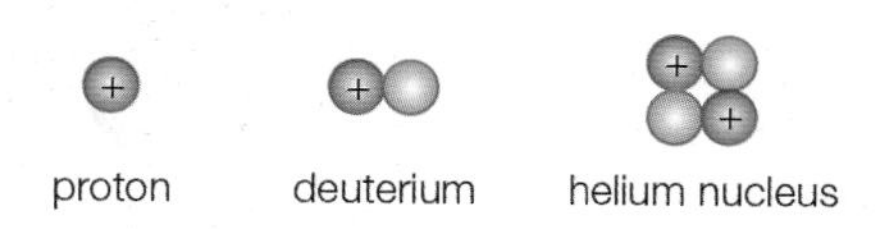

Figure 5.2(c) In the next half-hour nuclear fusion produced some deuterium and helium nuclei.

Figure 5.3 This gaseous nebula is known as the Pillars of Creation. Here clouds of hydrogen, helium and dust are collapsing to form new stars. This process began in the early Universe and new stars are being made today

- For the next 700 000 years the Universe continued to expand and to cool down. The next important step in the evolution of the Universe was that the temperature decreased to a point where nuclei could attract and keep orbiting electrons (at about 4000 K), so rather than protons, helium nuclei and electrons, the Universe filled with hydrogen and helium gases.
- Over the next billion years, gravity acted on the clouds of expanding gases. Gradually, the force of gravity pulled the expanding clouds of gases into large clumps, which condensed to form stars and galaxies. Some of the most distant objects seen in the Universe are galaxies which formed a few billion years after the Big Bang.

Figure 5.4 This giant cluster of galaxies in the constellation of Virgo shows how gravity had acted to clump together our expanding Universe. This photograph shows hundreds of galaxies, each of which contain hundreds of billions of stars.

- Now the Universe continues to expand and evolve. New stars are formed out of clouds of gas in the Milky Way and other galaxies. Our Sun is about 4.6 billion years old.

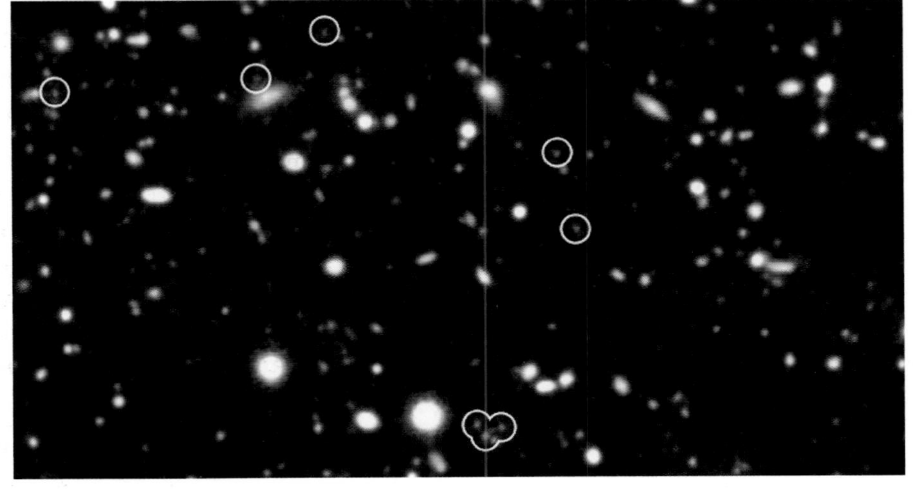

Figure 5.5 In the foreground of this picture, you can see stars in our galaxy and then some galaxies at a distance of hundreds of millions of light years. Behind those galaxies are some very distant galaxies, which are circled. These are over 10 billion years old, and some of the oldest and most distant objects in the Universe.

This chapter has briefly summarised the evolution of the Universe from its origin to the present state. In the next chapter, you will meet the evidence that allowed this theory to be developed.

STUDY QUESTIONS

1. A few seconds after the Big Bang, there was an abundance of protons and electrons. Explain how some further nuclei were also produced in the first few minutes of the Universe.
2. Explain why there was no hydrogen and helium gas until the Universe was about 700 000 years old.
3. Discuss why early stars and galaxies were not formed until a few billion years after the Big Bang.

8.6 The evidence for the Big Bang Theory

Figure 6.1 Edwin Hubble

This photograph shows Edwin Hubble in front of his telescope. No great theory in science has been produced without years of hard work. Hubble's study of the redshift of galaxies provided the evidence for an expanding universe.

The brief account in the last chapter of the evolution of the Universe produces a consistent theory which is widely accepted. The two most important pieces of evidence to support the Big Bang Theory are explained in this chapter.

Cosmic microwave background (CMB) radiation

When a hot body emits radiation, it does so with a range of wavelengths. The wavelength of the radiation depends on the temperature of the body. For example, our Sun, which has a surface temperature of 5800 K, emits radiation with the greatest power density at the wavelength of green light, as shown in Figure 6.2.

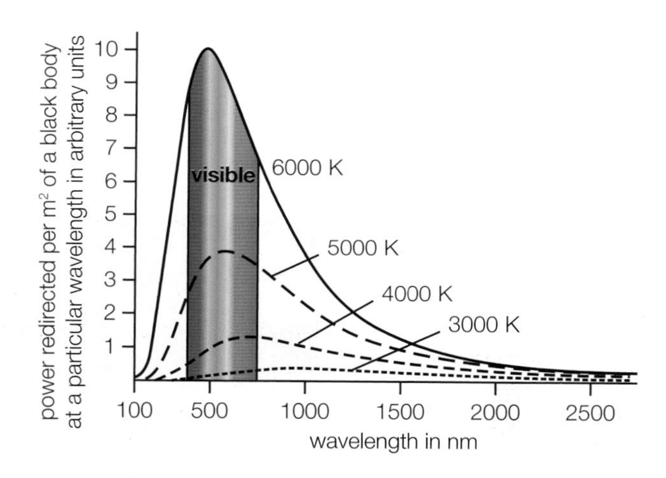

Figure 6.2 The hotter the body, the more energy is emitted; and the hotter bodies emit shorter wavelength radiation.

The Figure also shows that cooler objects emit radiation with the greatest power density at longer wavelengths.

When the Universe was young, it was very hot. At an age of 700 000 years, the temperature of the background radiation was about 4000 K. As the Universe expanded over billions of years, it cooled and the wavelength of the background radiation increased.

For the Big Bang Theory to make sense, astronomers needed to find this colder background radiation. In 1964, Penzias and Wilson detected background radiation, using a radio-receiver, which appeared to come from all directions of space. The peak wavelength of the radiation is about 1 mm, which is in the microwave region – hence the name: cosmic microwave background radiation. The CMB radiation corresponds to a background temperature of 2.7 K. This is the temperature of space between the stars and galaxies.

The CMB radiation is a distant echo which confirms that a long time ago, the Universe was very hot and full of short wavelength radiation produced by the Big Bang.

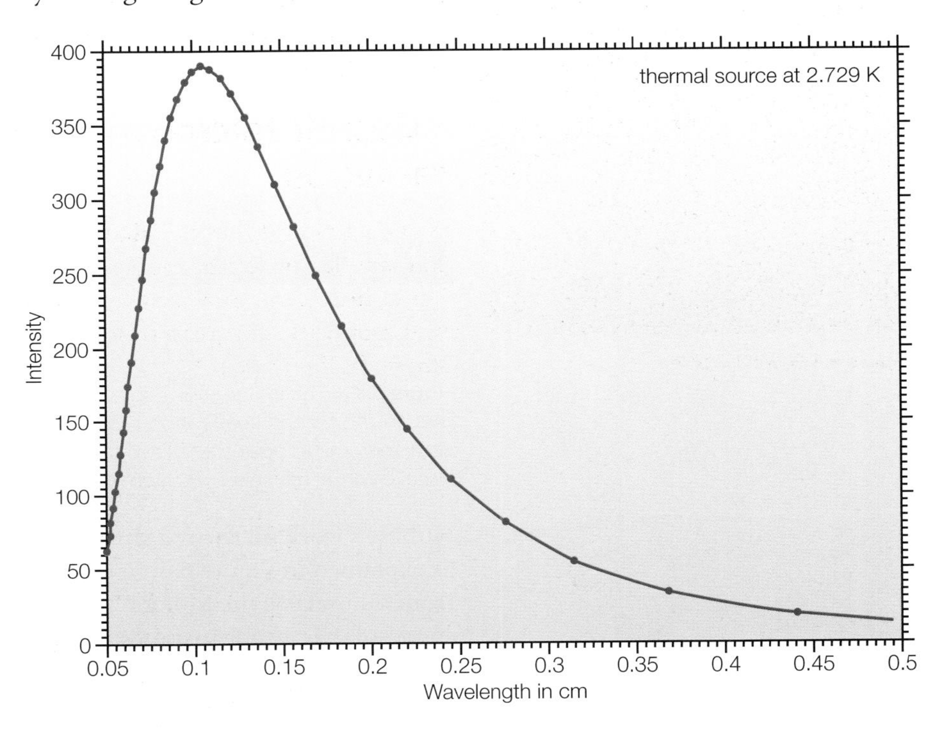

Figure 6.3 The graph shows the intensity of the cosmic microwave background radiation as a function of wavelength.

loudspeaker

frequency generator

Figure 6.4 A teacher uses a whirling loudspeaker to model the effect of red-shift. As the loudspeaker moves away from the students, they hear a decrease in pitch. A decrease in pitch means an increase in wavelength. So the loudspeaker moving away from the students produces the same effect as a light source moving away from us. The loudspeaker must be secured by string and not just swung relying on the electrical cables.

Red-shift

We have all heard an ambulance coming towards us and then speed past. As the ambulance goes past us and moves away, we hear the pitch of the sound from the siren become lower (fewer waves per second). The frequency of the sound waves we hear has decreased and the wavelength increased.

A similar effect happens with light waves, but it is only noticeable if the wave source is moving very quickly. When a source of light is moving away from us, the wavelength of the light that we see becomes longer. So the light moves towards the red end of the spectrum. We say the light has been **red-shifted**. If the source of the light moves even faster, the wavelength increases even more, producing a greater red-shift.

TIP

The recession of the galaxies shows us that the whole universe is expanding. Space expands as the galaxies move apart.

In 1929, Edwin Hubble began to study the light arriving from distant galaxies. He discovered three important facts about this light:

- The light emitted from distant galaxies is red-shifted – the wavelength is longer than expected. This tells us that distant galaxies are moving away from us.
- Galaxies appear to be moving away from us in all directions.
- The further away from us a galaxy is, the bigger the red-shift. This tells us that the further away a galaxy is, the faster it is moving.

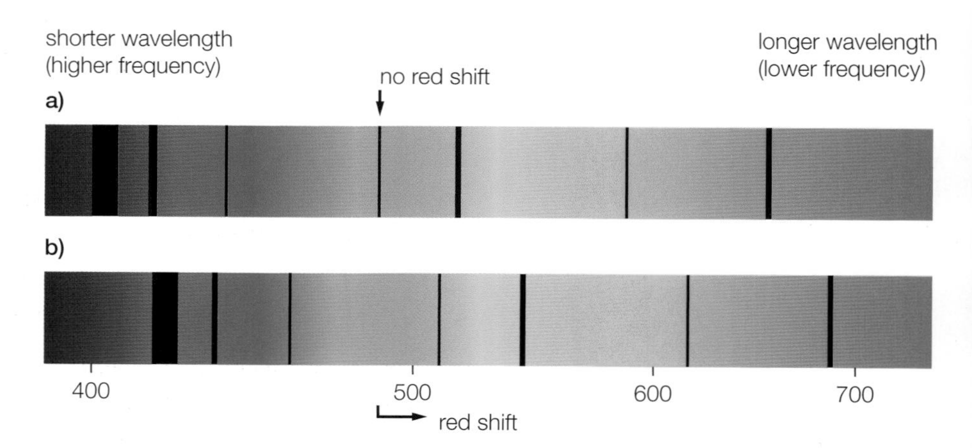

Figure 6.5 a) This image shows light which is emitted from the Sun. It is crossed by black lines. b) This second image shows light arriving at Earth from a distant galaxy. It has the same pattern has been shifted towards the red end of the spectrum. This is how Hubble first discovered the red-shift in the light of distant galaxies.

Hubble's work on the red shift also helped to support the Big Bang Theory, as explained in Figure 6.6. In this diagram, you can see galaxies moving away from our galaxy, the Milky Way. The galaxies further away move faster. This suggests that at one instance in the past, all the galaxies (or the particles in them) were in the same position. The outward movement and speeds of the galaxies are consistent with this theory.

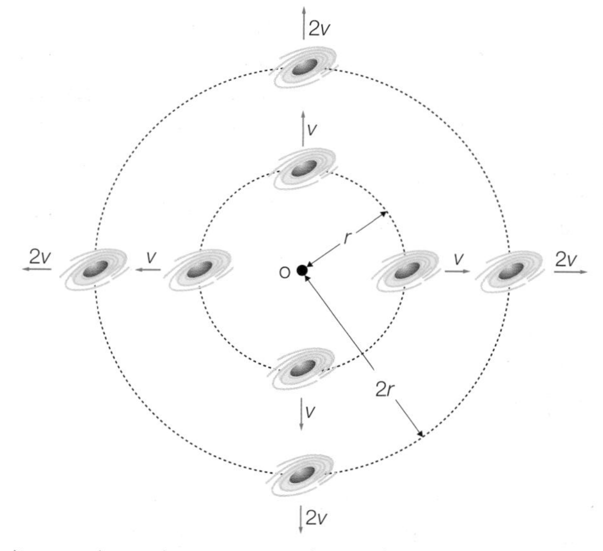

Figure 6.6 This diagram shows the expansion of space from our perspective – our galaxy is positioned at *o*. At a distance *r* galaxies are receding in all directions with speed *v*. At twice the distance, *2r*, the speed of recession has doubled to *2v*.

Calculating the speed and distance of galaxies

Figure 6.5 shows how light emitted by a galaxy moving away from us is red-shifted. The greater the speed, the greater the red shift or the change in wavelength. We can calculate the speed of a galaxy using the following equation:

$$\frac{\text{change in wavelength}}{\text{reference wavelength}} = \frac{\text{velocity of a galaxy}}{\text{speed of light}}$$

$$\frac{\lambda - \lambda_0}{\lambda_0} = \frac{\Delta\lambda}{\lambda_0} = \frac{v}{c}$$

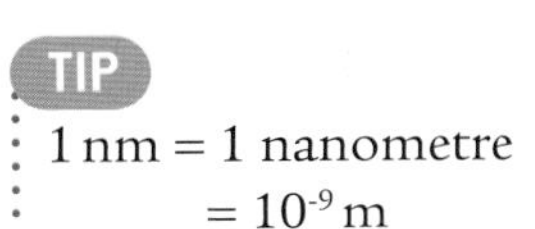

TIP

1 nm = 1 nanometre
= 10^{-9} m

Example

In Figure 6.5 (a), you can see light emitted by the Sun which is crossed by black lines. The same pattern is seen in a moving galaxy, Figure 6.5 (b).

The thickest black line (the reference) has a wavelength of about 400 nm in (a). For the moving galaxy, this wavelength has red-shifted to about 420 nm.

So: reference wavelength, $\lambda_0 = 400$ nm
change in wavelength, $\Delta\lambda = 420\text{ nm} - 400\text{ nm}$
$= 20$ nm

Therefore

$$\frac{\Delta\lambda}{\lambda_0} = \frac{v}{c}$$

$$\frac{20}{400} = \frac{v}{c}$$

$$v = 0.05\,c$$

Substitute in the value for $c = 3 \times 10^8$ m/s

$$v = 0.05 \times 3 \times 10^8$$
$$= 1.5 \times 10^7 \text{ m/s}$$

Figure 6.7 shows how we can link the speed of a galaxy to its distance. Hubble discovered that the speed of a galaxy is proportional to its distance from us. In this graph, the red dots show the distances and speeds of some galaxies. It is very difficult to determine the distance to very far away galaxies, but their red shift helps us.

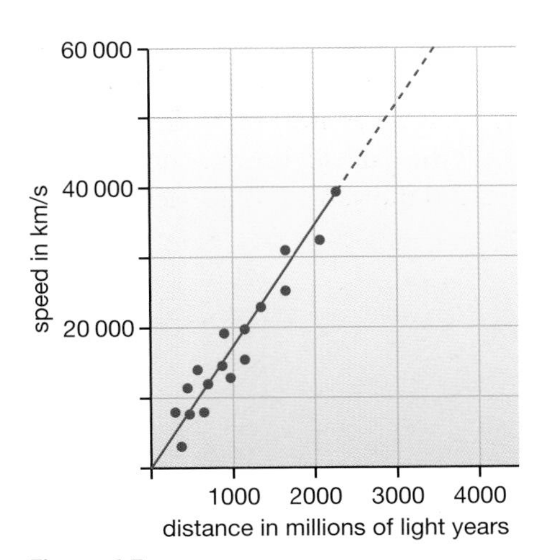

Figure 6.7

Example. A galaxy has a speed of 60 000 km/s. Calculate its distance from us.

Using the graph, we can show the galaxy is about 3600 million light years away.

STUDY QUESTIONS

1 Explain the terms:
   i) cosmic microwave background radiation
   ii) red shift.
2 Explain two pieces of evidence to support the Big Bang Theory. Treat both pieces as long answers earning 4 or 5 marks each, giving detailed explanations.
3 A spectral line seen in the Sun has a wavelength of 280 nm. The same line is seen in the spectrum of a galaxy and is measured to have a wavelength of 370 nm.
   Calculate the speed of the galaxy in km/s.

Summary

I am confident that:

✓ **I understand and can describe these facts about astronomy**

- The Universe contains billions of galaxies.
- A galaxy is a large collection of billions of stars.
- Our Solar System is in the Milky Way galaxy.
- Gravitational field strength is different on other planets and moons.
- Gravitational force causes moons to orbit planets, and planets and comets to orbit the Sun.
- There are differences between the orbits of planets, moons and comets.
- Stars can be classified according to colour.
- A star's colour is related to its temperature.
- The brightness of a star at a standard distance can be represented by an absolute magnitude.

✓ **I can recall and describe**

- The evolution of a star the mass of the Sun.
- The evolution of a star with a mass larger than the Sun.
- The Hertzsprung-Russell diagram.
- The past evolution of the Universe.
- The evidence for the Big Bang theory (red shift and cosmic microwave background radiation).
- The red shift in light received from galaxies at different distances from the Earth.

✓ **I have the mathematical skills and understanding to use these equations**

- $\text{orbital speed} = \dfrac{2 \times \pi \times \text{orbital radius}}{\text{time period}}$
- $\dfrac{\text{change in wavelength}}{\text{reference wavelength}} = \dfrac{\text{velocity of a galaxy}}{\text{speed of light}}$

Sample answers and expert's comments

1 Saturn has 62 moons. The time of orbit of a moon depends on its distance from Saturn.

The table shows radius of orbit of some of the moons of Saturn, and the time of orbit.

Moon	Radius of orbit in 1000 km	Time period of orbit in Earth days
Pandora	141	0.63
Aegaeon	168	0.80
Mimas	185	0.94
Enceladus	238	1.37
Tethys	295	1.89
Dione	377	2.74
Rhea	527	4.52

a) Plot a graph of the radius of orbit (y-axis) against the time period (x-axis) (5)

b) An astronomer discovers a new moon with a time of orbit 3.5 days. Use the graph to find the radius of the moon's orbit. Show your working on the graph. (2)

(Total for question = 7 marks)

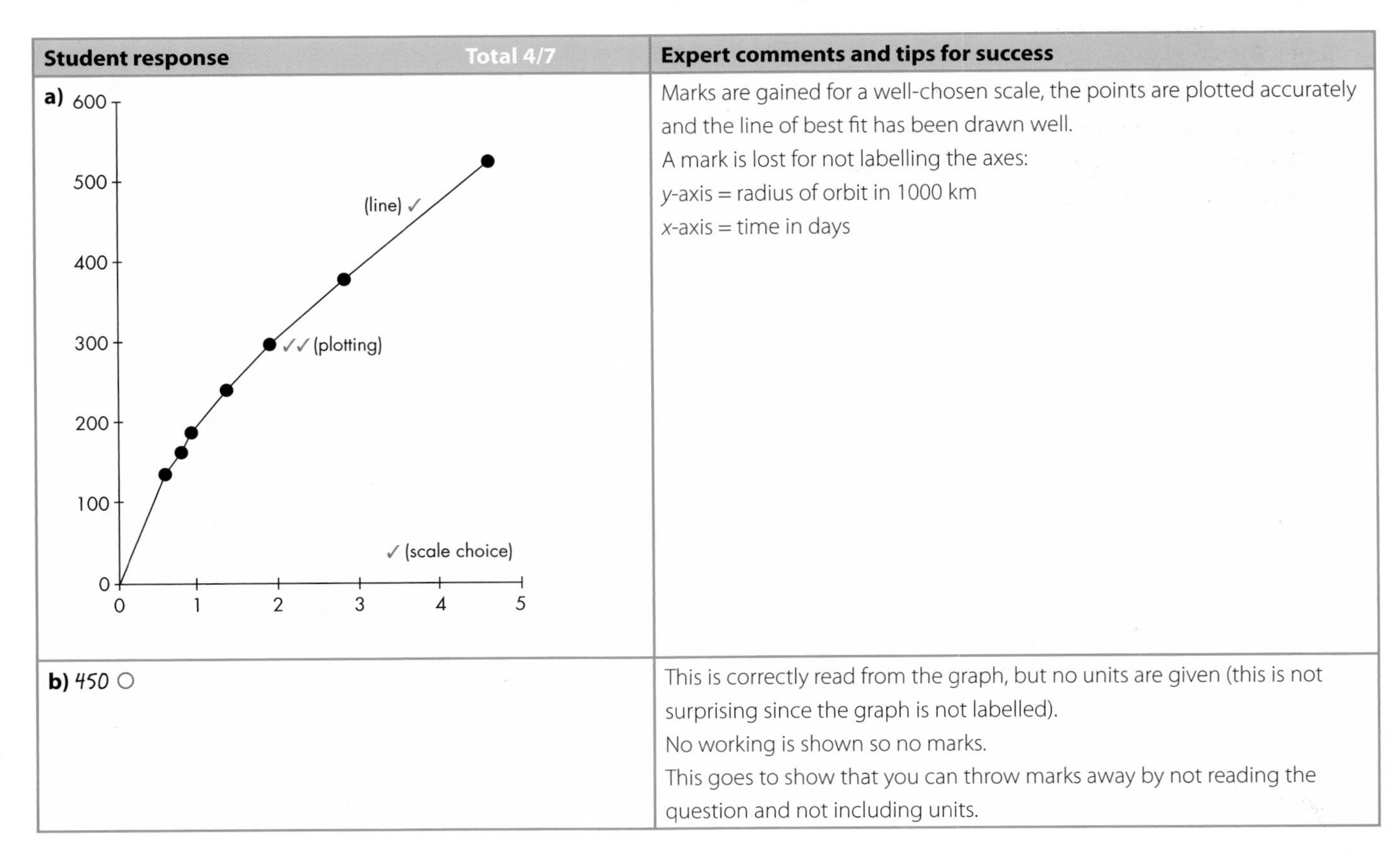

Student response
Total 4/7
Expert comments and tips for success
a)
600
500
400
300
200
100
0
0
1
2
3
4
5
(line) ✓
✓✓ (plotting)
✓ (scale choice)
Marks are gained for a well-chosen scale, the points are plotted accurately and the line of best fit has been drawn well.
A mark is lost for not labelling the axes:
y-axis = radius of orbit in 1000 km
x-axis = time in days
b) 450 ○
This is correctly read from the graph, but no units are given (this is not surprising since the graph is not labelled).
No working is shown so no marks.
This goes to show that you can throw marks away by not reading the question and not including units.

2 This satellite is in orbit round the Earth. The satellite is used for observation, and takes photographs of the Earth's surface.

a) Name the force that keeps the satellite in orbit round the Earth. (1)

b) The satellite moves in a circular orbit. The satellite is in orbit 1600 km above the Earth's surface.

i) The radius of the Earth is 6400 km. Calculate the radius of the orbit of the satellite. (1)

ii) The satellite completes one orbit in 120 minutes. Calculate the speed of the satellite in m/s. (3)

c) A second satellite also orbits the Earth. Its distance from the Earth changes in the orbit.

i) Name the shape of this orbit. (1)

ii) The satellite travels fastest when it is closest to the Earth. Use ideas about energy stores to explain why. (1)

(Total for question = 7 marks)

Student response Total 6/7	**Expert comments and tips for success**
a) *The pull of gravity* ✓	Correct
b) i) *8000 km* ✓	Correct
ii) $v = 2\pi r/T$ ✓ $= 2\pi \times 8000/120$ ○ $= 419$ m/s ✓	This is the correct formula. The radius is expressed in km and the time in minutes. The units are wrong, but the speed is correctly calculated in km/minute, so the answer gains a mark.
c) i) *An elipse* ✓	This is spelled incorrectly, but earns the mark (ellipse).
ii) *At a large distance there is energy in the gravitational potential store* ✓ *some of this is transferred to the satellite's kinetic store as it approaches the Earth.* ✓	The idea of travelling slowly at large distances is correct, and to say it then 'falls' is also correct. The student understands the idea of energy transfers very well.

3 Explain how the discovery of cosmic microwave background radiation supports the Big Bang Theory. (4)

Student response Total 2/4	**Expert comments and tips for success**
Electromagnetic radiation of microwave wavelength is found to come from all parts of the Universe. ✓ *This is an echo of the Big Bang which took place billions of years ago.* ✓	The student has made two good points, but they need to include the idea that the Universe was small and very hot billions of years ago. The Universe has expanded and cooled, and microwaves correspond to this cool temperature.

Exam-style questions

1 The name of our galaxy is:

A Andromeda

B The Great Cluster

C The Milky Way

D The Local Group [1]

2 Which of the following supports the Big Bang Theory?

A The number of stars in our galaxy

B Cosmic microwave background radiation

C The brightness of the Sun

D The diameter of the Universe [1]

3 At present the Sun is:

A A main sequence star

B A red giant

C A white dwarf

D A variable star [1]

4 Which of the following phrases correctly describes what happens to the pitch of a siren as a fire engine drives past you?

A The pitch decreases

B The pitch increases

C The pitch stays the same

D The pitch goes down and then up [1]

5 Which of the following statements is true about the Big Bang Theory?

A It has been proved correct by mathematical calculations.

B It is the only way to explain the origin of the Universe.

C It is based on scientific and religious facts

D It is the most satisfactory explanation of present scientific knowledge. [1]

6 An astronomer discovers that light from a galaxy has been shifted to the blue end of the spectrum. What can you deduce from this discovery.

A Blue stars have exploded as supernovas.

B The galaxy is moving towards us.

C The galaxy is moving away from us.

D The galaxy is not moving relative to us. [1]

7 A satellite is in orbit above a planet, with an orbital radius of 8000 km. The time of one orbit is 80 minutes. The orbital speed of the satellite is:

A 10 500 m/s

B 8000 m/s

C 1050 m/s

D 100 m/s [1]

8 A moon orbits a planet with an orbital speed of 430 m/s. The time of one orbit is exactly 15 earth days. Which of the following is the orbital radius of the moon?

A 44 000 km

B 89 000 km

C 6450 km

D 1000 km [1]

9 Which of the following best describes the numbers of stars and galaxies in the Universe?

A There are too many stars and galaxies to count

B There are thousands of galaxies and millions of stars

C There are millions of galaxies and billions of stars

D There are billions of galaxies and hundreds of billions of stars in each galaxy. [1]

10 Which of the following does not affect the absolute magnitude of a star?

A The star's colour

B The star's temperature

C The star's diameter

D The star's distance from us. [1]

11 Describe what these objects are:

a) A main sequence star [2]

b) The Milky Way [2]

c) A supernova [2]

d) A neutron star [2]

12 Stars go through a life cycle of change.

a) Explain how a star is formed. [4]

b) Explain why the amount of hydrogen in a star decrease as time goes on. [2]

c) Describe what happens to a star much larger than the Sun from the time it runs out of hydrogen. [5]

13 Our Solar System contains atoms of the heaviest elements.

a) State where these elements were formed. [1]

b) Explain how these elements are formed. [2]

c) Justify what this tells you about the age of our Solar System compared with the age of most stars in the Universe. [1]

14 The table shows information about the radius, surface temperature, colour and the luminosity of some stars relative to the Sun. The luminosity is a measure of the energy emitted per second by the star.

a) From the table above, identify a star in each of the following categories: main sequence, red dwarf, red giant, white dwarf. [4]

Star	Radius relative to the Sun	Surface temperature in K	Colour	Luminosity Sun = 1
Sun	1	5800	yellow	1
A	2	7300	yellow/green	10
B	0.2	3500	red	0.005
C	0.03	12 000	white	0.03
D	6	15 000	blue	1600
E	300	3300	red	9400
F	0.8	4800	orange	0.3

b) Stars A, D and E have lower absolute magnitudes than the Sun. Justify this statement. [2]

c) When viewed from the Earth, star F appears brighter than star A. Explain why. [2]

d) Star G has a green colour. Is its surface hotter than the Sun's? Explain your answer. [2]

15 The diagram shows the orbits of the Earth and two other bodies about the Sun.

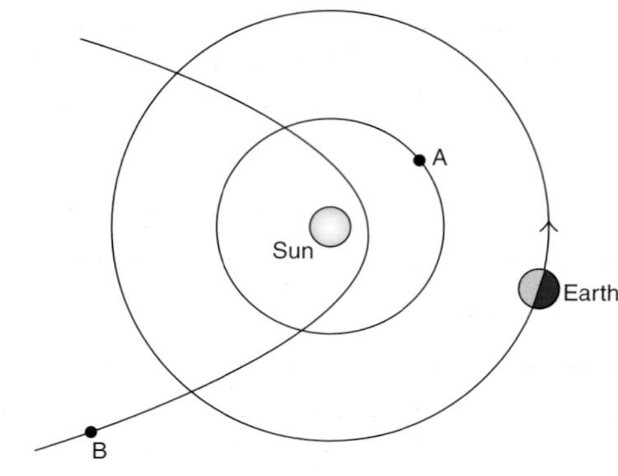

a) Which body could be:

i) the planet Mercury [1]

ii) a comet? [1]

b) i) Describe the shape of the orbit of B. [1]

ii) Describe how the speed of B changes in its orbit [2]

iii) Name the force that keeps comets and planets in orbit. [1]

16 The table below shows the gravitational field strength on the surface of some planets.

Planet	Gravitational field strength in N/kg
Earth	10
Mercury	3.7
Jupiter	24
Neptune	11

a) A man's mass is 80 kg. Calculate his weight on the surface of:

i) Mercury [1]

ii) Neptune. [1]

b) State two factors that affect the size of the gravitational field on the surface of a planet. [2]

17 a) The Moon orbits the Earth at a distance of 384 000 km. One orbit takes 27.3 days. Use this information to calculate the orbital speed of the Moon round the Earth in metres per second. [4]

b) A satellite orbits low over the Earth's surface in an orbit of radius 6500 km. Its orbital speed is 7.5 km/s. Calculate the time period of the orbit. [4]

18 The table shows the relationship between the speed of a galaxy and its distance away from us.

Galaxy	Distance of galaxy from Earth in 10^{18} km	Speed of galaxy in km/s
A	680	1200
B	3800	6700
C	8500	15 000
D	11 400	20 000
E	22 700	40 000
F	34 000	60 000

a) Plot a graph of the speed of the galaxies (y-axis) against the distance of the galaxy (x-axis). [5]

b) Does this graph help to support the Big Bang theory? Explain your answer. [2]

c) Two galaxies are discovered to be moving away from us. G is moving with a speed of 2000 km/s. H is moving with a speed of 4000 km/s. Use your graph to determine the distance of each galaxy from the Earth. [2]

d) A spectral line in the Sun has a wavelength of 580 nm. Calculate the red shift of this line for galaxy F. (The speed of light is 3×10^8 m/s). [4]

19 Make a sketch of the Hertzsprung-Russell diagram to show its main components. Mark the position of the Sun. Label your axes to show the range of star luminosities in comparison with the Sun, and the range of star surface temperatures. [6]

EXTEND AND CHALLENGE

1 A satellite is in circular orbit around the Earth.

a) The speed of the satellite remains constant. Explain why the velocity of the satellite changes during the orbit.

b) Explain why the satellite is accelerating constantly while it is in orbit around the Earth.

c) In which direction is the acceleration of the satellite, and what force is responsible for this acceleration?

2

The photograph shows the giant galaxy M87, which lies at a distance of 54 million light years from us. In the photograph, you can see some small star clusters which are in orbit around the giant galaxy.

a) Explain why the star clusters orbit the giant galaxy. Why can the clusters not remain stationary in space?

b) A star cluster orbits the galaxy with an orbital speed of 560 km/s with an orbital radius of 120 000 light years. Calculate the time of orbit in millions of years.

Index